Y0-CAN-850

Basic Concepts in Geometry

An Introduction to Proof

Basic Concepts in Geometry

An Introduction to Proof

Frank B. Allen
Elmhurst College

Betty Stine Guyer
Elmhurst College

Dickenson Publishing Company, Inc.
Encino, California and Belmont, California

ISBN-0-8221-0023-1
Library of Congress Catalog Card Number: 72-90241

Printed in the United States of America

9 8 7 6 5 4 3 2 1

Filmset by Typesetting Services Ltd, Glasgow, Scotland

Contents

Preface

This book attempts to help undergraduates understand the nature of proof through a study of the axiomatic method in the field of geometry. It is by no means a comprehensive treatment of the subject matter of Euclidean geometry. It emphasizes the methods that can be employed to construct valid arguments, rather than simply presenting a mass of information. An effort is made to vindicate the traditional view that geometry provides an appropriate setting for an intensive study of deductive proof. It is the authors' belief that this study is greatly facilitated by the use of elementary logic, set language, and the careful statement of basic incidence relationships through the use of appropriate symbolism. Accordingly, these topics provide the subject matter for the first three chapters.

The authors recognize that there are many advantages to developing geometry by coordinate methods as suggested by Birkhoff and Beatly in 1940 and later modified by the School Mathematics Study Group. The discussions of coordinate systems for a line in Chapter 4 and of ray-coordinates in Chapter 6 enable the student to utilize his knowledge of the real-number system in the study of proof in geometry. For example, betweenness of points and betweenness for rays can be simply stated in terms of betweenness for real numbers. This facile and accurate expression of incidence relationships provides a foundation for the construction of proofs without unwarranted assumptions that seem to be implied by a drawing.

Students who seek to learn the fundamentals of proof must have some criteria for determining the validity of an argument. Such criteria are provided in Chapter 2. Students must also be encouraged to give careful attention to the form in which a proof is written. Although three forms of proof are employed in the text (essay, ledger, and flow diagram), greatest emphasis is placed on the flow-diagram format. This type of proof is more appropriate for the beginner, because it more clearly delineates the structure of an argument and allows the student to see how the components of an argument fit together to form a proof. For these reasons, rules for writing a flow-diagram proof are developed in Chapter 2 and are systematically applied thereafter. Classroom experience indicates that playing the game of proof according to these rules is a challenging, instructive, and rewarding experience.

Believing that there is little need for another text in advanced plane geometry, the authors have attempted to provide a body of material in which the axiomatic structure of geometry is strikingly apparent. With this purpose, the geometric content has been selected in such a way that the significance of the parallel postulate is the prevailing theme. These limited objectives make it possible to tell the whole story in a one-semester course. The book is of reasonable length, and experience has shown that the entire text can be comfortably covered in one semester. Such a course will provide an important learning experience for three types of students: (1) liberal arts students who need to learn something of the nature of proof, (2) mathematics majors who need to acquire the mathematical maturity required in more advanced mathematics courses, and (3) prospective teachers of high school geometry who need to learn how to use logic as an expository tool. For the third group it is no longer sufficient to consider the facts of geometry from an intuitive viewpoint. This is now done very effectively in grades seven and eight. In high school geometry (usually taught in grade ten) the teacher is expected to deal effectively with proof. Study of this text should enable the prospective teacher of demonstrative geometry to face this task with confidence.

The authors wish to express their appreciation to Dr. Edwin Comfort for his careful reading of the manuscript and to Dr. Donald Mason for several helpful suggestions. We also wish to acknowledge our indebtedness to the School Mathematics Study Group for allowing us to use the basic postulational structure presented in the text *Geometry with Coordinates.*

F.B.A.
B.S.G.

List of Symbols

$\in$ and $\notin$	"is an element of" and "is not an element of"
$\{\ \}$	used to denote a set
$\{\ \mid\ \}$	set builder notation
$=$ and $\neq$	"is equal to" and "is not equal to"
$\subset, \subseteq, \not\subset, \not\subseteq$	proper subset, subset, not a proper subset, not a subset
$\emptyset$	used to denote the empty or null set
⌐	used in T ⌐ S to indicate the set of elements in T that are not in S
$>, <, \geqq, \leqq, \not>, \not<, \not\geqq, \not\leqq$	used throughout to indicate inequalities in the ordinary way
$\longleftrightarrow$ and $\nleftrightarrow$	The first is used to indicate that two statements are equivalent. The second that two statements are not equivalent.
$\leftarrow\!\!+$ and $+\!\!\rightarrow$	used when an equivalence must be broken because of spacing requirements
$\mid\ \mid$	used to denote the absolute value of a real number; also used to denote the distance between two points, as in $\lvert AB \rvert$
$\cap$	used to show the intersection of two sets
$\cup$	used to show the union of two sets
$'$	used like A' to indicate the complement of set A
$\wedge$	used to indicate the conjunction of two statements
$\vee$	denotes the disjunction of two statements
$\veebar$	denotes the exclusive "or"
$\sim$	indicates the contradiction of a statement
$\longrightarrow$	"implies"
$-\!\!\!/$ and $+\!\!\rightarrow$	used when an implication must be broken because of spacing requirements
$\times$	used like $L_1 \times L_2$ to denote that lines L_1 and L_2 intersect in a single point
$\cong$ and $\ncong$	"is congruent to" and "is not congruent to"
A-B-C	"point B is between points A and C"
a-b-c	"the real number b is between the real numbers a and c"
$\overleftrightarrow{AB}$	used to denote the line determined by points A and B
$\overline{AB}$, $\overset{\circ\!-\!\circ}{AB}$, $\overset{\circ\!-}{AB}$ or $\overset{-\!\circ}{AB}$, $\overrightarrow{AB}$, $\overset{\circ\!\rightarrow}{AB}$	segment AB, open segment AB, semi-open segments AB, ray AB, half-line AB

L/A	"the half plane with edge line L containing the point A" Can also be written $\overleftrightarrow{AB}/X$ to denote the half plane with edge $\overleftrightarrow{AB}$ containing the point X
$L/\sim A$	"the half plane opposite L/A"
$\overline{L/A}$ or $\overline{\overleftrightarrow{AB}/X}$	indicates a closed half plane
$\underline{L/A}$	the plane determined by line L and point A
M/A	"the half space with edge plane M and containing the point A"
$\overline{M/A}$	denotes a closed half space
$M/\sim A$	"the half space opposite M/A"
$-\!\!<$	used in flow proofs to show that a statement implies a disjunction
$A\underline{B}C$	"the union of ray BA and ray BC"
$\angle ABC$	"angle ABC"
$\langle a,b\rangle$ and $\rangle a,b\langle$	closed and open intervals
$\perp$	"perpendicular to"
$\triangle ABC$	"triangle ABC"
$\leftrightarrow$	used to denote a correspondence
$\parallel$ and $\nparallel$	"is parallel to" and "is not parallel to"
cop and $\sim$cop	"coplanar" and "noncoplanar"
$\downharpoonleft\upharpoonright$	"antiparallel"
$W\text{-}\underline{XY}\text{-}Z$	"the union of $\overline{\overleftrightarrow{XY}/W}$ and $\overline{\overleftrightarrow{XY}/Z}$"

Chapter 1
Sets and Logic

In our study of geometry we shall employ three principal tools:

1. Set theory.
2. A set of rules for reasoning called logic.
3. The knowledge of real numbers gained from the study of algebra.

In this chapter we introduce logic and set theory. In Chapter 2 we continue our study of logic and apply logic to set theory.

Set Language

A set is simply a collection of objects. The individual objects in a set are called *members* or *elements* of the set. We use a capital letter to name a set and use the symbol $\in$ for the phrase "is a member of" or "is an element of." The symbol $\notin$ is used for the phrase "is not an element of." If V is the set of vowels in the alphabet, then we can write $e \in V$ and $t \notin V$.

When we have indicated which objects belong to a set, we have *defined* the set. When we say that S is the set of elements that satisfy a given condition, we mean that S contains *all* elements that satisfy this condition and *no* others. For example, when we say "S is the set of counting numbers less than 6," we mean that S contains all the elements 1, 2, 3, 4, 5, and no others.

We often define a set by listing its members between braces. Thus $D = \{1, 2, 3, 4, \ldots, 37\}$ is the set consisting of the first 37 counting numbers. Listing the members of a set between braces in this way is called the *roster method* of defining the set. We use dots in either of the following ways to show that a set is infinite: $N = \{\ldots, -3, -2, -1, 0\}$ or $Z = \{\ldots, r, s, t, \ldots\}$.

If every element of set A is also an element of set B, then set A is a subset of set B; that is, set A is included in set B. To write "set A is a subset of set B" we write $A \subseteq B$. If every element of set A is an element of set B and there is at least one element of set B that is not in set A, we say that set A is a *proper subset* of set B and we indicate this by writing $A \subset B$. We use the symbols $A \nsubseteq C$ and $A \not\subset C$ to represent the phrases "A is not a subset of C" and "A is not a proper subset of C," respectively.

The set having no elements is called the *empty set* or the *null set* and is designated by the symbol $\emptyset$. We now assert that $\emptyset \subseteq A$ for any set A. In Chapter 2 this assertion is proved.

If in a certain discussion all of the sets under consideration are subsets of a given set, we may call the given set the *universal set* or the *universe* for that discussion.

You recall that R, the set of real numbers, has many subsets. Among these are

$C = \{1, 2, 3, \ldots\}$—the set of counting numbers,
$W = \{0, 1, 2, 3, \ldots\}$—the set of whole numbers,
$I = \{\ldots, -3, -2, -1, 0, 1, 2, 3, \ldots\}$—the set of integers,
Q, the set of rational numbers, such as $-\frac{4}{3}, \frac{5}{7}, \frac{21}{5}, \ldots$.

In this text all of the numbers we shall use will be real numbers. Thus we shall regard R as our universal set for numbers unless some other universe is specifically designated. Each of the sets C, W, I, and Q is a subset of R, as indicated by the *Venn diagram*. Observe that in a Venn diagram we use a rectangular region to represent the universal set.

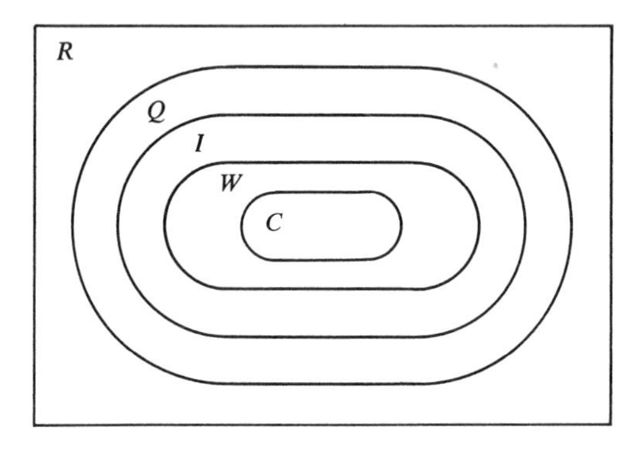

Equal sets are sets that have the same members. Thus $\{a, e, i, o, u\} = \{e, u, o, a, i\}$. Observe that the order in which we list the elements does not matter. The properties of equality for sets are the same as the properties of equality studied in algebra. (See E_1, E_2, E_3, and E_4 in Appendix A.)

The idea of equal sets has an applicaton to *definitions*. A definition is an exact description of a set of objects. Consider the

DEFINITION *A counting number is a nonzero whole number.*

This means that the set of counting numbers *is equal to* the set of nonzero whole numbers. Thus any member of the set of counting numbers is a member of the set of nonzero whole numbers and any member of the set of nonzero whole numbers is a member of the set of counting numbers. We accept either of the phrases *a counting number* and *a nonzero whole number* as a substitute of the other in any discussion.

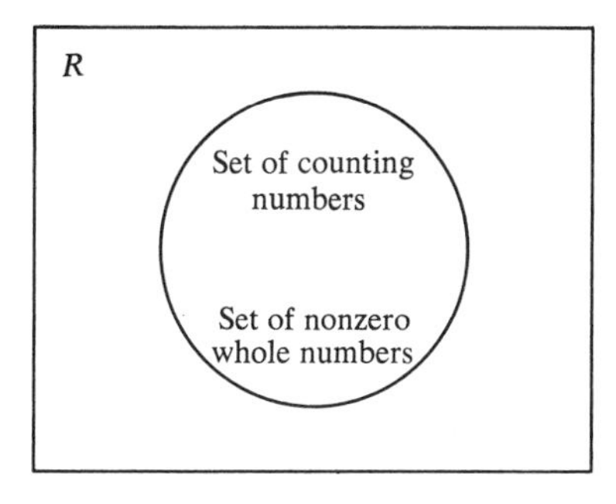

Notation: We sometimes indicate the set that consists of all members of T that are not in S by the symbol $T \sim S$. Thus if C is the set of counting numbers and W is the set of whole numbers, we have $C = W \sim \{0\}$ and $C \sim W = \emptyset$. Observe that $\{0\}$ is not the empty set because it has one element, namely 0.

If the members of set A are associated with the members of set B in such a way that each member of A is paired with (corresponds to) exactly one member of B, and each member of B is paired with exactly one member of A, we say that sets A and B are in *one-to-one correspondence.* If $A = \{1, 2, 3\}$ and $B = \{a, b, c\}$, then A and B can be put in one-to-one correspondence by pairing 1 with a, 2 with b, and 3 with c. Thus $\{(a, 1), (b, 2), (3, c)\}$ is a one-to-one correspondence between sets A and B. (There are, of course, other one-to-one correspondences between A and B.)

DEFINITION *Given two sets A and B. If S is a set of pairs of elements such that each pair contains one and only one element from each of the sets A and B and each element from each of the sets A and B is found in exactly one pair, then S is a* one-to-one correspondence *between sets A and B. Two one-to-one correspondences between sets A and B are equal if and only if they consist of the same pairs.*

Equivalent Statements and Equal Sets

If a sentence expresses an idea that can be labeled *true* or *false,* the sentence is a *statement.* The particular label "true" or "false" that we assign to a given statement is called the *truth value* of the statement. If two statements are both true or both false, they are said to have the *same truth value.* If one is true and the other false, they are said to have *opposite truth values.*

Note that not all sentences are statements. For example, such sentences as "Bring the coffee" or "Will you travel by train or plane?" cannot be labeled either true or false. In this text we shall use a small letter to represent a statement. Thus the letter s could represent the statement "George Washington was the first president of the United States" and the letter t could represent the statement "$(3 + 5)^2 > 3^2 + 5^2$."

DEFINITION *Two statements are equivalent if they have the same truth value.*

We shall use the symbol $\longleftrightarrow$ to indicate that two statements are equivalent to each other.

From the definition of equivalent statements it follows that

(a) Any statement is equivalent to itself. (*Reflexive property of equivalence*)
(b) If the first of two statements is equivalent to the second, then the second is equivalent to the first. (*Symmetric property of equivalence*)
(c) If the first of three statements is equivalent to the second, and the second is equivalent to the third, then the first is equivalent to the third. (*Transitive property of equivalence*)
(d) If we substitute one of two equivalent statements for the other in any third statement, we obtain an equivalent statement. (*Substitution property of equivalence*)

The *truth table* below indicates the conditions under which two statements are equivalent. The letter T written under s means that the statement s is true, and F written under s means that the statement s is false, and so on. The T's and F's under $s \longleftrightarrow t$ indicate that the statement "s is equivalent to t" is true when s and t have the same truth value and false when s and t have opposite truth values.

s	t	$s \longleftrightarrow t$
T	T	T
T	F	F
F	T	F
F	F	T

When we say that two statements are equivalent, we are not saying that they have the "same meaning" or that they are about the same subject. We are saying only that they have the same truth value. Thus the statements "Humphrey Bogart starred in *High Sierra*" and "$\frac{1}{2}(3 + 7) = 5$" are equivalent because both are true. The statements "Patrick Henry was the first president of the United States" and "7 is greater than 10" are equivalent because both are false. The statements "37 is divisible by 6" and "Washington, D.C., is the capital of the United States" are not equivalent because the first is false and the second is true.

If the letter q represents a statement, then when we write the letter q we are denoting a statement that is equivalent to the assertion that q is true. If we reserve the letter u to represent a statement that is known to be true, then the statement *q is true* is expressed by the equivalence $q \longleftrightarrow u$. From the truth table below we observe that $q \longleftrightarrow (q \longleftrightarrow u)$. In view of this discussion, we see that it is not necessary to write such phrases as "q is true" or "it is true that q."

q	u	$q \longleftrightarrow u$
T	T	T
F	T	F

same truth values

Our definition of equal sets provides a very important example of equivalent statements. Sets A and B are equal if they have the same members. This means that every member of A is a member of B ($A \subseteq B$) and that every member of B is a member of A ($B \subseteq A$). Thus, using our symbol for equivalence, we can express the definition of equal sets as follows:

DEFINITION $A = B \longleftrightarrow (A \subseteq B \text{ and } B \subseteq A)$.

If a sentence contains a variable, it is called an *open sentence.* Recall that a variable is a symbol that holds a place for the name of any element in a given nonempty set. We call the given set the *replacement set* or *domain of the variable.* Any member of the replacement set is called a *value* of the variable.

The following are examples of open sentences.

a: Student X is a member of the basketball team.
b: $2x + 7 = 5$.
c: $y > 15$.
d: $z^2 < 0$.

Consider an open sentence that contains one variable. If S is the set of values of the variable for which the truth value of the open sentence can be determined (and $S \neq \emptyset$), then the open sentence is an *open statement* and set S is the *replacement set for this open statement.* The members of the replacement set of an open statement that make the statement true when substituted for the variable make up the *truth set* or *solution set* of the open statement. Thus the truth set of statement a is A, the set of members of the basketball team, because if the name of any basketball player is substituted for X in the statement, it becomes a true statement. The replacement set for this statement is the set of students in college.

In this text we shall use a small letter to represent an open statement and the corresponding capital letter to represent its truth set. Each of the variables in open algebraic statements, such as statements b, c, and d above, will have R as its domain unless otherwise specified. Thus if C is the truth set of statement c, then C is the set of all real numbers greater than 15.

We shall often use the symbol $\{y \mid y > 15\}$ to represent the phrase "the set of all real numbers y such that y is greater than 15." This is called *set-builder notation* and can be used to denote the truth set of any open statement. For example, $B = \{-1\} = \{x \mid 2x + 7 = 5\}$.

DEFINITION *Two open statements are equivalent if they have the same truth set and the same replacement set.*

Open statements $3x = 12$ and $x - 4 = 0$ are equivalent, because each of these statements has R for its replacement set and $\{4\}$ for its solution set. Thus $3x = 12 \longleftrightarrow x - 4 = 0$.

Open statements $x > 4$ and $x^2 > 16$ are not equivalent, because -5 belongs to the solution set of the second statement but does not belong to the solution set of the first statement. Using the symbol $\neq$ for "is not equal to," we have $\{x \mid x > 4\} \neq \{x \mid x^2 > 16\}$. Using the symbol $\not\leftrightarrow$ for "is not equivalent to," we have $x > 4 \not\leftrightarrow x^2 > 16$.

If open statements a and b have the same replacement set, then our definition tells us that the statements $a \longleftrightarrow b$ and $A = B$ are equivalent. Thus for such statements we have

(i) $$(a \longleftrightarrow b) \longleftrightarrow (A = B).$$

EXAMPLE

Do open statements

$$a\colon \frac{x}{x-3} = \frac{4}{x-3} \quad \text{and} \quad b\colon 3x - 1 = 11$$

have the same truth set? the same replacement set? Are they equivalent?

DISCUSSION

Statements a and b have the same truth set, namely $\{4\}$. Thus

$$\left\{x \,\middle|\, \frac{x}{x-3} = \frac{4}{x-3}\right\} = \{4\} = \{x \mid 3x - 1 = 11\}.$$

Statements a and b do not have the same replacement set. The replacement set of b is R, while the replacement set of a is $R \sim \{3\}$, since division by 0 is not defined. Since a and b do not have the same replacement set, they are not equivalent.

Exercises

1. $U = \{1, 2, 3, 4, 5, 6, 7, 8\}$, $A = \{1, 2, 3, 4, 5\}$, $B = \{3, 4, 5\}$, and $C = \{5\}$.
   (a) Draw a Venn diagram showing sets U, A, B, and C.
   (b) $U \sim A =$ ____.
   (c) $A \sim B =$ ____.
2. $A = \{3, 4, 5\}$ and $B = \{3, 4, 5, 6\}$.
   (a) Find $B \sim A$. Is $(B \sim A) \subseteq B$? Is $(B \sim A) \subseteq A$?
   (b) Find $A \sim B$. Is $(A \sim B) \subseteq A$? Is $(A \sim B) \subseteq B$?
3. Which statements in each of the following columns are equivalent to the first statement in each column?

(a)	(b)	(c)
$2x \geqq -10$.	$x^2 > 9$.	$3(x+2) = 12$.
$x \not> -5$.	$x > 3$.	$3x = 6$.
$x - 5 \geqq -10$.	$\lvert x\rvert > 3$.	$x = 2$.
$-x \leqq 5$.	$\lvert x + 3\rvert > 6$.	$\frac{1}{x} = \frac{1}{2}$.

4. In each of the following, name the property of equivalence that justifies the statement.
   (a) $a \longleftrightarrow a$.
   (b) If $a \longleftrightarrow b$ and $b \longleftrightarrow c$, then $a \longleftrightarrow c$.
   (c) If $a \longleftrightarrow b$, then $b \longleftrightarrow a$.

5. Let p be an open statement involving the variable x. Let W be the domain of x, Y be the replacement set for p, and Z be the truth set for p.
   (a) Explain why $Y \subseteq W$ and $Z \subseteq Y$.
   (b) If we choose an element of $Y \frown Z$ as a value of the variable, what is the truth value for statement p for that value?
6. The domain of the variable is the set of real numbers. Find S, the replacement set, and T, the truth set, for each of the following open sentences.
   (a) $|x| > 0$.
   (b) $x(x-4) = 12$.
   (c) $\dfrac{1}{|x|-2} = \dfrac{1}{7}$.
   (d) $\dfrac{x^2}{x-3} = \dfrac{9}{x-3}$.
   (e) $\dfrac{x^2+1}{x^2-x} = \dfrac{1}{x} + \dfrac{2}{x-1}$.
7. How many subsets has a set of n elements ($n \in W$)?
8. Given sets X and Y each containing n elements ($n \in C$), how many different one-to-one correspondences are possible for X and Y?
9. (a) Use the roster method to write X, the set of numbers of the form $2k+1$, where $k \in \{0, 1, 2, 3, 4, \ldots\}$.
   (b) Use the roster method to write Y, the set of odd counting numbers.
   (c) Is $X \subseteq Y$? Is $Y \subseteq X$? Can you conclude that $X = Y$?
   (d) Write a definition of the set of odd counting numbers.

Operations with Sets

If we are given two sets, we may perform certain operations with them to obtain a set. Three very important operation with sets are: (1) finding the intersection of two given sets, (2) finding the union of the given sets, and (3) finding the complement of a set with respect to a given universal set that contains it.

The *intersection* of two sets A and B is the set of elements that belong to both A and B. This set is designated by the symbol $A \cap B$, which is read "A cap B," "A intersection B," or "the intersection of A and B." If $A = \{1, 2, 3, 4, 5, 6, 7, 8\}$ and $B = \{6, 7, 8, 9, 10\}$, then $A \cap B = \{6, 7, 8\}$. This set could also be denoted by the symbol $B \cap A$. In general

(ii*a*) $$A \cap B = B \cap A$$

and

(ii*b*) $$A \cap A = A.$$

In the diagram labeled $A \cap B$ the members of set A are represented by points in the region labeled A and the members of set B by points in the region labeled B. The shaded portion of the figure represents $A \cap B$ because it contains all points that are in both A and B and no others. In the diagram labeled $A \cap B = \emptyset$ we have two sets A and B that are *disjoint*; that is, two sets whose intersection is the null set. [*Note:* We can always speak of the "intersection of two sets" even though this intersection may be empty, as it is in the case of two disjoint sets. However, when we say, "Sets A and B intersect," we mean that there is at least one element common to these two sets. Thus "Sets A and B intersect" $\leftrightarrow$ $A \cap B \neq \emptyset$.]

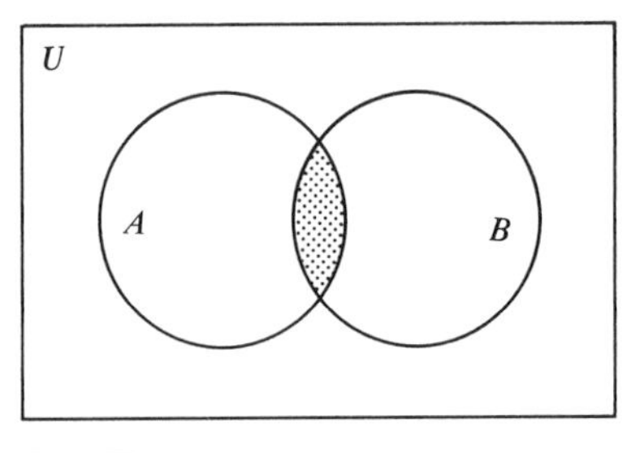

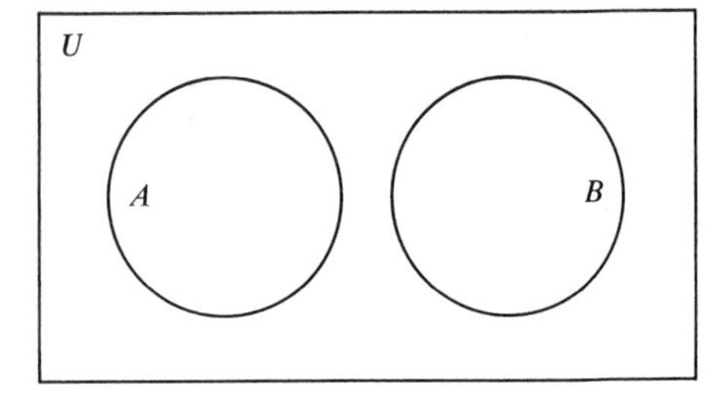

 $A \cap B$

$A \cap B = \phi$

We define the intersection of three or more sets as follows:

(iii)
$$A \cap B \cap C = (A \cap B) \cap C,$$
$$A \cap B \cap C \cap D = (A \cap B \cap C) \cap D.$$

To find the intersection of more than four sets we proceed in this pattern. There is an associative law for the intersection of sets—namely,

(iv)
$$(A \cap B) \cap C = A \cap (B \cap C).$$

The *union* of two sets A and B is the set of elements that belong to A or to B or to both A and B. This set is designated by the symbol $A \cup B$, which is read "A cup B," "A union B," or "the union of A and B." If $A = \{a, b, c, d\}$ and $B = \{c, d, e, f, g, h\}$, then $A \cup B = \{a, b, c, d, e, f, g, h\}$. This set could also be denoted by the symbol $B \cup A$. In general we have

(v*a*)
$$A \cup B = B \cup A$$

and

(v*b*)
$$A \cup A = A.$$

The union of three or more sets is defined as follows:

(vi)
$$A \cup B \cup C = (A \cup B) \cup C,$$
$$A \cup B \cup C \cup D = (A \cup B \cup C) \cup D.$$

To find the union of more than four sets we proceed in this pattern. We also have the associative property of union:

(vii)
$$(A \cup B) \cup C = A \cup (B \cup C).$$

It is also interesting to observe that there is a distributive property of intersection over union:

(viii)
$$A \cap (B \cup C) = (A \cap B) \cup (A \cap C)$$

and a distributive property of union over intersection:

(ix)
$$A \cup (B \cap C) = (A \cup B) \cap (A \cup C).$$

If A is any subset of the universal set U, then *the complement of A with respect to U* is the set of elements of U that are not elements of A. We use the symbol

A' to denote the complement of A. Thus $A' = U \sim A$. In the diagram A' is represented by the shaded area. A and A' are said to be complementary to each other with respect to the universe U.

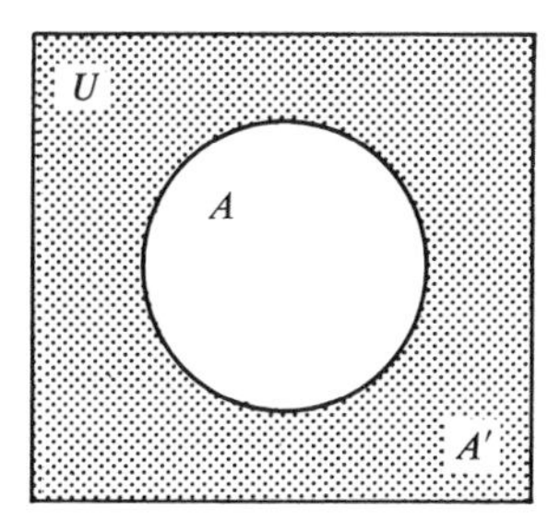

If A and B are subsets of the universe U, our definition tells us that A and B are complementary $\longleftrightarrow (A \cap B = \emptyset$ and $A \cup B = U)$.

Using A' as the complement of A with respect to U, we have the following:

(x*a*) $$A \cap A' = \emptyset,$$

(x*b*) $$A \cup A' = U,$$

(xi) $$(A')' = A,$$

(xii) $$A = B \longleftrightarrow A' = B'$$

(xiii*a*) $$\emptyset' = U,$$

and

(xiii*b*) $$U' = \emptyset.$$

Now that we have defined the complement of a set, we observe that for any two sets T and S, $T \sim S = T \cap S'$.

The following statements are readily verified.

(xiv*a*) $$U \cap A = A,$$

(xiv*b*) $$U \cup A = U,$$

(xiv*c*) $$\emptyset \cap A = \emptyset,$$

(xiv*d*) $$\emptyset \cup A = A.$$

Let us consider some examples that involve operations with sets.

EXAMPLE 1

Draw a Venn diagram to illustrate the statement $A \cap C' = \emptyset$. What relationship exists between sets A and C?

SOLUTION

A must be a subset of C, as shown in the diagram.

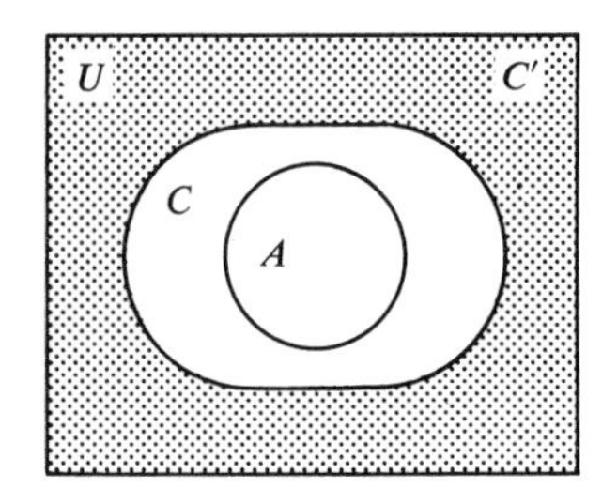

EXAMPLE 2

Draw a Venn diagram that shows the following situation for sets X, Y, and Z: $X \subset Y$ and $Y \cap Z = \emptyset$. What relationship exists between X and Z?

SOLUTION

$X \cap Z = \emptyset$, as shown in the diagram.

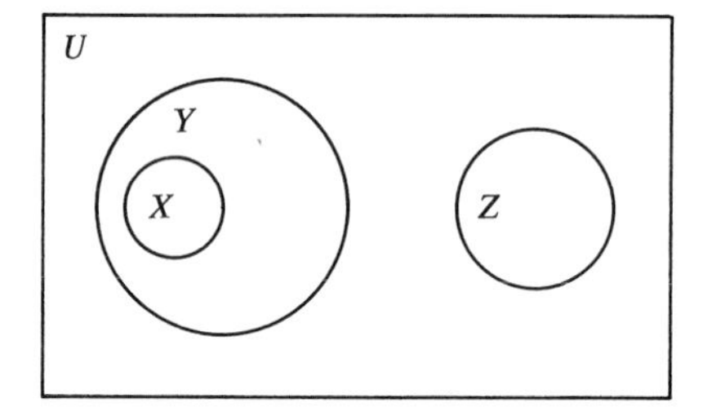

EXAMPLE 3

Tell which of the statements (ii) through (xiv) is used to justify each of the following:

1. $A' \cup (A \cap B) = (A' \cup A) \cap (A' \cup B)$.
2. $(A' \cup A) \cap (A' \cup B) = (A \cup A') \cap (A' \cup B)$.
3. $(A \cup A') \cap (A' \cup B) = U \cap (A' \cup B)$.
4. $U \cap (A' \cup B) = A' \cup B$.

SOLUTION

1: (ix), 2: (v*a*), 3: (x*b*), and 4: (xiv*a*).

Note that in part 2 the reason given justifies that $A' \cup A = A \cup A'$ and that in part 3 the reason justifies that $A \cup A' = U$. In each of these cases, it is understood that the substitution property of equality allows us to replace one of two equal sets with the other. In general, when the substitution property of equality is involved together with a reason that justifies the equality, we shall state the reason for the equality only.

Exercises

1. $U = \{1, 2, 3, \ldots, 10\}$, $A = \{2, 4, 6, 8\}$, $B = \{1, 2, 3, 5\}$, and $C = \{1, 3, 5, 7, 9\}$. Find
   (a) $B \cap C$. (b) $B \cap \emptyset$.

(c) $A \cup (B \cap C)$.
(d) $(A \cup B) \cap (A \cup C)$.
(e) $A' \cup B$.
(f) $A \cap B'$.
(g) $(A \cap B')'$.
(h) $A \sim B$.

2. A and B are subsets of universal set U. List the Roman numerals corresponding to the nonoverlapping regions that you would shade to show

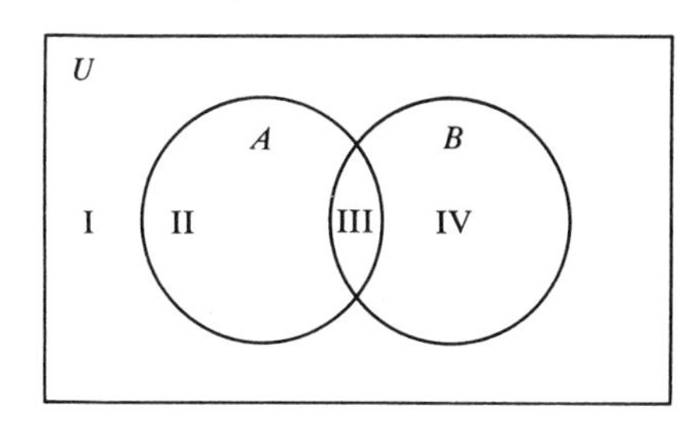

(a) A.
(b) $A \cup B$.
(c) $B \sim A$.
(d) A'.
(e) $(A \cap B)'$.
(f) $U \sim (B \cup A)$.
(g) $B \cap A'$.

3. If A and B are subsets of universal set U, draw Venn diagrams to show each of the following situations:
(a) A and B do not intersect.
(b) $A \subseteq B$.
(c) $B \cup A = B$.
(d) $B \subseteq A'$.
(e) $A \subseteq B$ and $B \subseteq A$.

4. (a) Make two copies of a Venn diagram showing any subsets A, B, and C of universe U. Verify the distributive property of intersection over union by shading the set $A \cap (B \cup C)$ in one diagram and the set $(A \cap B) \cup (A \cap C)$ in the other.
(b) In a similar manner verify the distributive property of union over intersection.

5. If set A consists of exactly five elements and B consists of exactly six elements, which of the following are true?
(a) $A \cap B$ has at most five elements.
(b) If $A \cap B$ has five elements, then $A \subset B$.
(c) If A and B have exactly two elements in common, then $A \cup B$ consists of exactly seven elements.
(d) If A and B are disjoint sets, then $A \cup B$ contains exactly eleven elements.
(e) If $B \sim A$ contains one element, then $A \cup B = A$.

6. X, Y, and Z are subsets of universe U. Draw a Venn diagram to represent the sets and state the relationship that exists between sets X and Z.
(a) $X \cap Y' = \emptyset$ and $Y \cup Z = Z$.
(b) $(X \cap Y) \subseteq Z'$ and $(X \cup Z) \subseteq Y$.

7. Use the distributive properties (viii) and (ix) to complete each of the following
(a) $X \cap (Y \cup Z) =$
(b) $A' \cup (B \cap C) =$
(c) $(X \cup Y) \cap (Z \cup Y) =$
(d) $(B \cap A') \cup (A' \cap C) =$
(e) $(X' \cap Y) \cup Z =$

8. Tell which of the statements (ii) through (xiv) is used to justify each of the following
(a) $A' \cap (A \cup B) = (A' \cap A) \cup (A' \cap B)$.
(b) $(A' \cap A) \cup (A' \cap B) = \emptyset \cup (A' \cap B)$.
(c) $\emptyset \cup (A' \cap B) = A' \cap B$.

9. Tell which of the statements (ii) through (xiv) is used to justify each of the following
(a) $(X' \cap Y) \cup X = X \cup (X' \cap Y)$.
(b) $X \cup (X' \cap Y) = (X \cup X') \cap (X \cup Y)$.
(c) $(X \cup X') \cap (X \cup Y) = U \cap (X \cup Y)$.
(d) $U \cap (X \cup Y) = X \cup Y$.

Operations with Statements

In operating with open statements we shall find that the words *conjunction, disjunction,* and *contradiction* correspond respectively to the words *intersection, union,* and *complement,* which we used to describe operations with sets.

Conjunction

If a and b represent statements, the compound sentence "a and b" is a *conjunctive sentence.* The statements a and b are called the *clauses* of this conjunctive sentence. Using the symbol $\wedge$ to represent the word "and," we can write a sentence that is the conjunction of a and b in the form $a \wedge b$.

Shown below is the truth table for the sentence $a \wedge b$. As the table indicates, a conjunctive sentence having two clauses is true when both of its clauses are true; otherwise it is false.

a	b	$a \wedge b$
T	T	T
T	F	F
F	T	F
F	F	F

A conjunctive sentence may, of course, have more than two clauses. In general, a conjunctive sentence is true if all of its clauses are true; otherwise it is false. Since this rule enables us to determine whether a conjunctive sentence is true or false, we can say that a conjunctive sentence is a statement.

We are often concerned with sentences whose clauses are open statements. Such sentences are called *open conjunctive statements.*

Suppose the universe for our discussion is the set of counting numbers, and we consider the truth set of the open conjunctive sentence $x > 2 \wedge x < 11$ for $x \in C$. The clause $x > 2$ has the truth set $\{3, 4, 5, \ldots\}$, and the clause $x < 11$ has the truth set $\{1, 2, 3, \ldots, 10\}$. We see that both of these clauses are true for those values of x and only those values which are common to these two truth sets. By definition, the set of elements common to two sets is the intersection of the two sets. Thus the truth set of the conjunctive statement $x > 2 \wedge x < 11$, where x is a counting number, is

$$\{3, 4, 5, \ldots\} \cap \{1, 2, 3, \ldots, 10\} = \{3, 4, 5, 6, 7, 8, 9, 10\}.$$

Thus

$$\{x \mid x > 2 \wedge x < 11\} = \{x \mid x > 2\} \cap \{x \mid x < 11\}.$$

In general, the truth set of an open conjunctive statement is the intersection of the truth sets of its clauses. Thus if a and b are open statements with truth sets A and B, respectively, we know that the truth set of the statement $a \wedge b$ is $A \cap B$.

Disjunction

If a and b represent statements, the compound sentence a or b is a *disjunctive sentence* with clauses a and b. Using the symbol $\vee$ to represent the word "or," we can write a sentence that is the disjunction of a and b in the form $a \vee b$.

Shown below is the truth table for the sentence $a \vee b$. As the table indicates, a disjunctive sentence having two clauses is false when both of its clauses are false; otherwise it is true. In general, a disjunctive sentence having two or more clauses is false when all of its clauses are false; otherwise it is true. Since this rule enables us to assign a truth value to any disjunctive sentence, we may now regard a disjunctive sentence as a statement.

a	b	$a \vee b$
T	T	T
T	F	T
F	T	T
F	F	F

Again using the counting numbers for our universe, we consider the truth set of the *open disjunctive statement* $x < 3 \vee x > 11$, where $x \in C$. The statement $x < 3$ has the truth set $\{1, 2\}$. The statement $x > 11$ has the truth set $\{12, 13, 14, \ldots\}$. Since a disjunctive statement is true when either clause is true or when both clauses are true, we see that our disjunctive statement is true for any number that can be found in either or both of the sets $\{1, 2\}$ and $\{12, 13, 14, \ldots\}$. By definition, the set of elements that belong to either or both of two sets is the union of the two sets. Thus the truth set of $x < 3 \vee x > 11$, where $x \in C$, is $\{1, 2\} \cup \{12, 13, 14, \ldots\}$. Thus

$$\{x \mid x < 3 \vee x > 11\} = \{x \mid x < 3\} \cup \{x \mid x > 11\}.$$

In general, the truth set of an open disjunctive sentence is the union of the truth sets of its clauses. That is, if a and b are open statements with truth sets A and B, respectively, the truth set of the statement $a \vee b$ is $A \cup B$.

We can verify by means of truth tables that the operations of conjunction and disjunction are commutative and associative and that there is a distributive property of each operation over the other. For a summary of these properties of disjunction and conjunction for open statements that correspond to set statements (ii) through (ix) see page 25.

Contradiction

Certain symbols convey the idea of contradiction. For example, we use $\neq$ for "is not equal to," $\not>$ for "is not greater than," and $\not\subset$ for "is not a proper subset of." Thus each of the following is a pair of contradictory statements:

$$a = b, \qquad a \neq b;$$
$$x > y, \qquad x \not> y;$$
$$S \subset T, \qquad S \not\subset T.$$

Two statements are contradictory statements if they have opposite truth values; that is, they are contradictory if each is true when and only when the other is false. If p is a statement, we denote a contradiction of p by the symbol $\sim p$. We read $\sim p$ as "not p," "p is false," or any statement that is equivalent. If two statements are contradictory, each is called a contradiction of the other. Two open statements having the same replacement set are said to be contradictory if their truth sets are complements of each other with respect to that set. Thus

(xv) a and b are contradictory statements $\longleftrightarrow$ A and B are complementary sets.

Alternate statements for (xv) are

$$(a \longleftrightarrow \sim b) \longleftrightarrow A = B',$$
$$(\sim a \longleftrightarrow b) \longleftrightarrow A' = B.$$

For example, let a and b represent the open statements $x^2 \geqq 9$ and $|x| < 3$, respectively, and let the replacement set for the statements be $\{-5, -4, -3, \ldots, 3, 4, 5\}$. Let us compare A and B, the truth sets of these two open statements. We see that $A = \{-5, -4, -3, 3, 4, 5\}$ and $B = \{-2, -1, 0, 1, 2\}$. Since the intersection of A and B is the empty set and their union is the replacement set, we conclude that A and B are complementary with respect to the replacement set. Accordingly, a and b are contradictory statements. Observe, too, that for any number we choose in the replacement set, a and b have opposite truth values.

A study of the table below shows that a contradiction of a contradiction of a statement is equivalent to the statement itself. Thus $\sim(\sim a) \longleftrightarrow a$. The table also indicates that two statements that are contradictions of the same statement are equivalent to each other. When a is an open statement with truth set A, the statement $\sim(\sim a) \longleftrightarrow a$ corresponds to set statement (xi), which asserts that $(A')' = A$.

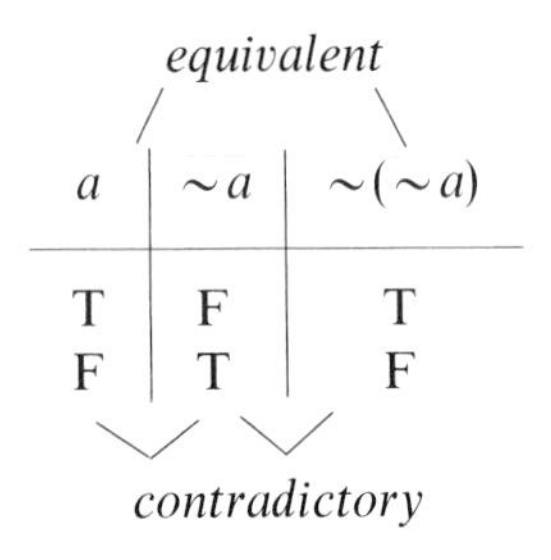

a	$\sim a$	$\sim(\sim a)$
T	F	T
F	T	F

It is also easily shown by a truth table that contradictions of equivalent statements are equivalent. Thus $(a \longleftrightarrow b) \longleftrightarrow (\sim a \longleftrightarrow \sim b)$. This statement is the "statement counterpart" of set statement (xii), $A = B \longleftrightarrow A' = B'$.

Exercises

1. If a and b represent statements that are true and p and q represent statements that are false, decide whether the following statements are true or false.
   (a) $a \vee p$. (b) $\sim a \vee q$.

(c) $\sim(a \wedge b)$. (d) $\sim a \wedge \sim q$.
(e) $a \vee \sim b$. (f) $\sim p \wedge \sim q$.

2. (a) If $a \wedge b$ is true, can you conclude that a is true? that b is true?
   (b) If $a \wedge b$ is false and a is true, what can be said of b?
   (c) If $a \wedge b$ is false and a is false, is b false?
   (d) If $a \vee b$ is true and b is true, is a true?
   (e) If $a \vee b$ is true and b is false, is a true?
3. Given statements a, b, c, and d such that $a \longleftrightarrow b$, $c \longleftrightarrow \sim a$, and $d \longleftrightarrow \sim b$.
   (a) Complete the truth table to show the truth values for b, c, and d.
   (b) What conclusions can you draw concerning c and d? b and c? d and a?

a	b	c	d
T			
F			

4. Decide whether or not the following pairs of statements are contradictory and explain your answers.
   (a) All complex numbers are real numbers.
   No complex numbers are real numbers.
   (b) Some students attend the class play.
   Some students do not attend the class play.
   (c) All undergraduates study Shakespeare.
   Some undergraduates do not study Shakespeare.
   (d) Lucy lives in Michigan.
   Lucy lives in California.
5. Construct an eight-line truth table to show $[(a \wedge b) \wedge c] \longleftrightarrow [a \wedge (b \wedge c)]$.
6. Construct an eight-line truth table to show $[a \wedge (b \vee c)] \longleftrightarrow [(a \wedge b) \vee (a \wedge c)]$.

Conditional Sentences—Implications

We have considered sentences formed by joining statements with the connective *and* and sentences formed by joining statements with the connective *or*. Now we consider sentences formed by combining statements with the words *if* and *then*. For example, given the two statements

p: I live in Michigan
q: I live in the United States

we can form the sentence: "*If* I live in Michigan, *then* I live in the United States." Observe that this sentence has the form: "If p, then q." To write "if p, then q," we often write $p \longrightarrow q$.

When p and q are statements, a sentence of the form "if p, then q" is called a *conditional sentence* or a *conditional*. The statements p and q are called the *terms* of the conditional. The statement p is called the *hypothesis* of the conditional, and the statement q is called the *conclusion* of the conditional. Observe that the word "if" is not a part of the hypothesis and the word "then" is not a part of the conclusion. The words "if" and "then" are merely connectives, which we combine with the hypothesis and conclusion to form a conditional sentence. In the conditional "If I live in Michigan, then I live in the United States" the hypothesis is "I live in Michigan" and the conclusion is "I live in the United States."

The conditional sentence "if p, then q" is a statement if we can find a rule by which we can always determine whether the sentence is true or false. Since we want our rule to be as consistent as possible with common usage, let us consider the sentence "If I have a driver's license, then I am at least 16 years old." We see that

1. If I have a driver's license *and* I am at least 16 years old, this conditional should be labeled true.
2. If I have a driver's license *and* I am not at least 16 years old, this conditional should be labeled false.

One might argue that if I do not have a driver's license, then the conditional sentence does not apply in my case. However, in order for a conditional sentence to be a statement, we must be able to assign truth values for all cases, even those for which the hypothesis is false. We adopt the following definition: The sentence "if p, then q" is false when p is true and q is false; otherwise it is true.

By using this definition we can always determine whether a given conditional is true or false. Thus we can regard a conditional sentence as a statement. The truth table below summarizes the above discussion.

p	q	$p \longrightarrow q$
T	T	T
T	F	F
F	T	T
F	F	T

In mathematics the conditional statement "if p, then q" is often replaced with "p implies q." For this reason a conditional statement is also called an *implication.* Thus each of the statements "if p, then q" and "$p \longrightarrow q$" is equivalent to the implication "p implies q." In our study of if-then statements we shall usually use the word "implication" instead of the word "conditional."

Observe that the implication $3x > 12 \longrightarrow x > 4$ is true for every real number x. *If p and q are open statements, the implication $p \longrightarrow q$ is true when and only when it is true for every member of its replacement set.*

There is an easy "set interpretation" of the sentence $a \longrightarrow b$, where a and b are open statements with truth sets A and B, respectively. Consider the implication

$$\underbrace{\text{point } X \text{ is in La Grange}}_{a} \longrightarrow \underbrace{\text{point } X \text{ is in Cook County}}_{b}$$

Clearly A is the set of points in La Grange and B is the set of points in Cook County. Our implication is true if A is a subset of B as shown in the left-hand drawing below and false if A is not a subset of B as shown in the right-hand drawing.

In general, when a and b are open statements with truth sets A and B we have the following important equivalence:

(xvi) $$(a \longrightarrow b) \longleftrightarrow A \subseteq B.$$

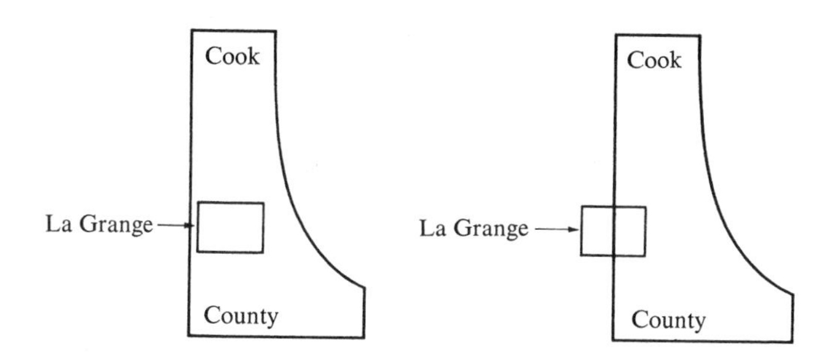

We have stated that the implication $a \longrightarrow b$ is false when a is true and b is false and that the implication is true for other truth values of a and b. The disjunction $\sim a \vee b$ is another statement that is false when a is true and b is false and true for other truth values of a and b. This indicates the following equivalence:

$$\underline{(a \longrightarrow b) \longleftrightarrow (\sim a \vee b).}$$

This equivalence may, of course, also be verified by means of a truth table.

In the diagrams below, the set A' is shaded with horizontal markings and the set B is shaded with vertical markings. These drawings suggest the following equivalence:

$$\underline{A \subseteq B \longleftrightarrow A' \cup B = U.}$$

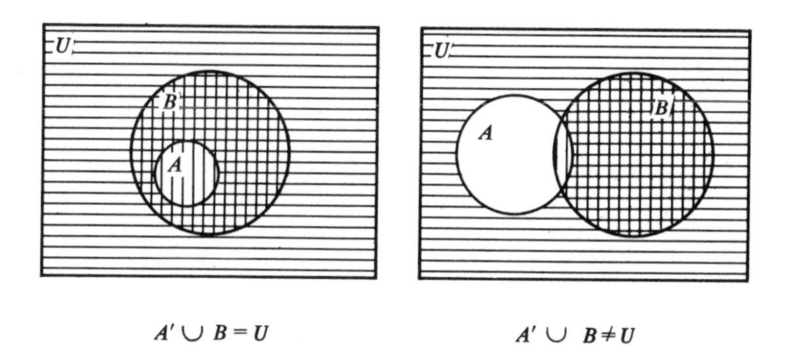

$A' \cup B = U$ $\qquad$ $A' \cup B \neq U$

From (xvi), the two underlined statements, and the transitive property of equivalence we have

(xvii) $$(a \longrightarrow b) \longleftrightarrow A \subseteq B \longleftrightarrow A' \cup B = U \longleftrightarrow (\sim a \vee b).$$

Variation in Stating an Implication

In making an if-then statement we frequently omit the word "then." For example, to say "If I own property, then I pay taxes" we may say "If I own property, I pay taxes."

The hypothesis of an implication need not be the first term. For example, instead of saying "if $x > 3$, then $x^2 > 9$," we may say "$x^2 > 9$ if $x > 3$." We may also say "$x^2 > 9$ provided $x > 3$."

The following are several verbal statements that can be regarded as equivalent to the statement "if p, then q."

p implies q.	p only if q.	p is sufficient for q.
q if p.	q provided p.	q is necessary for p.

Sometimes implications are so disguised that it is difficult to recognize them as implications. The column on the left below gives examples of implications that are somewhat disguised. Each is expressed in an equivalent if-then form in the column on the right.

All horses are animals.	If x is a horse, then x is an animal.
I shall make a B in this course only if I make at least a C on the final.	If I make a B in this course, then I shall have made at least a C on the final. *or* If I do not score at least a C on the final, then I shall not make a B in the course.
No insect is a mammal.	If x is an insect, then x is not a mammal. *or* If x is a mammal, then x is not an insect.
I will make the dean's list provided I get a B in geometry.	If I get a B in geometry, then I will make the dean's list.

More about Implication and Equivalence

By rearranging, contradicting, or rearranging and contradicting the hypothesis and conclusion of a given implication, we can obtain three other implications. These are shown below along with the name assigned to each.

Given implication:	If a then b	$a \longrightarrow b$
Converse:	If b then a	$b \longrightarrow a$
Inverse:	If $\sim a$ then $\sim b$	$\sim a \longrightarrow \sim b$
Contrapositive:	If $\sim b$ then $\sim a$	$\sim b \longrightarrow \sim a$

The table below has a column devoted to each of these implications.

a	b	$a \longrightarrow b$	$b \longrightarrow a$	$\sim a$	$\sim b$	$\sim a \longrightarrow \sim b$	$\sim b \longrightarrow \sim a$
T	T	T	T	F	F	T	T
T	F	F	T	F	T	T	F
F	T	T	F	T	F	F	T
F	F	T	T	T	T	T	T

converse ⟷ *inverse*

given implication ⟷ *contrapositive*

From our table we see that the contrapositive of a given implication is equivalent to the implication.

$$(a \longrightarrow b) \longleftrightarrow (\sim b \longrightarrow \sim a).$$

The table also tells us that the inverse of an implication is equivalent to the converse of the implication.

The equivalence of an implication and its contrapositive can also be obtained by considering open statements a and b whose truth sets are A and B, respectively. From a study of the Venn diagram below we see that if a point is in A, then it is in B, and if a point is not in B then it is not in A. Thus from the diagram we note that

$$\underline{A \subseteq B \longleftrightarrow B' \subseteq A'.}$$

Observe that the truth sets of $\sim b$ and $\sim a$ are B' and A', respectively.

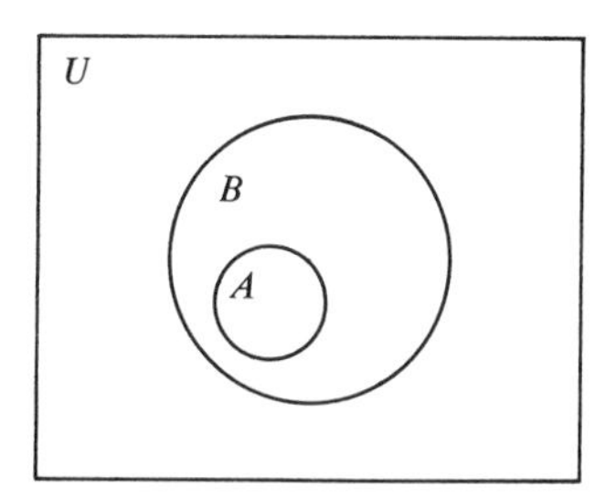

We obtain from (xvi) the following equivalences:

$$\underline{(a \longrightarrow b) \longleftrightarrow A \subseteq B} \qquad \text{and} \qquad \underline{(\sim b \longrightarrow \sim a) \longleftrightarrow B' \subseteq A'.}$$

From the underlined statements we have

(xviii) $$(a \longrightarrow b) \longleftrightarrow A \subseteq B \longleftrightarrow B' \subseteq A' \longleftrightarrow (\sim b \longrightarrow \sim a).$$

Applying the transitive property of equivalence, we see that

$$(a \longrightarrow b) \longleftrightarrow (\sim b \longrightarrow \sim a).$$

It is important to note that an implication is not necessarily equivalent to its converse. For example, the implication "If a person lives in Iowa, then he lives in the United States" is true because it is true for all people, while its converse "If a person lives in the United States, then he lives in Iowa" is false because it is false for some people. As a second example consider the implication $x > 3 \longrightarrow x^2 > 9$. This implication is true because it is true for all real numbers. However, the converse, $x^2 > 9 \longrightarrow x > 3$, is false because it is false for some real numbers. For example, it is false for $x = -5$, because for this value of x the hypothesis $x^2 > 9$ is true while the conclusion $x > 3$ is false. Since the truth sets for the statements $(x > 3 \longrightarrow x^2 > 9)$ and $(x^2 > 9 \longrightarrow x > 3)$ are not equal, we conclude that these statements are not equivalent.

Having noted that an implication $a \longrightarrow b$ and its converse $b \longrightarrow a$ are not equivalent, let us investigate the relation that exists between two statements a and b for which both $a \longrightarrow b$ and $b \longrightarrow a$ are true. The table below shows that

$$(a \longleftrightarrow b) \longleftrightarrow [(a \longrightarrow b) \wedge (b \longrightarrow a)].$$

a	b	$a \longleftrightarrow b$	$a \longrightarrow b$	$b \longrightarrow a$	$(a \longrightarrow b) \wedge (b \longrightarrow a)$
T	T	T	T	T	T
T	F	F	F	T	F
F	T	F	T	F	F
F	F	T	T	T	T

same truth values (columns $a \longleftrightarrow b$ and $(a \longrightarrow b) \wedge (b \longrightarrow a)$)

The statement $a \longleftrightarrow b$ is therefore equivalent to the conjunction of two implications, $b \longrightarrow a$ and $a \longrightarrow b$. The first of these implications can be written *a if b* and the second *a only if b*. The conjunction of these two statements is *a if b and a only if b*. This conjunctive statement is usually written *a if and only if b*. Sometimes the symbol "iff" is used for the phrase "if and only if." Thus we have

$$(a \longleftrightarrow b) \longleftrightarrow (a \text{ and } b \text{ are equivalent}) \longleftrightarrow a \text{ if and only if } b \leftrightarrow$$
$$\leftrightarrow b \text{ if and only if } a \longleftrightarrow a \text{ iff } b \longleftrightarrow b \text{ iff } a.$$

The statement $(a \longleftrightarrow b) \longleftrightarrow [(a \longrightarrow b) \wedge (b \longrightarrow a)]$, which was verified in the truth table, can also be obtained by considering A and B, the truth sets of statements a and b, respectively.

(xix) $(a \longleftrightarrow b) \longleftrightarrow A = B \longleftrightarrow (A \subseteq B \wedge B \subseteq A) \longleftrightarrow [(a \longrightarrow b) \wedge (b \longrightarrow a)].$

Using the transitive property of equivalence, we have

$$(a \longleftrightarrow b) \longleftrightarrow [(a \longrightarrow b) \wedge (b \longrightarrow a)].$$

We have previously described a definition as an assertion that two sets are equal. For example, the definition "A counting number is a nonzero whole number" is interpreted to mean $C = W \sim \{0\}$. It is now meaningful to restate this definition in "if-and-only-if" form as follows:

DEFINITION *A number is a counting number if and only if it is a nonzero whole number.*

Written in this form, the definition is clearly equivalent to the conjunction of an implication and its converse. We refer to this property of definitions when we say that "definitions are reversible." For convenience, some definitions will be stated in the if-then form. However, when an if-then statement is labeled "definition," we know that both the statement and its converse are true.

Exercises

1. Write each of the following statements in if-then form.
   (a) Every real number is a complex number.
   (b) A citizen of Chicago is a citizen of Illinois.
   (c) $x < 7$ provided $x + 1 < 8$.
   (d) No sophomore may attend the junior class meetings.

(e) A valid registration certificate is necessary for voting.
(f) When $a \neq b$, $a - b \neq 0$.
(g) $a \in X \cap Y$ provided $a \in X$ and $a \in Y$.
(h) Absence from two meetings is sufficient for dismissal.

2. Write the converse and contrapositive of each of the folowing statements in if-then form.
   (a) If $x > 4$, then $x > 3$.
   (b) If a number is an integer, then it is rational.
   (c) If a statement is true, then its contradiction is false.
   (d) If $A \subseteq B$, then $B' \subseteq A'$.
   (e) $x = 8$ provided $2x + 5 = 21$.
3. Which of the statements in Exercise 2 have true converses? Which have true contrapositives?
4. Which of the statements in Exercise 2 could be written as true "if-and-only-if" statements?
5. Write the converse, inverse, and contrapositive of each of the following implications.
   (a) $p \longrightarrow \sim q$.
   (b) $\sim p \longrightarrow \sim q$.
   (c) $\sim p \longrightarrow q$.
6. Write a disjunction that is equivalent to each of the implications in Exercise 5.
7. Write an implication that is equivalent to each of the following.
   (a) $\sim a \vee b$.
   (b) $\sim a \vee \sim b$.
   (c) $a \vee b$.
   (d) $a \vee \sim b$.
8. We have $A \subseteq B \longleftrightarrow A' \cup B = U$. Use this statement to write a statement equivalent to each of the following. [*Example:* $X' \subseteq Y \longleftrightarrow (X')' \cup Y = U \longleftrightarrow X \cup Y = U$.]
   (a) $X \subseteq Y' \longleftrightarrow$ ____.
   (b) $X' \subseteq Y' \longleftrightarrow$ ____.
   (c) $A' \cup B' = U \longleftrightarrow$ ____.
   (d) $A \cup B' = U \longleftrightarrow$ ____.
9. Supply a reason that justifies each of the following equivalences. You may use any of the statements (ii) through (xvii) on page 25 as reasons.
   (a) $X \subseteq Y \longleftrightarrow X' \cup Y = U$.
   (b) $X' \cup Y = U \longleftrightarrow Y \cup X' = U$.
   (c) $Y \cup X' = U \longleftrightarrow (Y')' \cup X' = U$.
   (d) $(Y')' \cup X' = U \longleftrightarrow Y' \subseteq X'$.
   (e) Which property of equivalence allows us to conclude that $X \subseteq Y \longleftrightarrow Y' \subseteq X'$?

Contradicting Conjunctions, Disjunctions, and Implications

The truth table below reveals that

$\sim(a \wedge b) \longleftrightarrow \sim a \vee \sim b$ (columns 6 and 7), [De Morgan's Law]
$\sim(a \vee b) \longleftrightarrow \sim a \wedge \sim b$ (columns 9 and 10). [De Morgan's Law]

a	b	$\sim a$	$\sim b$	$a \wedge b$	$\sim(a \wedge b)$	$\sim a \vee \sim b$	$a \vee b$	$\sim(a \vee b)$	$\sim a \wedge \sim b$
T	T	F	F	T	F	F	T	F	F
T	F	F	T	F	T	T	T	F	F
F	T	T	F	F	T	T	T	F	F
F	F	T	T	F	T	T	F	T	T
					↑ same	↑ truth values		↑ same	↑ truth values

EXAMPLE 1

A contradiction of the conjunctive statement "Today is Monday and the temperature is above forty degrees Fahrenheit" is "Today is not Monday or the temperature is not above forty degrees Fahrenheit."

EXAMPLE 2

A contradiction of the disjunctive statement "$x > 5 \vee x = 5$" is "$x \not> 5 \wedge x \neq 5$."

We can discover two important properties of sets by considering two open statements a and b whose truth sets are A and B, respectively. The statements $\sim a$ and $\sim b$ have truth sets A' and B', respectively. We have seen that the truth set of the conjunctive statement $a \wedge b$ is $A \cap B$ and the truth set of $\sim(a \wedge b)$ is $(A \cap B)'$. The truth set of $\sim a \vee \sim b$ is $A' \cup B'$. Since $\sim(a \wedge b) \longleftrightarrow \sim a \vee \sim b$, we conclude that $(A \cap B)' = A' \cup B'$. Thus we have the following corresponding statements:

(xx) $\qquad (A \cap B)' = A' \cup B', \qquad \sim(a \wedge b) \longleftrightarrow \sim a \vee \sim b.$

By a similar analysis we have

(xxi) $\qquad (A \cup B)' = A' \cap B', \qquad \sim(a \vee b) \longleftrightarrow \sim a \wedge \sim b.$

The truth table below indicates that

$$\sim(a \longrightarrow b) \longleftrightarrow (a \wedge \sim b).$$

a	b	$a \longrightarrow b$	$\sim(a \longrightarrow b)$	$\sim b$	$a \wedge \sim b$	$\sim(a \longrightarrow b) \longleftrightarrow (a \wedge \sim b)$
T	T	T	F	F	F	T
T	F	F	T	T	T	T
F	T	T	F	F	F	T
F	F	T	F	T	F	T

We can also obtain this equivalence from (xvii) by applying the fact that contradictions of equivalent statements are equivalent. Thus

$$[(a \longrightarrow b) \longleftrightarrow (\sim a \vee b)] \longleftrightarrow [\sim(a \longrightarrow b) \longleftrightarrow \sim(\sim a \vee b)] \leftrightarrow$$
$$\leftrightarrow [\sim(a \longrightarrow b) \longleftrightarrow \sim(\sim a) \wedge \sim b] \longleftrightarrow [\sim(a \longrightarrow b) \longleftrightarrow a \wedge \sim b].$$

EXAMPLE 3

Write a contradiction of the statement "If Harrington was late to work, then he missed the train."

SOLUTION

Let a represent the hypothesis "Harrington was late to work" and let b represent the conclusion "Harrington missed the train." Since a contradiction of "if a, then b" is "$a \wedge \sim b$," the contradiction we are seeking is "Harrington was late to work and did not miss the train."

Implications with Conjunctive Hypotheses

Each of the terms p and q of the implication $p \longrightarrow q$ may be a conjunctive statement. Consider the statement "If $a \not> b$ and $a \not< b$, then $a = b$." In this case the hypothesis is the conjunctive sentence $a \not> b$ and $a \not< b$. This implication can be written in the bracket form as follows:

$$\left.\begin{array}{r} a \not> b \\ a \not< b \end{array}\right\} \longrightarrow a = b.$$

When we write an implication in this form, we mean $(a \not> b \wedge a \not< b) \longrightarrow a = b$. Note that we do not write the "and." It is understood when the bracket form is used.

The statement: "I can reach Springfield by noon if (1) I start before 8:00 a.m., (2) I average 50 miles an hour, and (3) I do not have an accident" has the form $(a \wedge b \wedge c) \longrightarrow x$. We sometimes write such an implication in the form

$$\left.\begin{array}{r} a \\ b \\ c \end{array}\right\} \longrightarrow x.$$

This implication has a converse, which is written $x \longrightarrow (a \wedge b \wedge c)$ or

$$x \longrightarrow \left\{\begin{array}{l} a \\ b. \\ c \end{array}\right.$$

We shall seldom, if ever, be interested in this converse, which might be described as the *complete converse*. We shall often, however, be interested in forms known as *partial converses*. The implication $(a \wedge b \wedge c) \longrightarrow x$ has the partial converses $(x \wedge b \wedge c) \longrightarrow a$, $(a \wedge x \wedge c) \longrightarrow b$, and $(a \wedge b \wedge x) \longrightarrow c$. Any statement in the hypothesis of an implication is called a *premise*. If the conclusion of an implication is a single statement and if the hypothesis of the implication contains n premises, then any one of the n implications obtained by exchanging one of the premises with the conclusion is a *partial converse* of the original implication. Hereafter we shall drop the words partial and complete and refer to both partial and complete converses as converses.

Let us now consider the contrapositives of the implication $(a \wedge b \wedge c) \longrightarrow x$. This implication has a *complete contrapositive*, which is written as $\sim x \longrightarrow \sim (a \wedge b \wedge c)$. We shall have little use for this form. We shall, however, find the following *partial contrapositives* very useful:

$$\left.\begin{array}{r} \sim x \\ b \\ c \end{array}\right\} \longrightarrow \sim a, \qquad \left.\begin{array}{r} a \\ \sim x \\ c \end{array}\right\} \longrightarrow \sim b, \qquad \left.\begin{array}{r} a \\ b \\ \sim x \end{array}\right\} \longrightarrow \sim c.$$

If the conclusion of an implication is a single statement and if the hypothesis of the implication contains n premises, then any one of the n implications obtained by interchanging a contradiction of one of the premises with a contradiction of the conclusion is a partial contrapositive of the original implication. Hereafter, we shall make no distinction between complete and partial contrapositives. We shall refer to both as contrapositives. It is important to note that each of the contrapositives of a given implication is equivalent to the original implication. The truth table below verifies the equivalence of one of the contrapositives and the original implication for the case where there are two premises in the hypothesis.

a	b	x	$a \wedge b$	$(a \wedge b) \longrightarrow x$	$\sim x$	$\sim a$	$(\sim x \wedge b)$	$(\sim x \wedge b) \longrightarrow \sim a$
T	T	T	T	T	F	F	F	T
T	T	F	T	F	T	F	T	F
T	F	T	F	T	F	F	F	T
T	F	F	F	T	T	F	F	T
F	T	T	F	T	F	T	F	T
F	T	F	F	T	T	T	T	T
F	F	T	F	T	F	T	F	T
F	F	F	F	T	T	T	F	T

↑ *same truth values* ↑ (columns $(a \wedge b) \longrightarrow x$ and $(\sim x \wedge b) \longrightarrow \sim a$)

Exercises

1. Let $U = \{1, 2, 3, \ldots, 10\}$. Choose any two subsets A and B of U and show that
   (a) $(A \cap B)' = A' \cup B'$. (b) $(A \cup B)' = A' \cap B'$.
2. Write a contradiction for each of the following.
   (a) $a \wedge \sim b$. (b) $\sim p \vee \sim q$.
   (c) $p \longrightarrow q$. (d) $\sim (p \longrightarrow \sim q)$.
   (e) $p \vee \sim q$. (f) $p \longrightarrow \sim q$.
3. Given the statement $(a \wedge b \wedge c) \longrightarrow x$, determine whether each of the following is a *converse, contrapositive,* or *neither* a converse nor contrapositive of the given statement.
   (a) $(x \wedge b \wedge c) \longrightarrow a$. (b) $(a \wedge \sim x \wedge c) \longrightarrow \sim b$.
   (c) $(x \wedge b \wedge c) \longrightarrow \sim a$. (d) $\left.\begin{matrix} a \\ \sim x \\ c \end{matrix}\right\} \longrightarrow b$.
   (e) $\left.\begin{matrix} a \\ b \\ x \end{matrix}\right\} \longrightarrow c$. (f) $\left.\begin{matrix} a \\ b \\ \sim c \end{matrix}\right\} \longrightarrow \sim x$.
4. Write two converses and two contrapositives for each of the following statements.
   (a) If $a > b$ and $c > 0$, then $ac > bc$.
   (b) If $n = 2k + 1$ and k is an integer, then n is an odd number.
   (c) If $A \subseteq B$ and $A \neq B$, then $A \subset B$.
5. Which of the following statements are equivalent to $(a \wedge b) \longrightarrow c$?
   (a) $c \longrightarrow (a \wedge b)$. (b) $(a \wedge \sim c) \longrightarrow \sim b$.
   (c) $\sim c \longrightarrow (\sim a \vee \sim b)$. (d) $(a \wedge \sim b) \longrightarrow \sim c$.
6. Supply reasons for the following equivalences. You may select supporting reasons from those given on page 25.
   (a) $(b \longrightarrow a) \longleftrightarrow (\sim b \vee a)$. (b) $(\sim b \vee a) \longleftrightarrow (a \vee \sim b)$.

(c) $(a \vee \sim b) \longleftrightarrow [\sim(\sim a) \vee \sim b]$. (d) $[\sim(\sim a) \vee \sim b] \longleftrightarrow \sim(\sim a \wedge b)$.
(e) Which property of equivalence allows us to conclude that $(b \longrightarrow a) \longleftrightarrow \sim(\sim a \wedge b)$?

7. Supply reasons for the following equivalences. You may select supporting reasons from those given below. [For (b) supply two reasons.]
(a) $[(a \wedge \sim b) \longrightarrow \sim c] \longleftrightarrow [\sim(a \wedge \sim b) \vee \sim c]$.
(b) $[\sim(a \wedge \sim b) \vee \sim c] \longleftrightarrow [(\sim a \vee b) \vee \sim c]$.
(c) $[(\sim a \vee b) \vee \sim c] \longleftrightarrow [\sim a \vee (b \vee \sim c)]$.
(d) $[\sim a \vee (b \vee \sim c)] \longleftrightarrow [\sim a \vee (\sim c \vee b)]$.
(e) $[\sim a \vee (\sim c \vee b)] \longleftrightarrow [(\sim a \vee \sim c) \vee b]$.
(f) $[(\sim a \vee \sim c) \vee b] \longleftrightarrow [\sim(a \wedge c) \vee b]$.
(g) $[\sim(a \wedge c) \vee b] \longleftrightarrow [(a \wedge c) \longrightarrow b]$.
(h) $[(a \wedge \sim b) \longrightarrow \sim c] \longleftrightarrow [(a \wedge c) \longrightarrow b]$.

Statements and Their Truth Sets

In the preceding sections we have repeatedly observed that for every relationship involving open statements there is a corresponding relationship involving their truth sets. In the following summary of these corresponding relationships all sets are subsets of the universe U. In order to make certain comparisons we introduce the statement o, whose truth set is $\emptyset$, and the statement u, whose truth set is U. This summary will facilitate reference to these important relationships.

(i) $A = B$	(i′) $a \longleftrightarrow b$
(ii*a*) $A \cap B = B \cap A$	(ii′*a*) $a \wedge b \longleftrightarrow b \wedge a$
(ii*b*) $A \cap A = A$	(ii′*b*) $a \wedge a \longleftrightarrow a$
(iii) $A \cap B \cap C = (A \cap B) \cap C$	(iii′) $a \wedge b \wedge c \longleftrightarrow (a \wedge b) \wedge c$
(iv) $(A \cap B) \cap C = A \cap (B \cap C)$	(iv′) $(a \wedge b) \wedge c \longleftrightarrow a \wedge (b \wedge c)$
(v*a*) $A \cup B = B \cup A$	(v′*a*) $a \vee b \longleftrightarrow b \vee a$
(v*b*) $A \cup A = A$	(v′*b*) $a \vee a \longleftrightarrow a$
(vi) $A \cup B \cup C = (A \cup B) \cup C$	(vi′) $a \vee b \vee c \longleftrightarrow (a \vee b) \vee c$
(vii) $(A \cup B) \cup C = A \cup (B \cup C)$	(vii′) $(a \vee b) \vee c \longleftrightarrow a \vee (b \vee c)$
(viii) $A \cap (B \cup C) = (A \cap B) \cup (A \cap C)$	(viii′) $a \wedge (b \vee c) \longleftrightarrow (a \wedge b) \vee (a \wedge c)$
(ix) $A \cup (B \cap C) = (A \cup B) \cap (A \cup C)$	(ix′) $a \vee (b \wedge c) \longleftrightarrow (a \vee b) \wedge (a \vee c)$
(x*a*) $A \cap A' = \emptyset$	(x′*a*) $a \wedge \sim a \longleftrightarrow o$
(x*b*) $A \cup A' = U$	(x′*b*) $a \vee \sim a \longleftrightarrow u$
(xi) $(A')' = A$	(xi′) $\sim(\sim a) \longleftrightarrow a$
(xii) $A = B \longleftrightarrow A' = B'$	(xii′) $(a \longleftrightarrow b) \longleftrightarrow (\sim a \longleftrightarrow \sim b)$
(xiii*a*) $\emptyset' = U$	(xiii′*a*) $\sim o \longleftrightarrow u$
(xiii*b*) $U' = \emptyset$	(xiii′*b*) $\sim u \longleftrightarrow o$
(xiv*a*) $U \cap A = A$	(xiv′*a*) $u \wedge a \longleftrightarrow a$
(xiv*b*) $U \cup A = U$	(xiv′*b*) $u \vee a \longleftrightarrow u$
(xiv*c*) $\emptyset \cap A = \emptyset$	(xiv′*c*) $o \wedge a \longleftrightarrow o$
(xiv*d*) $\emptyset \cup A = A$	(xiv′*d*) $o \vee a \longleftrightarrow a$
(xv) A and B are complementary sets	(xv′) a and b are contradictory statements
(xvi) $A \subseteq B$	(xvi′) $a \longrightarrow b$
(xvii) $A \subseteq B \longleftrightarrow A' \cup B = U$	(xvii′) $(a \longrightarrow b) \longleftrightarrow \sim a \vee b$
(xviii) $A \subseteq B \longleftrightarrow B' \subseteq A'$	(xviii′) $(a \longrightarrow b) \longleftrightarrow (\sim b \longrightarrow \sim a)$
(xix) $A = B \longleftrightarrow (A \subseteq B \wedge B \subseteq A)$	(xix′) $(a \longleftrightarrow b) \longleftrightarrow [(a \longrightarrow b) \wedge (b \longrightarrow a)]$
(xx) $(A \cap B)' = A' \cup B'$	(xx′) $\sim(a \wedge b) \longleftrightarrow \sim a \vee \sim b$
(xxi) $(A \cup B)' = A' \cap B'$	(xxi′) $\sim(a \vee b) \longleftrightarrow \sim a \wedge \sim b$

We have also observed that the equality relationship for sets and the equivalence relationship for statements are reflexive, symmetric, and transitive and that each has a substitution property.

Exercises

In the following exercises supply reasons for statements of equality of sets or equivalence of statements from the summary on page 25. The numbers written above the symbols $=$ or $\longleftrightarrow$ are for your convenience in supplying reasons. Your reason number (1) should correspond to the equality or equivalence numbered (1), and so on.

1. $(A \cup B) \cup A' \overset{(1)}{=} (B \cup A) \cup A' \overset{(2)}{=} B \cup (A \cup A') \overset{(3)}{=} B \cup U \overset{(4)}{=} U \cup B \overset{(5)}{=} U.$

2. $B \cup (B' \cap A) \overset{(1)}{=} (B \cup B') \cap (B \cup A) \overset{(2)}{=} U \cap (B \cup A) \overset{(3)}{=} B \cup A.$

3. $(A \cap B) \cup (B \cap A') \overset{(1)}{=} (B \cap A) \cup (B \cap A') \overset{(2)}{=} B \cap (A \cup A') \overset{(3)}{=} B \cap U \overset{(4)}{=} B.$

4. $(A \cap B')' \cap A \overset{(1)}{=} [A' \cup (B')'] \cap A \overset{(2)}{=} (A' \cup B) \cap A \overset{(3)}{=} A \cap (A' \cup B)$

   $\overset{(4)}{=} (A \cap A') \cup (A \cap B) \overset{(5)}{=} \emptyset \cup (A \cap B) \overset{(6)}{=} A \cap B.$

5. $(A' \cap B)' \cup A' \overset{(1)}{=} [(A' \cap B) \cap A]' \overset{(2)}{=} [A \cap (A' \cap B)]' \overset{(3)}{=} [(A \cap A') \cap B]'$

   $\overset{(4)}{=} (\emptyset \cap B)' \overset{(5)}{=} \emptyset' \overset{(6)}{=} U.$

In the manner illustrated in Exercises 1 through 5, write simpler statements for each of the following.

6. $(A \cup B) \cap B'.$
7. $(A' \cap B) \cup A.$
8. $(A \cap B)' \cup B.$
9. $A' \cap (A \cup B')'.$
10. $A' \cap (A \cup B)'.$
11. $[A' \cap (A \cap B)]'.$

Supply reasons for the following equivalences.

12. $\sim(a \wedge b \wedge \sim c) \xleftrightarrow{(1)} \sim[(a \wedge b) \wedge \sim c] \xleftrightarrow{(2)} [\sim(a \wedge b) \vee \sim(\sim c)] \overset{(3)}{\leftrightarrow}$

    $\leftrightarrow [\sim(a \wedge b) \vee c] \xleftrightarrow{(4)} [(a \wedge b) \longrightarrow c].$

13. $[(a \wedge b) \vee \sim a] \xleftrightarrow{(1)} [\sim a \vee (a \wedge b)] \xleftrightarrow{(2)} [(\sim a \vee a) \wedge (\sim a \vee b)] \overset{(3)}{\leftrightarrow}$

    $\leftrightarrow [u \wedge (\sim a \vee b)] \xleftrightarrow{(4)} (\sim a \vee b).$

14. $[(a \vee b) \wedge (a \vee \sim b)] \xleftrightarrow{(1)} [a \vee (b \wedge \sim b)] \xleftrightarrow{(2)} (a \vee o) \xleftrightarrow{(3)} a.$

15. $[\sim a \vee \sim(a \wedge b)] \xleftrightarrow{(1)} \sim[a \wedge (a \wedge b)] \xleftrightarrow{(2)} \sim[(a \wedge a) \wedge b] \xleftrightarrow{(3)} \sim(a \wedge b) \overset{(4)}{\leftrightarrow}$

    $\leftrightarrow (\sim a \vee \sim b) \xleftrightarrow{(5)} (a \longrightarrow \sim b).$

16. $[(a \wedge b) \longrightarrow a] \xleftrightarrow{(1)} [\sim(a \wedge b) \vee a] \xleftrightarrow{(2)} [(\sim a \vee \sim b) \vee a] \xleftrightarrow{(3)} [a \vee (\sim a \vee \sim b)] \overset{(4)}{\leftrightarrow}$

    $\leftrightarrow [(a \vee \sim a) \vee \sim b] \xleftrightarrow{(5)} (u \vee \sim b) \xleftrightarrow{(6)} u.$

17. $[(a \vee b) \wedge \sim b] \longrightarrow a \xleftrightarrow{(1)} \sim[(a \vee b) \wedge \sim b] \vee a \xleftrightarrow{(2)} [\sim(a \vee b) \vee \sim(\sim b)] \vee a \overset{(3)}{\leftrightarrow}$

    $\leftrightarrow [\sim(a \vee b) \vee b] \vee a \xleftrightarrow{(4)} \sim(a \vee b) \vee (b \vee a) \xleftrightarrow{(5)} (a \vee b) \vee \sim(a \vee b) \xleftrightarrow{(6)} u.$

References

Adler, *Modern Geometry*
Courant and Robbins, *What Is Mathematics?*
Golos, *Foundations of Euclidean and Non-Euclidean Geometry*
Hemmerling, *Fundamentals of College Geometry*
Katz, *Axiomatic Analysis*

Chapter 2
Reasoning and Proof

Each of us has acquired a fund of geometric knowledge, which could be expressed in a series of statements such as:

The diagonals of a rectangle are congruent.
The base angles of an isosceles triangle are congruent.
Parallel lines do not intersect.

The principal purpose of this course in formal geometry is to organize these statements, along with many others, in such a way that we understand why many of them must be true if a few are assumed to be true. The statements that are assumed to be true are called *assumptions* or *postulates.* The more important statements that must be true if these assumptions are true are called *theorems.*

Formal geometry, and indeed all mathematics, is concerned with reasoning and proof. In the next two sections we illustrate two principal types of reasoning: inductive reasoning and deductive reasoning.

Inductive Reasoning

When we do inductive reasoning we make "educated guesses" based on incomplete evidence. Some of these guesses are quite plausible and some are true. Consider the following statements, each of which is denoted by a small letter.

a: All integers ending in 5 are multiples of 5.

$$65 = 13 \cdot 5, \qquad 135 = 27 \cdot 5,$$
$$485 = 97 \cdot 5, \qquad 795 = 159 \cdot 5.$$

b: Every product of two odd numbers is an odd number.

$$23 \cdot 11 = 453, \qquad 27 \cdot 31 = 837,$$
$$157 \cdot 83 = 13{,}031, \qquad 71 \cdot 71 = 5041.$$

c: No prime number is even.

$$3, 5, 7, 11, 13, 17, \ldots.$$

d: Any integer between twin primes is a multiple of 6.

$$5, \underline{6}, 7, \qquad 41, \underline{42}, 43,$$
$$11, \underline{12}, 13, \qquad 59, \underline{60}, 61,$$
$$17, \underline{18}, 19, \qquad 71, \underline{72}, 73,$$
$$29, \underline{30}, 31, \qquad 101, \underline{102}, 103.$$

e: If a number is a multiple of 9, then it is a multiple of 3.

$$45 = 5 \cdot 9, \qquad 45 = 15 \cdot 3,$$
$$72 = 8 \cdot 9, \qquad 72 = 24 \cdot 3,$$
$$369 = 41 \cdot 9, \qquad 369 = 123 \cdot 3.$$

Each of the statements a through e is a general statement or *generalization*, which is supposed to be true for every number in a certain set called the *subject set*. Observe that the numbers for which each of these general statements has actually been found true (verified) form a proper subset of its subject set. For example, the statement "If x is an integer ending in 5, then x is a multiple of 5," which is equivalent to statement a, has been verified for only four members of the set of integers ending in 5, which is the subject set for this statement. Again, the statement "If x is a prime number, then x is not an even number" has been verified for only six members of the set of prime numbers mentioned in statement c.

A general statement that has been verified for some, but not all, of the elements in its subject set is called a *conjecture*. The process of making a conjecture is called *inductive reasoning* or *induction*.

We can prove some conjectures by actually verifying them for every element in the subject set. This kind of proof is not practical if the subject set has many elements and impossible if it is infinite. Some conjectures can be proved by the use of deductive reasoning in ways to be described later. Some conjectures seem to be true although no one has been able to prove them, and some, of course, turn out to be false.

Consider statement d. We cannot prove this statement by merely extending our list of twin primes and observing that each "between integer" is a multiple of 6. On the other hand, if we find just one integer between twin primes that is *not* a multiple of 6, we shall have to admit that our general statement is false. The fact that 4 is between twin primes 3 and 5 and is not a multiple of 6 forces us to make this admission. At the same time it serves to remind us that *any exception* to a general statement will prove that the statement is false. Such an exception is called a *counterexample*. In this case 4 is a counterexample for the general statement "Any integer between twin primes is a multiple of 6." Find a counterexample for conjecture c.

Since some of the conjectures produced by inductive reasoning are false, we cannot use induction to prove statements. We must play the game of proof by the rules of logic soon to be described. When we do this, we are using deductive reasoning. We can use inductive reasoning to formulate conjectures, but when we try to prove these conjectures we must use deductive reasoning only. We should observe that the proofs by "mathematical induction" are forms of deductive reasoning, not to be confused with the inductive process described above.

Deductive Reasoning

When someone tries to convince us that a certain statement x is true, he usually lists several statements $r, s, t, \ldots, z$ and then asserts that statement x

"follows logically" from these. Such an assertion is an *argument* whose conclusion is the statement x and whose premises are the statements $r, s, t, \ldots, z$. Some arguments are sound (or, as we say, valid) and some are not. Let us consider a few examples. In each case we state an opinion on whether or not the argument is valid and we give the reason for this opinion. In a later section we shall be able to prove that these opinions are correct.

EXAMPLE 1

If a person is retired, then he is entitled to some tax exemption.
Mr. Hibbs is retired.
Therefore Mr. Hibbs is entitled to some tax exemptions.

DISCUSSION

The argument is valid. If we let one region represent the set of all people who are entitled to some tax exemption, then we see that the set of retired persons must be represented by some region inside this one. Since Mr. Hibbs is in the inner region, he must be in the outer region.

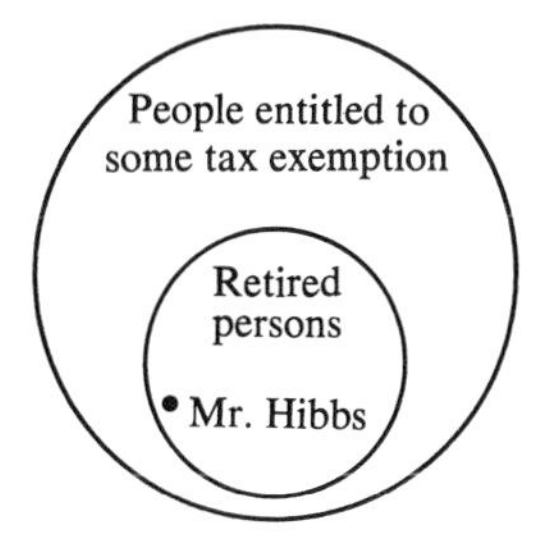

EXAMPLE 2

a: If a person lives in City X, then he lives in County Y.
b: If a person lives in County Y, then he lives in State Z.

ARGUMENT

If statements a and b are true, then the statement c, "If a person lives in City X, then he lives in State Z," is true.

DISCUSSION

The argument is valid as indicated by a study of the drawing below. Note that this argument could be written in the following "bracket form":

$$\left.\begin{matrix} a \\ b \end{matrix}\right\} \longrightarrow c,$$

which is read "If a and b are true, then c is true" or "If a and b, then c."

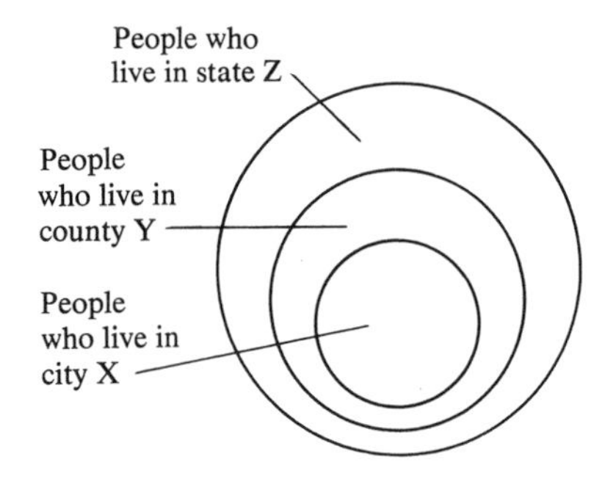

EXAMPLE 3

All home owners are taxpayers.
Mr. Jones is a taxpayer.
Therefore Mr. Jones is a home owner.

DISCUSSION

The argument is not valid. There may be many people who pay taxes who are not home owners. The statement "All home owners are taxpayers" tells us that the set of home owners is a subset of the set of taxpayers. The fact that Mr. Jones can be represented by a point in the outer region does not insure that this point will be in the inner region.

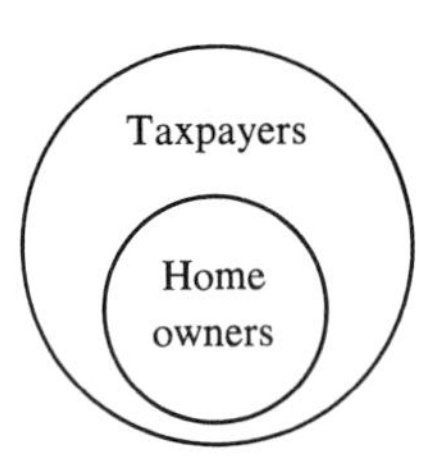

EXAMPLE 4

In trying to prove the defendant innocent a lawyer reasons as follows: "If the defendant committed the crime in Chicago at noon on February 11, he must have traveled 800 miles from New Orleans to Chicago in 30 minutes, because he was in New Orleans at 11:30 A.M. It is impossible to travel 800 miles in a half hour. Therefore my client did not commit this crime in Chicago at noon on February 11."

DISCUSSION

If we accept the facts as stated by the lawyer and assume that New Orleans and Chicago are in the same time zone, we must vote for acquittal.

Tautologies

A *composite statement* is a statement formed from other statements $a, b, c, \ldots$ by the use of some of the connectives *and, or, if-then.* The statements $a, b, c, \ldots$ are the *components* of this composite statement.

A composite statement that is true for all possible truth values of its composite statements is called a *tautology.*

We use truth tables to show that certain composite statements are tautologies. For example, the composite statement $p \vee \sim p$ is a tautology. Whether the statement p is true or false, the statement $p \vee \sim p$ is true. This is shown by the fact that both entries in the column headed $p \vee \sim p$ are T's. In general, a composite statement is a tautology if all entries under it in a truth table that shows all possible truth values of its components are T's.

p	$\sim p$	$p \vee \sim p$
T	F	T
F	T	T

The table below shows that $[(a \vee b) \wedge \sim a] \longrightarrow b$ is a tautology.

a	b	$a \vee b$	$\sim a$	$(a \vee b) \wedge \sim a$	$[(a \vee b) \wedge \sim a] \longrightarrow b$
T	T	T	F	F	T
T	F	T	F	F	T
F	T	T	T	T	T
F	F	F	T	F	T

We have seen other tautologies. For example $(p \longrightarrow q) \longleftrightarrow (\sim p \vee q)$ is a tautology. Our definition tells us that any composite statement that is equivalent to u is a tautology. (See Exercise 6 below.) Thus any composite statement that is equivalent to a known tautology is a tautology.

If a statement can be transformed into another statement by the use of the properties of equivalence, then the assertion that these two statements are equivalent is a tautology. (See Exercise 7 below.)

Exercises

1. In each of the drawings points of a circle are connected to form nonoverlapping regions. Make a conjecture as to the number of regions formed by connecting six points of a circle. Check your conjecture by drawings.

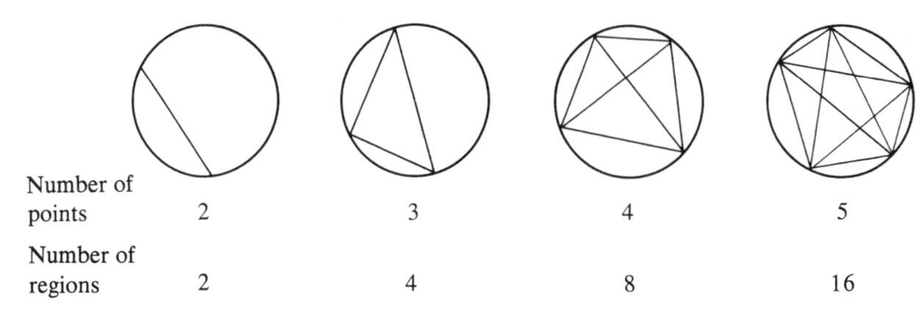

2. Consider numbers of the form $n^2 - n + 41$, where n represents a counting number. We see that

$$\begin{array}{ll} \text{if } n = 1, & \text{then } n^2 - n + 41 = 41, \\ \text{if } n = 2, & \text{then } n^2 - n + 41 = 43, \\ \text{if } n = 3, & \text{then } n^2 - n + 41 = 47, \\ \text{if } n = 4, & \text{then } n^2 - n + 41 = 53. \end{array}$$

Each of the numbers 41, 42, 47, and 53 is a prime number. Can you conclude that $n^2 - n + 41$ represents a prime number for any counting number n? Can you find a counterexample?

3. Examine each of the following statements.

$$\begin{array}{l} 1 = 1^2 \\ 1 + 3 = 2^2 \\ 1 + 3 + 5 = 3^2 \\ 1 + 3 + 5 + 7 = 4^2 \\ 1 + 3 + 5 + 7 + 9 = ? \\ 1 + 3 + 5 + 7 + 9 + 11 = ? \end{array}$$

Can you conclude that the sum of the first n odd counting numbers will equal n^2? Can you find a counterexample?

4. After studying the examples below,

$$\begin{array}{lll} 4 = 2 + 2, & 10 = 3 + 7, & 16 = 3 + 13, \\ 6 = 3 + 3, & 12 = 5 + 7, & 18 = 5 + 13, \\ 8 = 3 + 5, & 14 = 3 + 11, & 20 = 7 + 13, \end{array}$$

we may employ inductive reasoning to make the conjecture that any even number greater than two can be expressed as the sum of two prime numbers. This is the famous Goldbach conjecture, which seems to be true but has never been proved true. Verify the Goldbach conjecture for the following numbers: 80, 100, 200, and 400.

5. Show by means of truth tables that the following statements are tautologies.
(a) $(a \wedge b) \longrightarrow a$. (b) $b \longrightarrow (a \vee b)$.
(c) $(p \longrightarrow q) \vee (q \longrightarrow p)$.

6. We can show that each statement in Exercise 5 is a tautology by verifying that it is equivalent to u. This has already been done for the tautology $(a \wedge b) \longrightarrow a$ (see Exercise 16, page 26). Supply reasons for the following:

(a) $[b \longrightarrow (a \vee b)] \overset{(1)}{\longleftrightarrow} [\sim b \vee (a \vee b)] \overset{(2)}{\longleftrightarrow} [(b \vee \sim b) \vee a] \overset{(3)}{\longleftrightarrow} (u \vee a) \overset{(4)}{\longleftrightarrow} u$.
Thus $[b \longrightarrow (a \vee b)] \longleftrightarrow u$. Why?

(b) $[(p \longrightarrow q) \vee (q \longrightarrow p)] \overset{(1)}{\longleftrightarrow} [(\sim p \vee q) \vee (\sim q \vee p)] \overset{(2)}{\longleftrightarrow} [(p \vee \sim p) \vee (q \vee \sim q)] \overset{(3)}{\leftrightarrow}$
$\leftrightarrow (u \vee u) \overset{(4)}{\longleftrightarrow} u$. Thus $[(p \longrightarrow q) \vee (q \longrightarrow p)] \longleftrightarrow u$. Why?

Note that in some cases steps involving properties of operations with statements have been combined. In such cases you may supply more than one reason for an equivalence. This will continue to be our practice in future exercises.

7. Verify that $[(x \wedge y) \longrightarrow z] \longleftrightarrow [x \longrightarrow (y \longrightarrow z)]$ is a tautology by supplying reasons for the following:

$$[(x \wedge y) \longrightarrow z] \overset{(1)}{\longleftrightarrow} [\sim(x \wedge y) \vee z] \overset{(2)}{\longleftrightarrow} [(\sim x \vee \sim y) \vee z] \overset{(3)}{\leftrightarrow}$$
$$\leftrightarrow [\sim x \vee (\sim y \vee z)] \overset{(4)}{\longleftrightarrow} [x \longrightarrow (\sim y \vee z)] \overset{(5)}{\longleftrightarrow} [x \longrightarrow (y \longrightarrow z)].$$

Thus $[(x \wedge y) \longrightarrow z] \overset{(6)}{\longleftrightarrow} [x \longrightarrow (y \longrightarrow z)]$. Why?

8. Verify that $[x \longrightarrow (y \longleftrightarrow z)] \longleftrightarrow ([(x \wedge y) \longrightarrow z] \wedge [(x \wedge z) \longrightarrow y])$ is a tautology.
9. Verify that $[x \longrightarrow (a \vee b)] \longleftrightarrow [(x \wedge \sim a) \longrightarrow b]$ is a tautology.
10. Verify that $(a \longrightarrow b) \longrightarrow [(a \wedge c) \longrightarrow b]$ is a tautology.
11. Verify that the following statement is a tautology:

$$\left.\begin{array}{l} (x \longrightarrow z) \\ (y \longrightarrow z) \end{array}\right\} \longrightarrow [(x \vee y) \longrightarrow z].$$

12. Verify that the following statement is a tautology:

$$[(a \longrightarrow x) \wedge (b \longrightarrow y)] \longrightarrow [(a \wedge b) \longrightarrow (x \wedge y)].$$

13. Verify that the following statement is a tautology:

$$\left.\begin{matrix}[x \longrightarrow (a \vee b)] \\ x \\ \sim a\end{matrix}\right\} \longrightarrow b.$$

14. Write an argument to establish the following statement: In a group consisting of twenty people, there are at least two people whose birthdays fall in the same month.
15. Suppose that each of the following statements is true.

    a: The number of trees in Illinois is larger than the number of leaves on any one tree.
    b: Every tree has at least one leaf.

    Write an argument to support the statement: There are at least two trees in Illinois that have the same number of leaves.
16. The checkerboard shown has 64 squares. As you know, it can be covered by 32 dominoes, each of which is the size of two squares. Suppose that the two squares marked with a cross are removed. Can the remaining part of the board be covered by 31 dominoes? Explain.

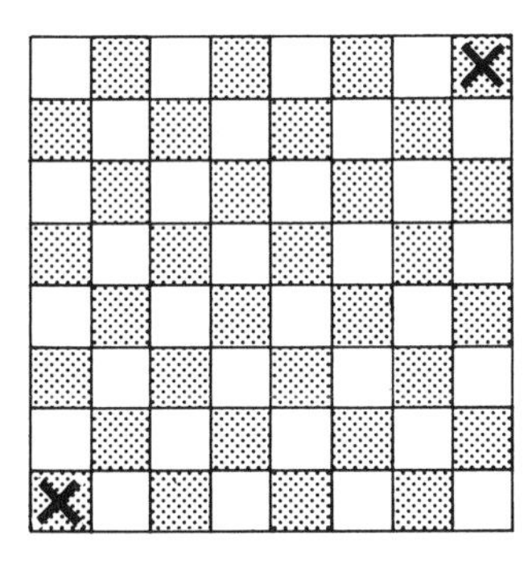

Argument and Proof

Let us consider how tautologies are involved in argument and proof. We have noted that an *argument* is an assertion that a certain statement q, called the conclusion of the argument, is true because certain other statements $a, b, c, \ldots, x$, called the *premises* of the argument, are true. Thus an argument is an implication of the form $(a \wedge b \wedge c \wedge \cdots \wedge x) \longrightarrow q$.

DEFINITION *An* argument *is valid if the truth of its premises (hypothesis) guarantees the truth of its conclusion.*

If an implication is a tautology, then the truth of its hypothesis guarantees the truth of its conclusion because an implication with a true hypothesis and a false conclusion is false, and hence, is not a tautology. Thus we have a criterion for determining the validity of an argument:

An argument is valid if it is a tautology.

An argument is a *proof* of its conclusion if it is valid and all of its premises are true. Since valid arguments are used to prove statements, they are called proof patterns.

Let us review our examples of deductive reasoning in terms of our criterion for judging validity.

Since, in Example 1, Mr. Hibbs is a retired person, the statement "If a person is retired, then he is entitled to some tax exemption" becomes "If Mr. Hibbs is a retired person, then Mr. Hibbs is entitled to some tax exemption." Our argument can be displayed as follows:

Premises		Conclusion
Mr. Hibbs is a retired person. If Mr. Hibbs is a retired person, then he is entitled to some tax exemption.	$\longrightarrow$	Mr. Hibbs is entitled to some tax exemption.

If we let p represent the statement "Mr. Hibbs is a retired person" and q the statement "Mr. Hibbs is entitled to some tax exemption," we see that this argument has the form

$$A_1: \quad \left.\begin{matrix} p \\ (p \longrightarrow q) \end{matrix}\right\} \longrightarrow q.$$

This argument is valid because it is a tautology, as shown by the truth table below.

p	q	$p \longrightarrow q$	$p \wedge (p \longrightarrow q)$	$[p \wedge (p \longrightarrow q)] \longrightarrow q$
T	T	T	T	T
T	F	F	F	T
F	T	T	F	T
F	F	T	F	T

This tautology is important enough to deserve a special name. It is called a *syllogism.* Statement p is called the *minor premise* of the syllogism, the statement $p \longrightarrow q$ is called the *major premise,* and the statement q is called the *conclusion.* If we accept the statement p and the statement $p \longrightarrow q$ in this tautology, we must accept the statement q. In other words, if we accept the truth of the premises of a syllogism, we must accept the truth of its conclusion. Thus A_1 is a proof of statement q, provided its premises are true.

In Example 2 let $p \longrightarrow q$ represent the statement "If a person lives in City X, then he lives in County Y" and let $q \longrightarrow r$ represent the statement "If a person lives in County Y, then he lives in State Z." Then $p \longrightarrow r$ represents the statement "If a person lives in City X, then he lives in State Z," and our argument takes the following form:

$$A_2: \quad \left.\begin{matrix} (p \longrightarrow q) \\ (q \longrightarrow r) \end{matrix}\right\} \longrightarrow (p \longrightarrow r).$$

The fact that this is a tautology is verified by the truth table below.

p	q	r	$p \longrightarrow q$	$q \longrightarrow r$	$(p \longrightarrow q) \wedge (q \longrightarrow r)$	$p \longrightarrow r$	$[(p \longrightarrow q) \wedge (q \longrightarrow r)] \longrightarrow (p \longrightarrow r)$
T	T	T	T	T	T	T	T
T	T	F	T	F	F	F	T
T	F	T	F	T	F	T	T
F	T	T	T	T	T	T	T
F	F	T	T	T	T	T	T
F	T	F	T	F	F	T	T
T	F	F	F	T	F	F	T
F	F	F	T	T	T	T	T

If we accept the statements $p \longrightarrow q$ and $q \longrightarrow r$ we must accept the statement $p \longrightarrow r$ because a valid argument with true premises is a proof of its conclusion.

Valid argument A_2 also has a special name. It is called the *transitive property of implication.* The transitive property of implication is, of course, readily extended to include three, four, or any finite number of implications. For three implications we have

A_3: $$[(p \longrightarrow q) \wedge (q \longrightarrow r) \wedge (r \longrightarrow s)] \longrightarrow (p \longrightarrow s).$$

If we accept the truth of $p \longrightarrow q$, $q \longrightarrow r$, and $r \longrightarrow s$, we must accept the truth of $p \longrightarrow s$. We observe that, in this sequence of implications, each implication after the first has the conclusion of the preceding implication for its hypothesis. This suggests that these implications can be "hooked together" and written in the following more compact form:

$$p \longrightarrow q \longrightarrow r \longrightarrow s.$$

Thus we can write valid argument A_3 more compactly as

A_3: $$(p \longrightarrow q \longrightarrow r \longrightarrow s) \longrightarrow (p \longrightarrow s)$$

with the understanding that $p \longrightarrow q \longrightarrow r \longrightarrow s$ is merely a short way to write $(p \longrightarrow q) \wedge (q \longrightarrow r) \wedge (r \longrightarrow s)$.

The expression $p \longrightarrow q \longrightarrow r \longrightarrow s$ is an example of what we call a *flow-diagram arrangement* of implications. A flow-diagram arrangement of implications is also referred to as a *deductive sequence.* The components of this particular deductive sequence are the implications $p \longrightarrow q$, $q \longrightarrow r$, and $r \longrightarrow s$. Other deductive sequences are considered later.

Using the conjunction of the statement p and the deductive sequence $p \longrightarrow q \longrightarrow r \longrightarrow s$, we can form another tautology.

A_4: $$[p \wedge (p \longrightarrow q \longrightarrow r \longrightarrow s)] \longrightarrow s.$$

This valid argument is a proof pattern that can be used to prove the statement s. The truth table that shows that A_4 is a tautology has sixteen lines.

In Example 3 the statement "All home owners are taxpayers" is equivalent to the following implication: "If x is a home owner, then x is a taxpayer." Substituting Mr. Jones for x, we have "If Mr. Jones is a home owner, then Mr. Jones is a taxpayer." Thus our argument can be presented as follows:

Mr. Jones is a taxpayer.
If Mr. Jones is a home owner, then he is a taxpayer. } ⟶ Mr. Jones is a home owner.

If we let q represent the statement "Mr. Jones is a taxpayer" and p represent the statement "Mr. Jones is a home owner," we see that this argument has the form

$$\left.\begin{array}{l} q \\ (p \longrightarrow q) \end{array}\right\} \longrightarrow p.$$

A truth table will show that this is *not a tautology* and hence is not a valid argument. We know that the truth of the conclusion of an implication does not guarantee the truth of the hypothesis.

The argument in Example 4 can be arranged as follows:

If the defendant committed the crime in Chicago at noon on February 11, then *he traveled 800 miles from New Orleans to Chicago in 30 minutes.*

He cannot travel 800 miles from New Orleans to Chicago in 30 minutes.

} ⟶ The defendant did not commit the crime in Chicago at noon on February 11.

If we let p represent the statement "The defendant committed the crime in Chicago at Noon on February 11" and q the statement "He traveled 800 miles from New Orleans to Chicago in 30 minutes," this argument takes the form

$$A_5: \quad \left.\begin{array}{l} (p \longrightarrow q) \\ \sim q \end{array}\right\} \longrightarrow \sim p.$$

A_5 is clearly a tautology, because it is equivalent to its second contrapositive —namely,

$$\left.\begin{array}{l} (p \longrightarrow q) \\ p \end{array}\right\} \longrightarrow q,$$

which we recognize as tautology A_1.

In using A_5 it is sometimes convenient to consider a statement a such that $\sim a \longleftrightarrow p$. Substituting $\sim a$ for p in A_5, we obtain

$$\left.\begin{array}{l} (\sim a \longrightarrow q) \\ \sim q \end{array}\right\} \longrightarrow \sim(\sim a)$$

or

$$A_5: \quad \left.\begin{array}{l} (\sim a \longrightarrow q) \\ \sim q \end{array}\right\} \longrightarrow a.$$

Stated thus, A_5 is a proof pattern we use for *indirect proof.* It can be stated as follows: *If the contradiction of a given statement implies a false statement, then the given statement is true.* We refer to this statement as *the indirect proof rule.*

We summarize this section in the following statements:

An argument is valid if the implication which has the premise(s) of the argument for its hypothesis and the conclusion of the argument for its conclusion is a tautology.

A proof is a valid argument having true premises.

Each of the following tautologies becomes a proof of its conclusion if we know that each statement in its hypothesis is true:

A_1: $$[p \wedge (p \longrightarrow q)] \longrightarrow q.$$

A_2: $$(p \longrightarrow q \longrightarrow r) \longrightarrow (p \longrightarrow r).$$

A_3: $$(p \longrightarrow q \longrightarrow r \longrightarrow s) \longrightarrow (p \longrightarrow s).$$

A_4: $$[p \wedge (p \longrightarrow q \longrightarrow r \longrightarrow s)] \longrightarrow s.$$

A_5: $$[(\sim a \longrightarrow q) \wedge \sim q] \longrightarrow a.$$

Three Forms of Proof

Often the statement we are to prove is an *implication*—that is, a statement such as $p \longrightarrow s$.

Suppose we want to prove the implication

$$\frac{3x + b}{2} = c \longrightarrow x = \frac{2c - b}{3}.$$

This implication has the form $p \longrightarrow s$, where p is the statement $(3x + b)/2 = c$ and s is the statement $x = (2c - b)/3$. To construct a proof of this statement we first try to construct a deductive sequence from the hypothesis p to the conclusion s. The following flow-diagram arrangement will serve.

$$\frac{3x + b}{2} = c \longrightarrow 3x + b = 2c \longrightarrow 3x = 2c - b \longrightarrow x = \frac{2c - b}{3}.$$

We now use this deductive sequence to construct the following valid argument.

A: $$\left[\frac{3x + b}{2} = c \xrightarrow{(1)} 3x + b = 2c \xrightarrow{(2)} 3x = 2c - b \xrightarrow{(3)} x = \frac{2c - b}{3}\right] \longrightarrow \left(\frac{3x + b}{2} = c \longrightarrow x = \frac{2c - b}{3}\right).$$

We know that A is a valid argument because it has the form

$$[p \longrightarrow q \longrightarrow r \longrightarrow s] \longrightarrow (p \longrightarrow s),$$

which is the same as the valid argument A_3 we considered in the last section.

This valid argument becomes a *proof* of its conclusion if we can assure ourselves that each of its premises is true. The three premises of this argument are

$$\frac{3x + b}{2} = c \xrightarrow{(1)} 3x + b = 2c,$$

$$3x + b = 2c \xrightarrow{(2)} 3x = 2c - b,$$

$$3x = 2c - b \xrightarrow{(3)} x = \frac{2c - b}{3}.$$

These are clearly identified by the symbols (1), (2), and (3) that appear in the hypothesis of our argument. We must supply correspondingly numbered reasons for the truth of each implication. The reasons are as follows:

(1) Multiplication property of equality (E_6).
(2) Addition property of equality (E_5).
(3) Multiplication property of equality (E_6).

Now we know that A is a proof of

$$\frac{3x+b}{2} = c \longrightarrow x = \frac{2c-b}{3}$$

because (a) A is a valid argument, (b) each of its premises is true for the reasons indicated.

In writing a proof of the form $[p \xrightarrow{(1)} q \xrightarrow{(2)} r \xrightarrow{(3)} s] \longrightarrow (p \longrightarrow s)$ you will be expected to write only the part that is in the brackets together with appropriate reasons for the numbered implications. The deductive sequence $p \longrightarrow q \longrightarrow r \longrightarrow s$ together with its indicated reasons is a *flow-diagram proof* of the statement $p \longrightarrow q$.

The reason we supply for a statement in a flow-diagram proof must meet two qualifications:

1. It must fit the statement. Generally, the statement is a specific application of the general principle stated in the reason.
2. It must be a statement that we accepted as true *before* using it in the proof.

The flow-diagram proof is one of the three principal forms of proof used in mathematics. The other two forms are known as *ledger* and *essay*. In a ledger arrangement we make a list of numbered statements in the left-hand column and a list of correspondingly numbered reasons in the right-hand column. Thus a ledger proof is merely a ledger arrangement of the statements and reasons in a flow-diagram proof.

We illustrate the process of constructing a ledger proof by putting the flow-diagram proof above in ledger form.

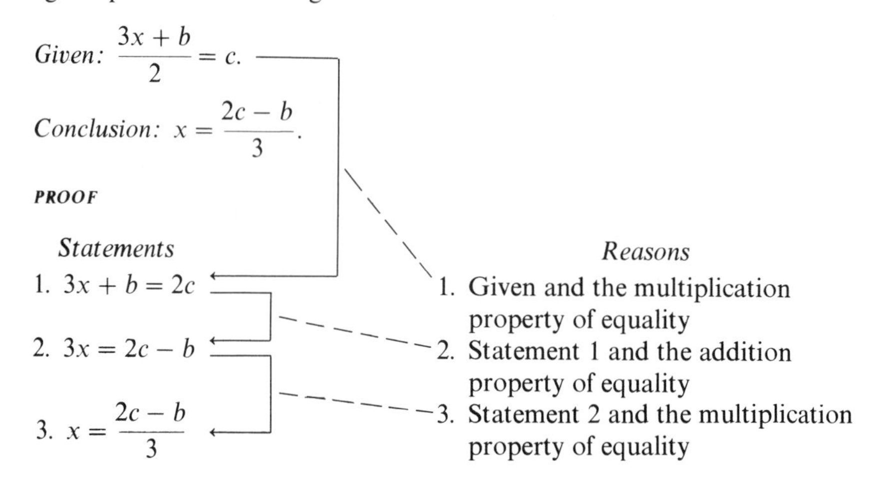

In a ledger proof we write

$$\text{Given: } \frac{3x+b}{2} = c \quad \text{and} \quad \text{Conclusion: } x = \frac{2c-b}{3}$$

to indicate that we must prove the implication

$$\frac{3x+b}{2} = c \longrightarrow x = \frac{2c-b}{3}.$$

A ledger proof does not clearly indicate how each statement follows from previous statements. In order to show this we indicate which previous statements are involved by listing them just before each reason, as shown above. Each reason really supports an *implication,* as indicated by the broken lines in the proof above. Since these lines are ordinarily not shown in the ledger arrangement, we must rely on the statement number(s) that precede each reason to convey this idea.

An essay proof of $\frac{3x+b}{2} = c \longrightarrow x = \frac{2c-b}{3}$ can be presented as follows:

If $(3x + b)/2 = c$, then $3x + b = 2c$ because of the multiplication property of equality. By the addition property of equality (adding $-b$ to each member of our equation) it follows that $3x = 2c - b$. Again applying the multiplication property of equality by multiplying each member of this equation by $\frac{1}{3}$, we obtain $x = (2c - b)/3$, and our proof is complete.

You have, no doubt, observed that most of the proofs in college mathematics texts are of the essay type. Such proofs have the advantage of seeming quite "readable," but they may actually be harder to understand than either flow-diagram or ledger type proofs. The difficulty arises because essay proofs sometimes show only the principal ideas in the argument, omitting some of the reasons and some of the more "obvious" statements. If an essay proof is difficult to understand, we may find it helpful to analyze it by putting it in flow-diagram or ledger form.

Exercises

1. Write an example of a valid argument patterned after each of the tautologies A_1 through A_5.
2. Given the implications $a \longrightarrow \sim b$, $d \longrightarrow c$, and $\sim b \longrightarrow d$, we can arrange the statements in a deductive sequence as follows: $a \longrightarrow \sim b \longrightarrow d \longrightarrow c$. In each of the following, arrange the implications in a deductive sequence.
   (a) $d \longrightarrow a$, $a \longrightarrow b$, $c \longrightarrow d$.
   (b) $\sim a \longrightarrow b$, $s \longrightarrow r$, $t \longrightarrow \sim a$, $b \longrightarrow s$.
   (c) $x \longrightarrow \sim a$, $y \longrightarrow a$, $\sim x \longrightarrow b$.
   (d) $d \longrightarrow c$, $\sim d \longrightarrow b$, $a \longrightarrow \sim b$, $x \longrightarrow \sim c$.
3. In an election speech for office in the Student Senate, a candidate made the following statement: "In order to perform the duties of this office, a candidate must be a senior. My opponent is a junior. Therefore, I conclude that he is unfit for this office." Has this candidate given us a valid argument for his conclusion? Has he proved to us that his opponent is unfit for office? Explain.

4. Suppose that each of the following statements is accepted as true.

   If a person is unsuccessful, then he is lazy.
   If a person is successful, then he is not in debt.
   If a person is lazy, then he is foolish.

   (a) Write a flow-diagram proof of the statement: If a person is in debt, then he is foolish.
   (b) Write a ledger proof of the statement: If a person is in debt, then he is foolish.
5. Write a flow-diagram, ledger, and essay proof for the statement:

$$\frac{2a-b}{3}=c\longrightarrow a=\frac{3c+b}{2}.$$

More about Proving If-Then Statements

When implications such as $a\longrightarrow x$, $w\longrightarrow t$, $t\longrightarrow h$, and $x\longrightarrow w$ are given, it is easy to "hook them together" to form the flow-diagram sequence $a\longrightarrow x\longrightarrow w\longrightarrow t\longrightarrow h$, which will serve as the basis for a flow-diagram proof of the theorem $a\longrightarrow h$. In this section we consider how we can form a flow-diagram sequence when implications having conjunctive terms are involved. First we give some examples of valid arguments that contain such implications arranged in flow-diagram form. In each case the if-then statement to be proved is the conclusion of the argument. (Recall that u is a statement that is known to be true.)

A_6: $$\left[\left.\begin{matrix} u \\ \underline{a}\end{matrix}\right\}\longrightarrow x\longrightarrow y\right]\longrightarrow(\underline{a}\longrightarrow y).$$

A_7: $$\left[\left.\begin{matrix} \underline{a}\longrightarrow x \\ \underline{c}\end{matrix}\right\}\longrightarrow d\right]\longrightarrow\left(\left.\begin{matrix} \underline{a} \\ \underline{c}\end{matrix}\right\}\longrightarrow d\right).$$

A_8: $$\left[\left.\begin{matrix} \left.\begin{matrix}\underline{a} \\ \underline{b}\end{matrix}\right\}\longrightarrow x \\ \left.\begin{matrix}\underline{c} \\ \underline{d}\end{matrix}\right\}\longrightarrow y\end{matrix}\right\}\longrightarrow z\right]\longrightarrow\left(\left.\begin{matrix} \underline{a} \\ \underline{b} \\ \underline{c} \\ \underline{d}\end{matrix}\right\}\longrightarrow z\right).$$

A_9: $$\left[\begin{matrix} a \\ b \\ \left.\begin{matrix} a \\ b \\ \underline{c}\end{matrix}\right\}\longrightarrow x\longrightarrow y\end{matrix}\right]\longrightarrow(\underline{c}\longrightarrow y).$$

A_{10}: $$\left[\left.\begin{matrix} \underline{a}\longrightarrow\left\{\begin{matrix} b \\ c\end{matrix}\right. \\ \left.\begin{matrix}\underline{d} \\ \underline{e}\end{matrix}\right\}\longrightarrow f\end{matrix}\right\}\longrightarrow w\right]\longrightarrow\left(\left.\begin{matrix} \underline{a} \\ \underline{d} \\ \underline{e}\end{matrix}\right\}\longrightarrow w\right).$$

A_{11}: $$\left[\left.\begin{matrix} \underline{e}\longrightarrow f\longrightarrow g \\ \underline{x}\longrightarrow y\longrightarrow w\end{matrix}\right\}\longrightarrow h\right]\longrightarrow\left(\left.\begin{matrix} \underline{e} \\ \underline{x}\end{matrix}\right\}\longrightarrow h\right).$$

$$A_{12}: \quad \left[\underline{a} \longrightarrow b \longrightarrow \left.\begin{matrix} c \\ d \\ u \end{matrix}\right\} \longrightarrow x\right] \longrightarrow (\underline{a} \longrightarrow x).$$

A study of the way in which the implications in each of these arguments have been "hooked together" suggests the following simple rule.

Rule: To construct a flow-diagram (deductive) sequence using the premises of a given argument, arrange the implications involved in such a way that the hypothesis of each contains only statements from one or more of the following categories:

(1) Statements from the hypothesis of the if-then statement to be proved.
(2) Statements that have been previously accepted as true.
(3) Statements that are conclusions of preceding implications.

If such a flow-diagram sequence can be constructed using the premises of a given argument and if the conclusion of the if-then statement to be proved appears as the conclusion of the last implication in the sequence, then we can be sure that the argument is valid without resorting to a truth table. We accept this statement as a second criterion for determining the validity of an argument. For example, we can see that A_8 is a valid argument without constructing the 128-line truth table that shows it is a tautology. Thus

$$\left.\begin{matrix} \left.\begin{matrix} a \\ b \end{matrix}\right\} \xrightarrow{(1)} x \\ \left.\begin{matrix} c \\ d \end{matrix}\right\} \xrightarrow{(2)} y \end{matrix}\right\} \xrightarrow{(3)} z$$

is a proof of the implication

$$\left.\begin{matrix} a \\ b \\ c \\ d \end{matrix}\right\} \longrightarrow z,$$

provided we supply a suitable reason for each of its three premises. While these premises have been "hooked together," they are clearly indicated by means of the numbers (1), (2), and (3) as the implications

$$\left.\begin{matrix} a \\ b \end{matrix}\right\} \longrightarrow x, \quad \left.\begin{matrix} c \\ d \end{matrix}\right\} \longrightarrow y, \quad \text{and} \quad \left.\begin{matrix} x \\ y \end{matrix}\right\} \longrightarrow z,$$

respectively.

Valid argument A_9 has four premises, namely

$$a, \quad b, \quad \left.\begin{matrix} a \\ b \\ c \end{matrix}\right\} \longrightarrow x, \quad \text{and} \quad x \longrightarrow y.$$

It can be written compactly as follows:

$$\left.\begin{array}{ll}(1) & a\\ (2) & b\\ & \underline{c}\end{array}\right\}\xrightarrow{(3)} x \xrightarrow{(4)} y.$$

This argument is a proof of the implication $c \longrightarrow y$ if suitable reasons are supplied for each of the numbered statements.

Observe that in each argument the statements in the hypothesis of the if-then statement to be proved are underlined wherever they appear, and no other statements are underlined. We shall continue to use this plan.

The structure of a deductive sequence such as the one below

$$\left.\begin{array}{l} \left.\begin{array}{l} a \longrightarrow t \\ \left.\begin{array}{l} c\\ d\end{array}\right\}\longrightarrow x\end{array}\right\}\longrightarrow m \\ \left.\begin{array}{l} e\\ f\end{array}\right\}\longrightarrow y \longrightarrow z\end{array}\right\}\longrightarrow k \tag{1}$$

gives us some idea of the nature of a deductive system. As we continue our study of geometry, each deductive sequence will be extended. For example, the following deductive sequence

$$\left.\begin{array}{r} k \\ \left.\begin{array}{l} h\\ i\end{array}\right\}\longrightarrow l \longrightarrow \left\{\begin{array}{l} x\\ y\end{array}\right.\end{array}\right\}\longrightarrow w$$

can be "hooked" to (1) to form the following more extensive sequence:

$$\left.\begin{array}{l} \left.\begin{array}{l} \left.\begin{array}{l} a \longrightarrow t \\ \left.\begin{array}{l} c\\ d\end{array}\right\}\longrightarrow x\end{array}\right\}\longrightarrow m \\ \left.\begin{array}{l} e\\ f\end{array}\right\}\longrightarrow y \longrightarrow z\end{array}\right\}\longrightarrow k \\ \left.\begin{array}{l} h\\ i\end{array}\right\}\longrightarrow l \longrightarrow \left\{\begin{array}{l} x\\ y\end{array}\right.\end{array}\right\}\longrightarrow w. \tag{2}$$

Eventually all of the theorems in this text can be included in one vast "web of implications."

When we arrange the principal facts of a subject so that each takes its place in one deductive sequence, we have constructed what is known as a *deductive system.* Clearly every statement in a deductive system cannot be the conclusion of an implication. When we start to prove our first theorem, we must have accepted some statements, such as statements *a, c, d, e, f, h,* and *i* in (2), that can serve as hypotheses. Such statements are called assumptions or postulates. These postulates provide the "starting points" that we must have if we are to draw any conclusions at all. In building a deductive system it is absolutely essential that these basic assumptions (postulates) exist and that they be clearly stated. In this text our basic assumptions are

properties of equality,
properties of real numbers, including the order properties (see Appendix A),
postulates about sets,
postulates about geometry.

When we begin constructing a set of definitions, we encounter the same situation as when we start to prove the first theorem in a deductive system. Since nothing has yet been defined, we have nothing to refer to when we try to describe any idea, thing, or term. If we are to get started with our set of definitions, therefore, some terms must be left undefined. Thus *undefined* terms are just as essential in the building of a set of definitions as postulates are in the building of a deductive system.

Exercises

1. In each of the following, rewrite the argument and indicate—by placing numbers (1), (2), (3), and so on—the points where you would need to supply reasons to prove the conclusion of the argument. For example, in the argument A_7 we would place the numbers (1) and (2) as shown:

$$\left[\left.\begin{array}{r}\underline{a}\xrightarrow{(1)}x\\ \underline{c}\end{array}\right\}\xrightarrow{(2)}d\right]\longrightarrow\left(\left.\begin{array}{c}\underline{a}\\ \underline{c}\end{array}\right\}\longrightarrow d\right)$$

to indicate that the implications $a\longrightarrow x$ and $\left.\begin{array}{c}x\\ c\end{array}\right\}\longrightarrow d$ need supporting reasons.

(a) $\left[\left.\begin{array}{r}\underline{a}\longrightarrow b\\ \underline{\ }\longrightarrow d\end{array}\right\}\longrightarrow f\right]\longrightarrow\left(\left.\begin{array}{c}\underline{a}\\ \underline{c}\end{array}\right\}\longrightarrow f\right).$

(b) $\left[\left.\begin{array}{r}\left.\begin{array}{r}\underline{a}\longrightarrow x\\ \underline{b}\end{array}\right\}\longrightarrow c\\ \underline{d}\end{array}\right\}\longrightarrow f\right]\longrightarrow\left(\left.\begin{array}{c}\underline{a}\\ \underline{b}\\ \underline{d}\end{array}\right\}\longrightarrow f\right)$

(c) $\left[\left.\begin{array}{c}a\\ \underline{b}\end{array}\right\}\longrightarrow d\right]\longrightarrow(\underline{b}\longrightarrow d).$

(d) $\left[\left.\begin{array}{r}\underline{x}\longrightarrow\left\{\begin{array}{l}a\\ b\end{array}\right.\\ \underline{y}\end{array}\right\}\longrightarrow z\right]\longrightarrow\left(\left.\begin{array}{c}\underline{x}\\ \underline{y}\end{array}\right\}\longrightarrow z\right).$

(e) $\left[\underline{a}\longrightarrow b\longrightarrow\left\{\left.\begin{array}{c}c\\ d\\ e\end{array}\right\}\longrightarrow f\right.\right]\longrightarrow(\underline{a}\longrightarrow f).$

2. Supply the best conclusion to each deductive sequence so that the resulting argument is valid.

(a) $\left[\left.\begin{array}{r}\underline{a}\longrightarrow b\\ \underline{c}\end{array}\right\}\longrightarrow x\longrightarrow y\right]\longrightarrow ?$

(b) $\left[\left.\begin{array}{l}\left.\begin{array}{l}\underline{a}\\ \underline{b}\end{array}\right\}\longrightarrow x\\ \qquad\quad \underline{c}\end{array}\right\}\longrightarrow y\right]\longrightarrow ?$

(c) $\left[\left.\begin{array}{l}\left.\begin{array}{l}a\\ \underline{x}\end{array}\right\}\longrightarrow b\\ \qquad\quad \underline{y}\end{array}\right\}\longrightarrow c\right]\longrightarrow ?$

(d) $\left[\left.\begin{array}{l}\left.\begin{array}{l}\left.\begin{array}{l}\underline{x}\\ \underline{y}\end{array}\right\}\longrightarrow a\\ \qquad\quad \underline{z}\end{array}\right\}\longrightarrow c\\ \qquad\qquad\qquad \underline{d}\end{array}\right\}\longrightarrow e\right]\longrightarrow ?$

(e) $\left[\left.\begin{array}{l}\qquad\qquad \underline{a}\longrightarrow k\\ \left.\begin{array}{l}\left.\begin{array}{l}\underline{b}\\ \underline{c}\end{array}\right\}\longrightarrow l\\ \qquad\quad d\end{array}\right\}\longrightarrow m\end{array}\right\}\longrightarrow n\right]\longrightarrow ?$

3. Show how the given statements may be arranged to form a deductive sequence. *Example:* given statements:

$$e\longrightarrow h,\quad a\longrightarrow x,\quad h\longrightarrow y,\quad x\longrightarrow m,\quad \left.\begin{array}{l}y\\ m\end{array}\right\}\longrightarrow p,$$

deductive sequence:

$$\left.\begin{array}{l}e\longrightarrow h\longrightarrow y\\ a\longrightarrow x\longrightarrow m\end{array}\right\}\longrightarrow p.$$

(a) $x\longrightarrow\left\{\begin{array}{l}a\\ b\end{array}\right.,\ \left.\begin{array}{l}w\\ f\end{array}\right\}\longrightarrow g,\ (a \wedge b \wedge g)\longrightarrow y.$

(b) $a\longrightarrow\left\{\begin{array}{l}b\\ c\end{array}\right.,\ d\longrightarrow\left\{\begin{array}{l}e\\ f\end{array}\right.,\ b\longrightarrow g,\ c\longrightarrow h,\ e\longrightarrow i,\ f\longrightarrow j,$
$(g \wedge h \wedge i \wedge j)\longrightarrow k.$

(c) $x\longrightarrow(c \wedge y),\ a\longrightarrow w,\ (w \wedge c)\longrightarrow d,\ (d \wedge y)\longrightarrow z.$

(d) $f\longrightarrow(y \wedge c),\ x,\ a\longrightarrow(b \wedge d),\ (m \wedge q)\longrightarrow l,\ y\longrightarrow i,\ (b \wedge d \wedge x)\longrightarrow m,\ c\longrightarrow k,$
$(i \wedge k)\longrightarrow q.$

(e) $s,\ q\longrightarrow h,\ w\longrightarrow b,\ (x \wedge h)\longrightarrow k,\ (m \wedge s)\longrightarrow q,\ b\longrightarrow(x \wedge m).$

(f) $a\longrightarrow \sim s,\ b,\ (q \wedge b)\longrightarrow \sim m,\ t,\ (h \wedge x)\longrightarrow m,\ x,\ (t \wedge \sim s)\longrightarrow q.$

Proving Theorems about Sets

On page 25 we listed some definitions and assumptions about sets. In this section we gain some experience with proof, using these definitions and assumptions to prove some theorems about sets that we will need in later chapters. The definitions and assumptions previously stated are restated here for reference. Set definitions are indicated by SD, set postulates by SP, and set theorems by ST.

SD_1: $A = B\longleftrightarrow(A \subseteq B \wedge B \subseteq A).$

SD_2: (a) $A \cap B \cap C = (A \cap B) \cap C.$

(b) $A \cup B \cup C = (A \cup B) \cup C.$

SP_1: (a) $A \cap B = B \cap A.$
(b) $A \cup B = B \cup A.$

SP_2: (a) $A \cap A = A.$
(b) $A \cup A = A.$

SP_3: (a) $(A \cap B) \cap C = A \cap (B \cap C).$
(b) $(A \cup B) \cup C = A \cup (B \cup C).$

SP_4: (a) $A \cap (B \cup C) = (A \cap B) \cup (A \cap C).$
(b) $A \cup (B \cap C) = (A \cup B) \cap (A \cup C).$

SP_5: $B \cap B' = \emptyset.$

SP_6: $(A')' = A.$

SP_7: $A = B \longleftrightarrow A' = B'.$

SP_8: $\emptyset' = U.$

SP_9: (a) $U \cap A = A.$
(b) $U \cup A = U.$

SP_{10}: $A \subseteq B \longleftrightarrow A' \cup B = U.$

SP_{11}: $(A \cap B)' = A' \cup B'.$

SP_{12}: $(A \cup B)' = A' \cap B'.$

Proofs for the following theorems are considered in the examples and exercises below. Some of these theorems are illustrated by specific statements or by diagrams. All are fairly "obvious." However, it is more difficult to construct a flow-diagram proof for one of these theorems than it is to "see" that it is true by studying a diagram.

ST_1: (a) $A = B \longrightarrow A \cap C = B \cap C.$
(b) $A = B \longrightarrow A \cup C = B \cup C.$

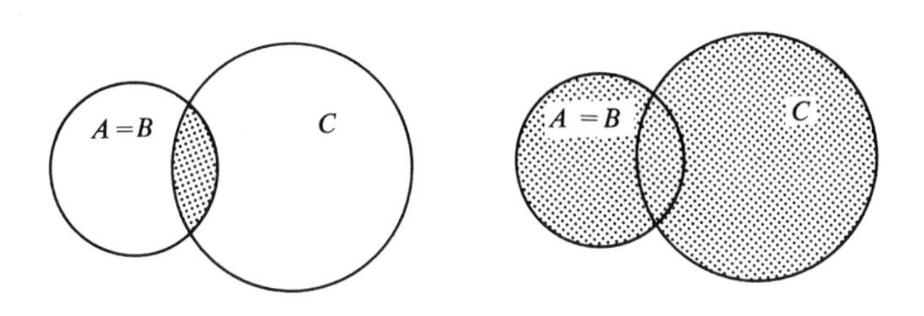

ST_2: $\left.\begin{matrix} x \in B \\ y \notin B \end{matrix}\right\} \longrightarrow x \neq y.$

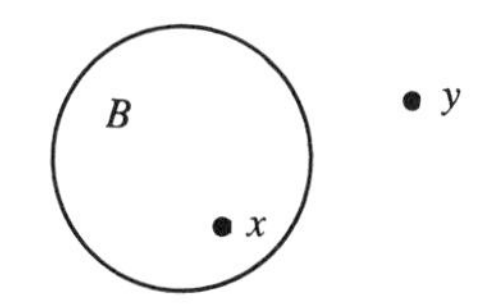

ST_3: $\left.\begin{array}{l} x \in A \\ x \notin B \end{array}\right\} \longrightarrow A \neq B.$

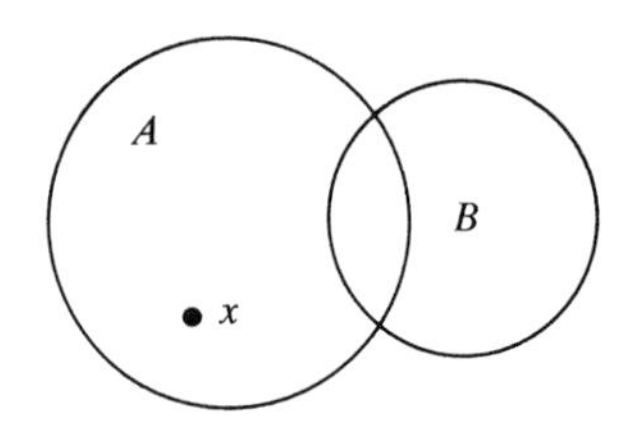

ST_4: $B \cup B' = U.$

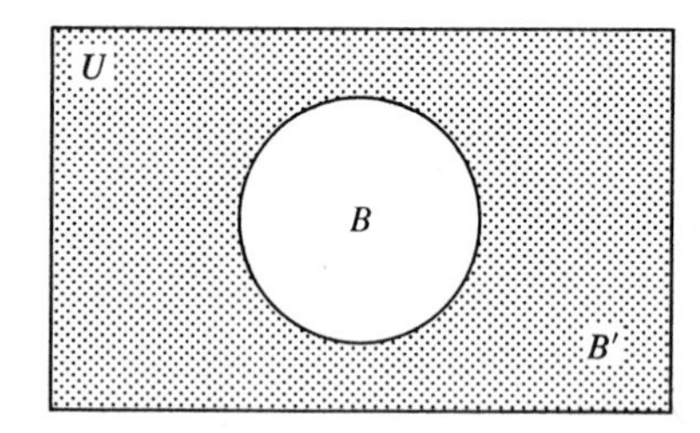

ST_5: $U' = \emptyset.$
ST_6: $\emptyset \cap A = \emptyset.$
ST_7: $\emptyset \cup A = A.$
ST_8: $A \subseteq B \longleftrightarrow A \cap B' = \emptyset.$

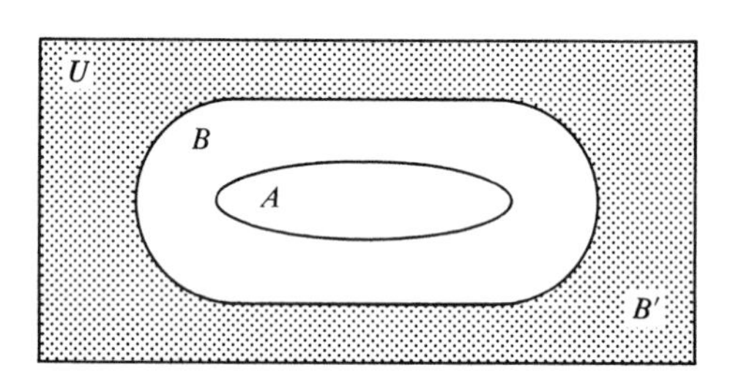

ST_9: $A \subseteq B \longleftrightarrow B' \subseteq A'.$

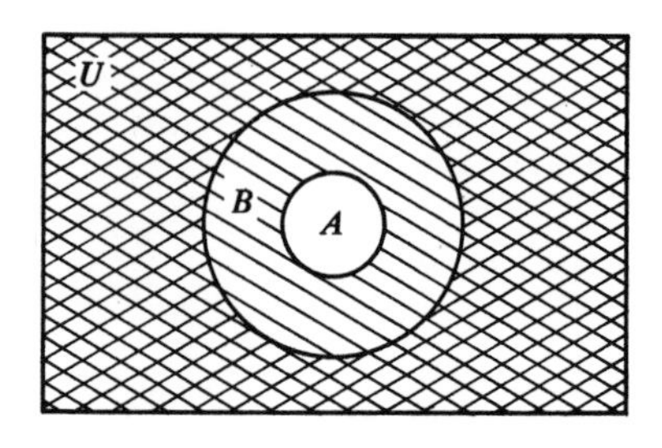

ST_{10}: $A \subseteq B \longleftrightarrow A \cup B = B.$

When a theorem leads easily to the proof of another statement, the statement is often called a *corollary*. ST_{10} has the following corollary.

Corollary to ST_{10}: $\emptyset \subseteq B$

ST_{11}: $A \subseteq B \longleftrightarrow A \cap B = A.$

ST_{12}: $A \subseteq A \cup B.$

ST_{13}: $A \cap B \subseteq A.$

ST_{14}: $\left.\begin{array}{r} A \subseteq B \\ B \cap C = \emptyset \end{array}\right\} \longrightarrow A \cap C = \emptyset.$

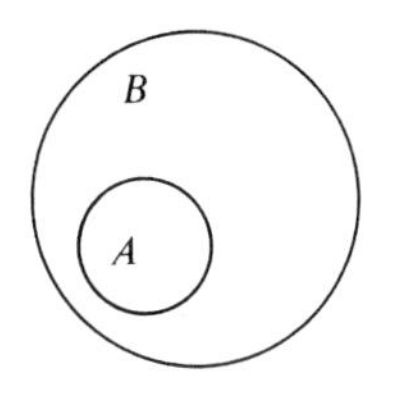

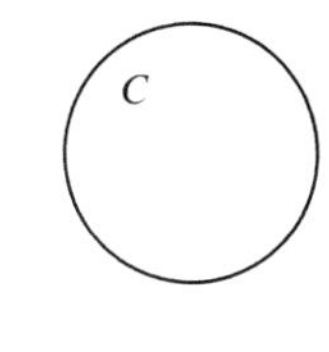

ST_{15}: $\left.\begin{array}{r} A \subseteq B \\ B \subseteq C \end{array}\right\} \longrightarrow A \subseteq C.$

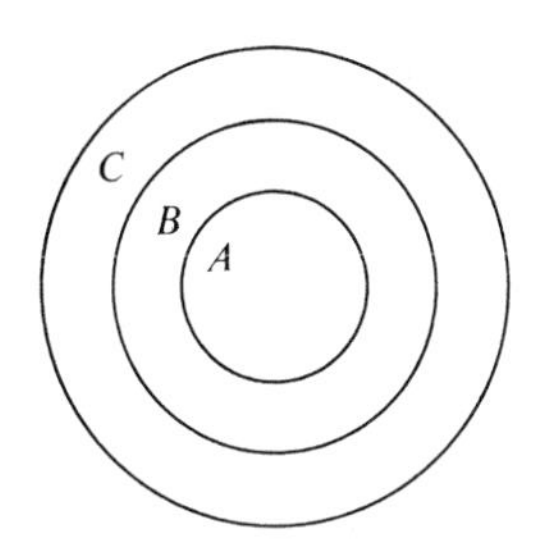

Corollary to ST_{15}: $\left.\begin{array}{r} x \in A \\ A \subseteq B \end{array}\right\} \longrightarrow x \in B.$

ST_{16}: $A \subseteq B \longrightarrow A \cap C \subseteq B \cap C.$

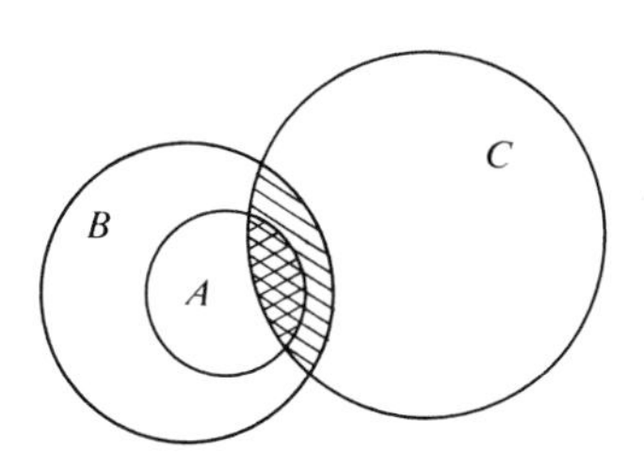

ST_{17}: $\left.\begin{array}{r} A \subseteq B \\ C \subseteq D \end{array}\right\} \longrightarrow A \cap C \subseteq B \cap D.$

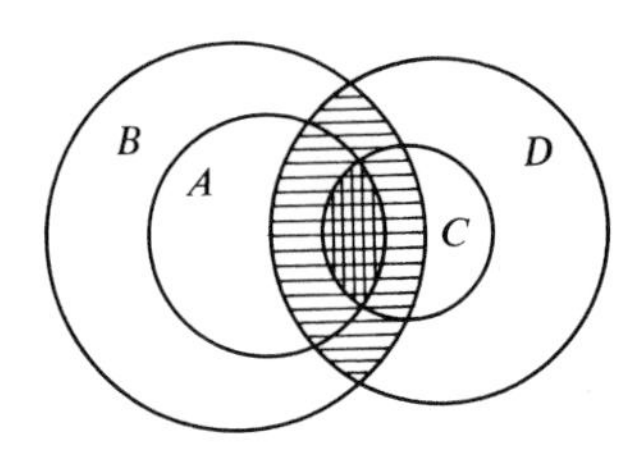

ST_{18}: $\left.\begin{array}{r} A \subseteq B \\ C \subseteq D \end{array}\right\} \longrightarrow A \cup C \subseteq B \cup D.$

ST_{19}: $\left.\begin{array}{r} A \subseteq B \cup C \\ A \cap B = \emptyset \end{array}\right\} \longrightarrow A \subseteq C.$

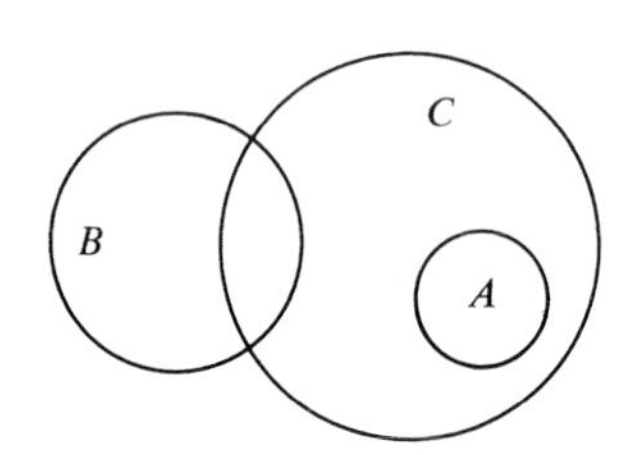

Exercises

In proving any given set theorem you may use set postulates, set definitions, properties of equality of sets, and any theorems that precede the given theorem in the sequence of set theorems as reasons.

1. Supply reasons to the following proof of ST_4: $B \cup B' = U$.

   (1) $B \cap B' = \emptyset \xrightarrow{(2)} (B \cap B')' = \emptyset' \xrightarrow{(3)} B' \cup (B')' = \emptyset' \xrightarrow{(4)} B' \cup B = \emptyset' \xrightarrow{(5)} B \cup B' = U.$

2. Complete the following proof of ST_5: $U' = \emptyset$.

   (1) $\emptyset' = U \xrightarrow{(2)} (\emptyset')' = U' \xrightarrow{(3)} ?$

3. (a) Complete the following proof of ST_6: $\emptyset \cap A = \emptyset$.

   (1) $U \cup A' = U \xrightarrow{(2)} (U \cup A')' = U' \xrightarrow{(3)} ?$

   (b) Prove ST_7: $\emptyset \cup A = A$.

4. Prove ST_8: $A \subseteq B \longleftrightarrow A \cap B' = \emptyset$.

5. Supply reasons for the following proof of ST_9: $A \subseteq B \longleftrightarrow B' \subseteq A'$.

   $A \subseteq B \xleftrightarrow{(1)} A \cap B' = \emptyset \xleftrightarrow{(2)} B' \cap A = \emptyset \xleftrightarrow{(3)} B' \cap (A')' = \emptyset \xleftrightarrow{(4)} B' \subseteq A'$. Therefore

   $A \subseteq B \xleftrightarrow{(5)} B' \subseteq A'$.

6. (a) Supply reasons for the following proof of the left-to-right implication of ST_{10}: $A \subseteq B \longleftrightarrow A \cup B = B$.

$$\underline{A \subseteq B} \xrightarrow{(1)} A \cap B' = \emptyset \xrightarrow{(2)} (A \cap B') \cup B = \emptyset \cup B \xrightarrow{(3)} (A \cap B') \cup B = B \overset{(4)}{\dashv}$$

$$\leftrightarrow (A \cup B) \cap (B \cup B') = B \xrightarrow{(5)} (A \cup B) \cap U = B \xrightarrow{(6)} A \cup B = B.$$

(b) Supply reasons for the right-to-left implication of ST_{10}.

$$\underline{A \cup B = B} \xrightarrow{(1)} (A \cup B) \cap B' = B \cap B' \xrightarrow{(2)} (A \cup B) \cap B' = \emptyset \overset{(3)}{\dashv}$$

$$\leftrightarrow (A \cap B') \cup (B \cap B') = \emptyset \xrightarrow{(4)} (A \cap B') \cup \emptyset = \emptyset \xrightarrow{(5)} A \cap B' = \emptyset \xrightarrow{(6)} A \subseteq B.$$

7. Prove the Corollary to ST_{10}: $\emptyset \subseteq B$.
8. Prove ST_{11}: $A \subseteq B \longleftrightarrow A \cap B = A$.
9. (a) Supply reasons for the following proof of ST_{12}: $A \subseteq A \cup B$.

$$\left.\begin{array}{l}(1)\ A \cup B = A \cup B \\ (2)\ A \cup A = A\end{array}\right\} \xrightarrow{(3)} (A \cup A) \cup B = A \cup B \xrightarrow{(4)} A \cup (A \cup B) = A \cup B \overset{(5)}{\dashv}$$

$$\leftrightarrow A \subseteq A \cup B$$

(b) Prove ST_{13}: $A \cap B \subseteq A$.

10. Prove ST_{14}:

$$\left.\begin{array}{l}A \subseteq B \\ B \cap C = \emptyset\end{array}\right\} \longrightarrow A \cap C = \emptyset.$$

11. Prove ST_{15}:

$$\left.\begin{array}{l}A \subseteq B \\ B \subseteq C\end{array}\right\} \longrightarrow A \subseteq C.$$

12. Prove ST_{16}: $A \subseteq B \longrightarrow A \cap C \subseteq B \cap C$.
13. Prove ST_{17}:

$$\left.\begin{array}{l}A \subseteq B \\ C \subseteq D\end{array}\right\} \longrightarrow A \cap C \subseteq B \cap D.$$

14. Prove ST_{18}:

$$\left.\begin{array}{l}A \subseteq B \\ C \subseteq D\end{array}\right\} \longrightarrow A \cup C \subseteq B \cup D.$$

15. Prove ST_{19}:

$$\left.\begin{array}{l}A \subseteq B \cup C \\ A \cap B = \emptyset\end{array}\right\} \longrightarrow A \subseteq C.$$

References

Adler, *Modern Geometry*
Carroll, *Pillow Problems and a Tangled Tale*
Carroll, *Symbolic Logic and Game of Logic*
Courant and Robbins, *What Is Mathematics?*
Golos, *Foundations of Euclidean and Non-Euclidean Geometry*
Hemmerling, *Fundamentals of College Geometry*
Katz, *Axiomatic Analysis*
Stabler, *An Introduction to Mathematical Thought*
Stoll, *Sets, Logic and Axiomatic Theories*
Suppes, *Introduction to Logic*

Chapter 3

Incidence Geometry—Points, Lines, and Planes

We have noted the need for undefined terms in any deductive system. The fundamental undefined terms in our system are *point, line,* and *plane.* The fact that we cannot define or describe points, lines, and planes does not mean that we cannot think effectively using these terms. We can play chess without having a complete description of each piece in a chess set. We don't need to know whether a piece is made of wood or ivory, whether it is carved or molded, large or small. The only features that really matter are its properties—that is, the things it can do, the way it moves, the way it can capture other pieces. These are prescribed by the *rules* of the game, which are accepted without proof or argument by those who play the game. Points, lines, and planes are somewhat like the pieces in a game of chess. Their real significance comes from the rules of the game—the postulates we agree to accept when we study geometry.

Incidence Relationships

If we say that two sets of points are equal or that one set of points is a subset of another, we are describing incidence relationships for these sets of points. Thus statements such as $E \in S$, $E \notin T$, $W \subseteq T$, $K = S \cap T$, or $\{P\} = S \cap T$ describe incidence relationships. When the intersection of two sets is a single point, we may omit the braces. Thus our last statement, which applies to the second drawing, could be written $P = S \cap T$.

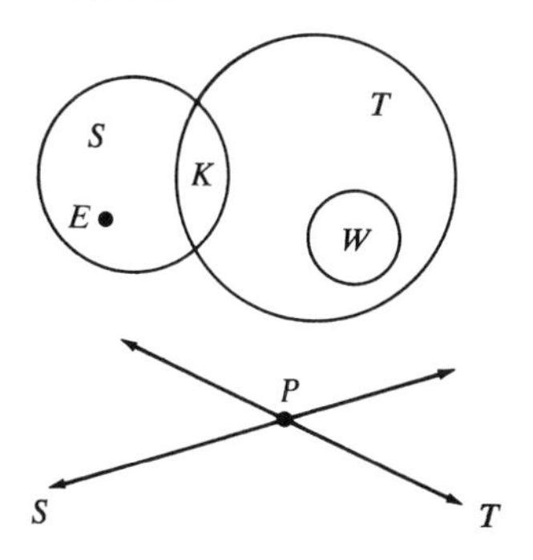

We state a definition and four incidence postulates for points and lines.

DEFINITION *Space is the set of all points. (We shall denote space by the symbol S.)*

POSTULATE 1 *Space contains at least two distinct points.*

POSTULATE 2 *If P and Q are distinct points, then there is exactly one line that contains them.*

POSTULATE 3 *Every line is a set of points and contains at least two points.*

POSTULATE 4 *No line contains all the points in space.*

There are two symbols we shall use to denote a line.

(a) We may use a capital letter to denote a line. When there are several lines, we may use a capital letter with subscripts, such as L_1, L_2, L_3, and so on.
(b) We shall use the symbol $\overleftrightarrow{PQ}$ to denote the line that contains the distinct points P and Q.

We know that given any line L there are at least two distinct points P and Q in it (Postulate 3), and at least one point X is not in it (Postulate 4). To distinguish between a set of points that is contained in a line and a set that is not, we introduce the terms collinear and noncollinear.

DEFINITION *The points in set T are* collinear *if and only if there is a line that contains all the points in set T.*

Since contradictions of equivalent statements are equivalent, we also have:

The points in a set are noncollinear if and only if there is no line that contains all of them.

Looking at the diagram, we see that $\{P, S, R, Q\}$ appears to be a set of collinear points, while $\{P, S, X, R\}$ is a set of noncollinear points.

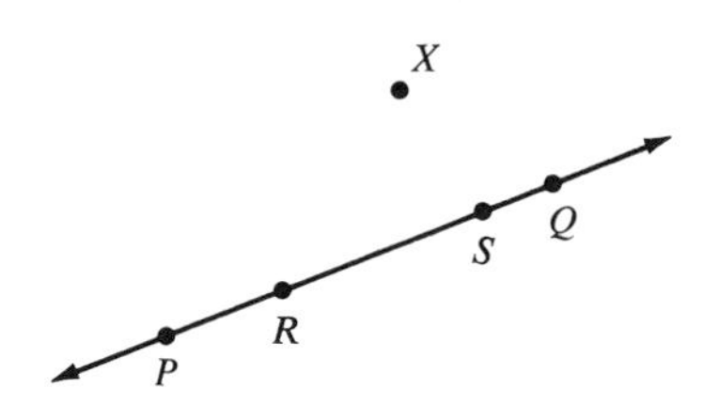

DEFINITION *Two sets of points are collinear if their union is collinear.*

Deductions from the Postulates

We continue our discussion of the incidence postulates by considering Postulate 2, "If P and Q are distinct points, then there is exactly one line that contains them." In order to understand Postulate 2 we must consider precisely what we mean by such expressions as "the set T has exactly k elements" and "the set T has k elements," where $k \in W$. We shall see that these two expressions are not equivalent.

When we say a set contains exactly k elements, we mean that the number of distinct elements is not more than k *and* not less than k. If T is a set, we shall use the symbol $n(T)$ to denote the number of elements in the set. Thus for $k \in W$ we have

$$T \text{ contains exactly } k \text{ elements} \longleftrightarrow n(T) = k \longleftrightarrow [n(T) \not> k \wedge n(T) \not< k].$$

When we say that a set T contains k elements, we mean that k distinct elements can be found among the elements of T. This is true if and only if $n(T) \geqq k$. Thus we agree that

$$T \text{ contains } k \text{ elements} \longleftrightarrow n(T) \geqq k \longleftrightarrow T \text{ contains at least } k \text{ distinct elements.}$$

Now that we understand the meaning of "exactly one," we see that the following statement is equivalent to Postulate 2.

> If P and Q are distinct points, then (1) there is at least one line that contains them *and* (2) there is not more than one line that contains them.

The second part of the postulate tells us that if two distinct points are in each of two lines, then the two lines are, in fact, the same line. In other words, Postulate 2 tells us that two points determine a line, as indicated by the following corollary.

COROLLARY 1 TO POSTULATE 2

(i)
$$\left.\begin{array}{l} A \in L_1 \wedge A \in L_2 \\ B \in L_1 \wedge B \in L_2 \\ A \neq B \end{array}\right\} \longrightarrow L_1 = L_2,$$

or equivalently,

(ii)
$$[A, B \in (L_1 \cap L_2) \wedge A \neq B] \longrightarrow L_1 = L_2.$$

The corresponding verbal statement is:

> If the intersection of lines L_1 and L_2 contains two distinct points, then $L_1 = L_2$.

On the basis of our agreement about the meaning of the statement that a set contains two distinct elements, we see that the hypothesis of (ii) can be written $n(L_1 \cap L_2) \geqq 2$. Therefore, (ii) is equivalent to

(iii)
$$n(L_1 \cap L_2) \geqq 2 \longrightarrow L_1 = L_2.$$

However, we know that for $x \in C$, $x \geqq 2 \longleftrightarrow x > 1$, and we have the following restatement of Corollary 1 to Postulate 2:

(iv) $$n(L_1 \cap L_2) > 1 \longrightarrow L_1 = L_2.$$

Corollary 2 to Postulate 2 justifies the common practice of naming a line by naming *any two* points in it.

COROLLARY 2 TO POSTULATE 2 *For points A, B, C, D*

$$\left.\begin{array}{l} A \neq B \\ C \neq D \\ A, B, C, D \text{ are in line } L \end{array}\right\} \longrightarrow \overleftrightarrow{AB} = \overleftrightarrow{CD} = L.$$

PROOF

$$\begin{array}{lll}
\underline{A \neq B} \xrightarrow{(1)} \text{There is a line } AB \text{ such that} & \left.\begin{array}{l} A, B \in \overleftrightarrow{AB} \\ \underline{A, B \in L} \\ \underline{A \neq B} \end{array}\right\} \xrightarrow{(2)} \overleftrightarrow{AB} = L & \\
\underline{C \neq D} \xrightarrow{(1)} \text{There is a line } CD \text{ such that} & \left.\begin{array}{l} C, D \in \overleftrightarrow{CD} \\ \underline{C, D \in L} \\ \underline{C \neq D} \end{array}\right\} \xrightarrow{(2)} \overleftrightarrow{CD} = L & \Bigg\} \xrightarrow{(3)} \dashv
\end{array}$$

$$\leftrightarrow \overleftrightarrow{AB} = \overleftrightarrow{CD} = L.$$

Reasons:

(1) Postulate 2.
(2) Corollary 1 to Postulate 2.
(3) Transitive property of equality.

We state two other corollaries to Postulate 2 and consider their proofs in the exercises.

COROLLARY 3 TO POSTULATE 2 *If points P and Q are in each of two distinct lines, then $P = Q$.*

COROLLARY 4 TO POSTULATE 2 *If two distinct lines have a point in common, then any* other point *of one is not in the other.*

The following set theorem is often useful:

ST_{20}: If two sets (lines) have a single element (point) in common and subsets of these sets (lines) intersect, then this element (point) is the intersection of these subsets.

We state the following theorem:

THEOREM 3-1 *If two distinct lines intersect, then their intersection contains exactly one point.*

In order to prove Theorem 3-1, we must show:

$$\left.\begin{array}{l} L_1 \neq L_2 \\ L_1 \text{ and } L_2 \text{ intersect} \end{array}\right\} \longrightarrow n(L \cap L) = 1.$$

PROOF

$$\left.\begin{array}{l} \underline{L_1 \neq L_2 \xrightarrow{(1)} n(L_1 \cap L_2) \ngtr 1} \\ \underline{L_1 \text{ and } L_2 \text{ intersect}} \xrightarrow{(2)} n(L_1 \cap L_2 \nless 1 \end{array}\right\} \xrightarrow{(3)} n(L_1 \cap L_2) = 1.$$

Reasons:

(1) Contrapositive of Corollary 1 to Postulate 2 as stated in (iv).
(2) Definition of "intersecting sets."
(3) Definition of "exactly one."

COROLLARY TO THEOREM 3-1 *If subsets of distinct lines have a point in common, this point is their intersection.*

We shall sometimes find it convenient to use the symbol $L_1 \times L_2$ to represent such expressions as "lines L_1 and L_2 intersect in a single point" or "the intersection of lines L_1 and L_2 contains exactly one point." When we write $L_1 \times L_2(P)$, we mean that $L_1 \cap L_2 = P$. We shall use the symbol $\sim(L_1 \times L_2)$ for the statement, "The intersection of L_1 and L_2 is not a single point."

Suppose that lines L_1 and L_2 intersect in a single point $P[L_1 \times L_2(P)]$. We know by Postulate 3 that L_1 contains at least one other point, Q. Now Q is not in L_2, because, if it were, the intersection of L_1 and L_2 would contain at least two points, namely P and Q. Therefore, $Q \in L_1$ and $Q \notin L_2$. By ST_3 it follows that $L_1 \neq L_2$. Thus we have the following corollary.

COROLLARY TO POSTULATE 3 *If two lines intersect in a single point, then the lines are distinct.*

If A and B are distinct points and $C \in \overleftrightarrow{AB}$, then by definition A, B, and C are collinear. Thus

(v)
$$\left.\begin{array}{l} A \neq B \\ C \in \overleftrightarrow{AB} \end{array}\right\} \longrightarrow A, B, C \text{ are collinear.}$$

In the exercises you are asked to prove the second converse of statement (v):

(vi)
$$\left.\begin{array}{l} A \neq B \\ A, B, C \text{ are collinear} \end{array}\right\} \longrightarrow C \in \overleftrightarrow{AB}.$$

Combining (v) and (vi), we have the following theorem.

THEOREM 3-2 *If A and B are distinct points and C is any point, then $C \in \overleftrightarrow{AB} \longleftrightarrow A, B, C$ are collinear.*

Observe that Theorem 3-2 can be expressed as follows: *If A and B are distinct points and C is any point, then*

(1) $C \in \overleftrightarrow{AB} \longrightarrow A, B, C$ *are collinear,*
(2) A, B, C *are collinear* $\longrightarrow C \in \overleftrightarrow{AB}$,
(3) A, B, C *are noncollinear* $\longrightarrow C \notin \overleftrightarrow{AB}$,
(4) $C \notin \overleftrightarrow{AB} \longrightarrow A, B, C$ *are noncollinear.*

We conclude this section by stating two additional theorems. The proofs are requested in the exercises.

THEOREM 3-3 *Space contains at least three distinct points that are noncollinear.*

THEOREM 3-4 *Space contains at least three distinct lines.*

Exercises

1. (a) If point P is in line L and point Q is not in line L, what can be said of points P and Q?
   (b) Complete the following statement.
$$\left.\begin{matrix} P \in L \\ Q \notin L \end{matrix}\right\} \longrightarrow ?$$
   (c) On which set theorem do you base your conclusions in parts (a) and (b)?
2. Explain why every line is a proper subset of space.
3. We agreed that "line" is to be regarded as an undefined term. Explain why Postulate 3 is not a definition of a line.
4. If A, B, and C are points, is the following implication true?
$$\left.\begin{matrix} A \neq B \\ B \neq C \end{matrix}\right\} \longrightarrow A \neq C.$$
5. (a) Write a verbal statement of the following implication about lines L_1 and L_2 and points P and Q.
$$\left.\begin{matrix} L_1 \cap L_2 = \emptyset \\ P \in L_1 \\ Q \in L_2 \end{matrix}\right\} \longrightarrow P \neq Q.$$
   (b) Use set properties to supply reasons to the following proof of the implication in part (a).
$$\left.\begin{matrix} \underline{L_1 \cap L_2 = \emptyset} \\ \underline{P \in L_1} \end{matrix}\right\} \xrightarrow{(1)} \left.\begin{matrix} P \notin L_2 \\ \underline{Q \in L_2} \end{matrix}\right\} \xrightarrow{(2)} P \neq Q.$$
6. We state Corollary 1 to Postulate 2 in the following bracket form:
$$\left.\begin{matrix} (1)\ P \neq Q \\ (2)\ P \in L_1 \\ (3)\ P \in L_2 \\ (4)\ Q \in L_1 \\ (5)\ Q \in L_2 \end{matrix}\right\} \longrightarrow L_1 = L_2.$$

(a) Form the first contrapositive and verbalize it.
(b) Form the second contrapositive and verbalize it.
(c) Explain why parts (a) and (b) of this exercise prove, respectively, Corollaries 3 and 4 to Postulate 2.

7. Supply reasons for the following proof of ST_{20}:

$$\left.\begin{array}{l} S_1 \subseteq L_1 \\ S_2 \subseteq L_2 \\ L_1 \cap L_2 = \{P\} \\ S_1 \cap S_2 \neq \emptyset \end{array}\right\} \longrightarrow S_1 \cap S_2 = \{P\}.$$

$$\left.\begin{array}{l} S_1 \subseteq L_1 \\ \underline{S_2 \subseteq L_2} \end{array}\right\} \xrightarrow{(1)} \left.\begin{array}{l} S_1 \cap S_2 \subseteq L_1 \cap L_2 \\ \underline{L_1 \cap L_2 = \{P\}} \end{array}\right\} \xrightarrow{(2)} S_1 \cap S_2 \subseteq \{P\} \overset{(3)}{\dashv}$$

$$\mapsto \left.\begin{array}{r} (S_1 \cap S_2 = \emptyset) \vee (S_1 \cap S_2 = \{P\}) \\ \underline{S_1 \cap S_2 \neq \emptyset} \end{array}\right\} \xrightarrow{(4)} S_1 \cap S_2 = \{P\}.$$

8. Prove the Corollary to Theorem 3-1.
9. Supply reasons for the following proof of statement (vi):

$$\left.\begin{array}{l} A \neq B \\ A, B, C \text{ are collinear} \end{array}\right\} \longrightarrow C \in \overleftrightarrow{AB}.$$

$$\underline{A, B, C \text{ are collinear}} \xrightarrow{(1)} \text{There is a line } L \text{ such that} \begin{cases} C \in L \\ B \in L \\ A \in L \end{cases}$$

$$\underline{A \neq B} \xrightarrow{(2)} \text{There is a line } AB \wedge A, B \in \overleftrightarrow{AB}$$

$$\underline{A \neq B} \quad \overset{(3)}{\dashv}$$

$$\mapsto \left.\begin{array}{r} \overleftrightarrow{AB} = L \\ C \in L \end{array}\right\} \xrightarrow{(4)} C \in \overleftrightarrow{AB}.$$

*L 10. (a) If L_1 and L_2 are lines, explain why the following implication is true:

$$L_1 \cap L_2 \neq \emptyset \longrightarrow (L_1 = L_2 \vee L_1 \times L_2).$$

(b) Verbalize each of the following statements and show each is equivalent to the implication in part (a).
(1) $[L_1 \cap L_2 \neq \emptyset \wedge L_1 \neq L_2] \longrightarrow L_1 \times L_2$.
(2) $[L_1 \cap L_2 \neq \emptyset \wedge \sim(L_1 \times L_2)] \longrightarrow L_1 = L_2$.
(3) $[L_1 \neq L_2 \wedge \sim(L_1 \times L_2)] \longrightarrow L_1 \cap L_2 = \emptyset$.

11. How many distinct lines are determined by points in set T if T consists of distinct points no three of which are collinear and $n(T) = 2$? $n(T) = 3$? $n(T) = 4$? $n(T) = 5$? $n(T) = k$ $(k \in C \wedge k \geqq 2)$?
12. Prove Theorem 3-3.
13. Prove Theorem 3-4.

Incidence Postulates for Points, Lines, and Planes

We next consider planes and their relationships with points, lines, and other planes. Our discussion is based on the following postulates.

* We use an **L** to indicate an exercise that may be used in other proofs.

POSTULATE 5 *If P, Q, and R are three distinct noncollinear points, there is exactly one plane that contains them.*

POSTULATE 6 *Every plane is a set of points and contains at least three noncollinear points.*

POSTULATE 7 *No plane contains all the points in space.*

Notation: There are two symbols we shall use to denote a plane.

(a) We may use a capital letter, or, when there are several planes, we may use a capital letter with subscripts: $M_1, M_2, M_3, \ldots$.
(b) We shall use the symbol pl PQR to denote the plane that contains the distinct noncollinear points P, Q, and R.

Since there is exactly one plane that contains three distinct noncollinear points, we say that three distinct noncollinear points *determine* a plane.
To distinguish between a set of points that is contained in a plane and a set of points that is not, we introduce the terms *coplanar* and *noncoplanar*.

DEFINITION *The points in a set T are* coplanar *if and only if there is a plane that contains all the points in set T.*

Since contradictions of equivalent statements are equivalent, this definition is equivalent to the statement:

The points in set T are noncoplanar $\longleftrightarrow$ There is no plane that contains all the points in set T.

DEFINITION *Two sets of points are coplanar if their union is coplanar.*

By an analysis similar to that applied to Postulate 2 we see that Postulate 5 has two parts and can be stated as follows:

POSTULATE 5 *If P, Q, and R are three distinct noncollinear points, then* (1) *there is at least one plane that contains them* and (2) *there is not more than one plane that contains them.*

The second part of the postulate tells us that if three distinct noncollinear points are in each of two planes (and thus in their intersection), then the two planes are, in fact, the same plane. In other words, Postulate 5 tells us that the following implication is true.

COROLLARY 1 TO POSTULATE 5

$(P, Q, R$ *are distinct noncollinear points* $\wedge\ P, Q, R \in M_1 \cap M_2) \longrightarrow M_1 = M_2$.

We can now state Corollary 2 to Postulate 5. This corollary justifies the common practice of naming a plane by naming any three noncollinear points in the plane.

COROLLARY 2 TO POSTULATE 5

$$\left.\begin{array}{l} X, Y, Z \text{ are distinct noncollinear points} \\ A, B, C \text{ are distinct noncollinear points} \\ A, B, C, X, Y, Z \text{ are in plane } M \end{array}\right\} \longrightarrow \text{pl } ABC = \text{pl } XYZ = M.$$

We can use Corollary 1 to Postulate 5 to prove the following theorem. The proof is considered in the exercises.

THEOREM 3-5 *If B, C, D are distinct noncollinear points and A is any point, then*

(1) $A \in \text{pl } BCD \longrightarrow A, B, C, D$ *are coplanar,*
(2) A, B, C, D *are coplanar* $\longrightarrow A \in \text{pl } BCD,$
(3) A, B, C, D *are noncoplanar* $\longrightarrow A \notin \text{pl } BCD,$
(4) $A \notin \text{pl } BCD \longrightarrow A, B, C, D$ *are noncoplanar.*

We are now ready to prove

THEOREM 3-6 *Space contains four distinct points that are noncoplanar.*

PROOF

According to Theorem 3-3, space contains three distinct points W, X, Y that are noncollinear. ($\underline{W \in S}$, $\underline{X \in S}$, $\underline{Y \in S}$, $\underline{W \neq X}$, $\underline{X \neq Y}$, $\underline{W \neq Y}$, and W, X, Y are noncollinear.) Postulate 5 tells us that these three points are contained in a plane, namely pl WXY. We know from Postulate 7 that there is a point Z in space that is not in pl WXY ($\underline{Z \in S} \wedge Z \notin \text{pl } WXY$). It follows that $\underline{Z \neq W}$ because $(W \in \text{pl } WXY \wedge Z \notin \text{pl } WXY) \longrightarrow W \neq Z$. Why? Similarly $\underline{Z \neq X}$ and $\underline{Z \neq Y}$. We know that $\underline{W, X, Y, Z \text{ are noncoplanar}}$ because by Theorem 3-5(4) $Z \notin \text{pl } WXY \longrightarrow W, X, Y, Z$ are noncoplanar. The underlined statements tell us that S contains four distinct points that are noncoplanar, and our proof is complete.

We now state two theorems whose proofs are considered in the exercises.

THEOREM 3-7 *If P is a point, then there is a plane that contains it.*

THEOREM 3-8 *Space contains at least two planes.*

We now state two postulates that describe relationships between lines and planes.

POSTULATE 8 *If two distinct planes intersect, their intersection is a line.*

POSTULATE 9 *If two distinct points in a line lie in a plane, then every point in the line lies in the plane.*

Postulate 8 can be expressed as follows (M_1 and M_2 represent planes):

$$\left.\begin{array}{l}(1)\ M_1 \cap M_2 \neq \emptyset \\ (2)\ M_1 \neq M_2\end{array}\right\} \longrightarrow M_1 \cap M_2 \text{ is a line.}$$

Consideration of the contrapositives will give us a deeper understanding of this postulate. Consider the first contrapositive.

Symbolic statement	*Verbal statement*
$\left.\begin{array}{l}M_1 \cap M_2 \text{ is not a line} \\ M_1 \neq M_2\end{array}\right\} \longrightarrow M_1 \cap M_2 = \emptyset.$	If the intersection of two distinct planes is not a line, then the planes do not intersect.

We use the word "parallel" to describe two planes that do not intersect.

DEFINITION *If two planes do not intersect, then they are* parallel.

We indicate that plane M_1 is parallel to plane M_2 by writing pl $M_1 \parallel$ pl M_2. Then pl $M_1 \parallel$ pl $M_2 \longleftrightarrow M_1 \cap M_2 = \emptyset$.

We now consider the second contrapositive of Postulate 8.

$\left.\begin{array}{l}M_1 \cap M_2 \neq \emptyset \\ M_1 \cap M_2 \text{ is not a line}\end{array}\right\} \longrightarrow M_1 = M_2.$	If two planes intersect and their intersection is not a line, then they are equal (they are the same plane).

If S and T are points, L is a line, and M is a plane, Postulate 9 can be expressed as follows:

(vii) $$\left.\begin{array}{l}(1)\ T \neq S \\ (2)\ T \in L \\ (3)\ T \in M \\ (4)\ S \in L \\ (5)\ S \in M\end{array}\right\} \longrightarrow L \subseteq M.$$

The conjunction of statements (2), (3), (4), and (5) in the symbolic statement of Postulate 9 is equivalent to the statement $T, S \in L \cap M$. Why? Therefore Postulate 9 also has the following symbolic expression:

$$\left.\begin{array}{l}T \neq S \\ T, S \in L \cap M\end{array}\right\} \longrightarrow L \subseteq M.$$

Thus if the intersection of a line and a plane contains at least two points, then Postulate 9 tells us that the line lies in the plane—that is,

(viii) $$n(L \cap M) \geqq 2 \longrightarrow L \subseteq M.$$

The contrapositive of this statement is

(ix) $$L \nsubseteq M \longrightarrow n(L \cap M) \ngtr 1.$$

We are now ready to prove

THEOREM 3-9 *If a line intersects a plane not containing it, the intersection is a single point.*

PROOF

Let L be a line and M be a plane.

$$\left.\begin{array}{l}\underline{L \nsubseteq M} \xrightarrow{(1)} n(L \cap M) \ngtr 1 \\ \underline{L \text{ intersects } M} \xrightarrow{(2)} n(L \cap M) \nless 1\end{array}\right\} \xrightarrow{(3)} n(L \cap M) = 1.$$

You may supply the reasons.

Suppose that plane M intersects parallel planes M_1 and M_2 in lines L_1 and L_2, respectively. We know that $\underline{L_1 \text{ and } L_2 \text{ are coplanar}}$ because they are both in plane M. Moreover,

$$\left.\begin{array}{l}L_1 \subseteq M_1 \\ L_2 \subseteq M_2 \\ M_1 \cap M_2 = \emptyset\end{array}\right\} \longrightarrow \underline{L_1 \cap L_2 = \emptyset}. \qquad \text{Why?}$$

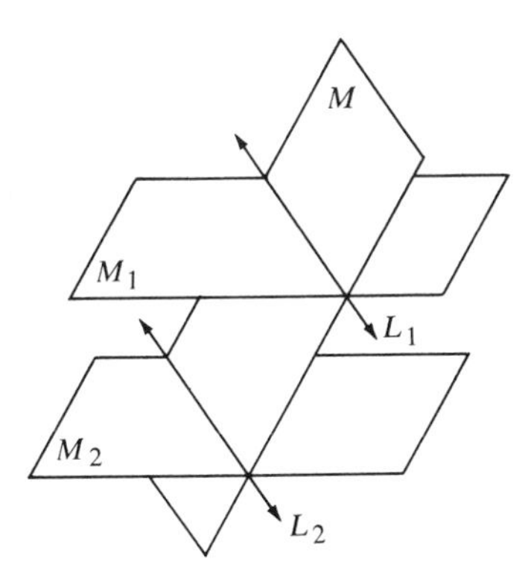

Since each of the underlined statements follows from the fact that L_1 and L_2 are the lines of intersection of plane M with the parallel planes M_1 and M_2, we have proved the following theorem.

THEOREM 3-10 *If a plane intersects two parallel planes, then the lines of intersection are coplanar and nonintersecting.*

Exercises

1. Form the first three contrapositives of Postulate 9 as expressed in statement (vii). For each contrapositive, write an equivalent verbal statement and draw a diagram illustrating it.
2. Given that distinct points A, B, C lie in planes M_1 and M_2, can you conclude that $M_1 = M_2$? Explain.
3. Let A be the intersection of planes M_1 and M_2. Sketch a figure illustrating the planes if
   (a) A is the empty set.
   (b) A contains at least two distinct points.
   (c) A contains three noncollinear points.

4. Draw a figure showing distinct planes M_1 and M_2 intersecting in line L_1 and a plane M_3 such that $L_1 \cap M_3$ is the point P. If $L_2 = M_3 \cap M_1$ and $L_3 = M_3 \cap M_2$, what is true about L_2 and L_3?
5. The proof of Theorem 3-5(1) follows at once from our definition of coplanar points.
   (a) Prove part (2) of the theorem:

$$\left.\begin{array}{l} B, C, D \text{ are distinct noncollinear points} \\ A, B, C, D \text{ are coplanar} \end{array}\right\} \longrightarrow A \in \text{pl } BCD.$$

   (b) Now that parts (1) and (2) have been proved, explain why parts (3) and (4) are true.
6. (a) If P is a given point, prove that there exist points Q and R such that P, Q, and R are noncollinear.
   (b) Use part (a) to prove Theorem 3-7: If P is a point, then there is a plane that contains it.
7. (a) After considering the following statements, write an essay proof of Theorem 3-8: Space contains at least two planes.
      (1) Space contains at least three noncollinear points.
      (2) Space contains at least one plane, M_1.
      (3) Space contains a point P such that $P \notin M_1$.
      (4) Space contains a plane M_2 such that $P \in M_2$ and $M_1 \neq M_2$.
   (b) Given the postulates and definitions introduced at this time, how many distinct planes can you prove that space contains?
8. With our present definitions, postulates, and theorems,
   (a) How many different lines must contain a given point?
   (b) How many different planes must contain a given point?
   (c) How many different planes must contain a given set of two distinct points?
9. How many distinct planes are determined in set T if T consists of distinct points, no four of which are coplanar and no three of which are collinear, and $n(T) = 3$? $n(T) = 4$? $n(T) = 5$? $n(T) = k$ $(k \in C \wedge k \geqq 3)$?
10. Is the following implication true?

$$\left.\begin{array}{l} S \text{ is a set of coplanar points} \\ T \text{ is a set of coplanar points} \\ n(S \cap T) \geqq 3 \end{array}\right\} \longrightarrow \left\{\begin{array}{l} S \cup T \text{ is a set of} \\ \text{coplanar points.} \end{array}\right.$$

    If you believe it is true, prove it. If you believe it is false, show a counterexample.

Some Applications of Indirect Proof

If A and B are distinct points in plane M and point C is not in plane M, we readily conjecture that A, B, C are noncollinear. We should be able to prove this conjecture. We wish to prove

$$\text{(I)} \qquad \left.\begin{array}{l} (1)\ A \neq B \\ (2)\ A, B \in M \\ (3)\ C \notin M \end{array}\right\} \longrightarrow A, B, C \text{ are noncollinear.}$$

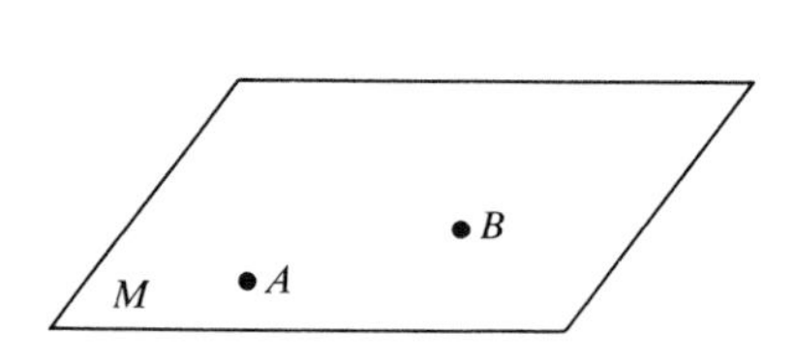

Reviewing our theorems, we find none that help us to prove that three points are noncollinear under the conditions given.

In a situation of this kind, two basic strategies are open to us: (a) try to use the indirect proof pattern A_5 described in Chapter 2,

$$\left.\begin{array}{l}\sim q \longrightarrow t \\ \sim t\end{array}\right\} \longrightarrow q,$$

or (b) try to prove an equivalent contrapositive form. While these two strategies are logically equivalent, they do involve different procedures.

If we elect to try an indirect proof, we must try to show that statement q (A, B, C are noncollinear) is true by showing that its contradiction $\sim q$ (A, B, C are collinear) implies a false statement. We argue as follows:

If A, B, C are collinear, then $C \in \overleftrightarrow{AB}$. According to Postulate 9, $\overleftrightarrow{AB} \subseteq M$, because two distinct points in the line are in M. Therefore, by the corollary to ST_{15}, C is an element of M. [$(C \in \overleftrightarrow{AB} \wedge \overleftrightarrow{AB} \subseteq M) \longrightarrow C \in M$.] But according to our hypothesis $C \notin M$. Therefore the supposition that A, B, C are collinear has led to a false statement, namely $C \in M$. It follows by our indirect-proof rule that A, B, C are noncollinear, and our indirect proof is complete.

The following analysis shows how the indirect-proof pattern has been applied.

$$\left.\begin{array}{l}\sim q \longrightarrow t \\ \sim t\end{array}\right\} \longrightarrow q,$$

$$\left.\begin{array}{l}\sim(A, B, C \text{ are noncollinear}) \longrightarrow C \in M \\ \sim(C \in M)\end{array}\right\} \longrightarrow A, B, C \text{ are noncollinear.}$$

Observe that q is the statement "A, B, C are noncollinear" and t is the statement "$C \in M$."

If we elect to prove an equivalent contrapositive form, we must decide which of the three contrapositives to choose. In view of the foregoing discussion, the third contrapositive looks promising. A flow-diagram proof of

$$\text{(II)} \qquad \left.\begin{array}{l}A \neq B \\ A, B \in M \\ A, B, C \text{ are collinear}\end{array}\right\} \longrightarrow C \in M$$

is easily constructed.

$$\left.\begin{array}{l}\left.\begin{array}{l}\underline{A, B \in M} \\ \underline{A \neq B}\end{array}\right\} \\ \underline{A, B, C \text{ are collinear}}\end{array}\right\} \begin{array}{l}\xrightarrow{(1)} \overleftrightarrow{AB} \subseteq M \\ \xrightarrow{(2)} C \in \overleftrightarrow{AB}\end{array}\Bigg\} \xrightarrow{(3)} C \in M.$$

The reader may supply the reasons.

We have proved the following theorem.

THEOREM 3-11 *If two distinct points are in a plane and a third point is not in that plane, then the three points are noncollinear.*

Conditions that Determine a Plane

Postulate 5 tells us that a set of three distinct noncollinear points determines a plane. There are two other sets of points that we can show determine a plane. These are (1) a line and a point not in the line, and (2) two distinct intersecting lines. Accordingly we now consider the following theorems.

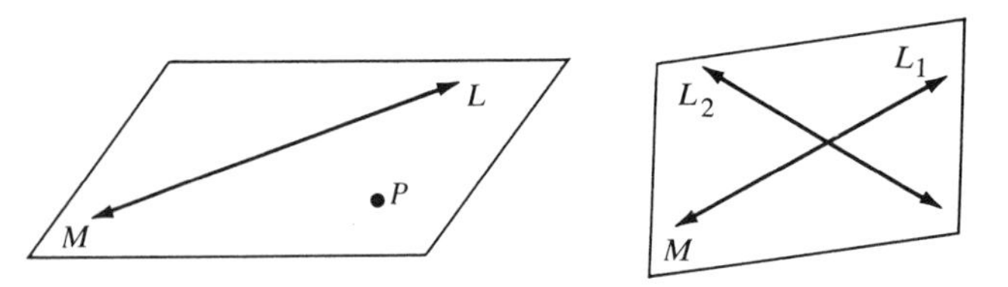

THEOREM 3-12 *If a point P is not in line L, then there is exactly one plane that contains $L \cup P$.*

THEOREM 3-13 *If two distinct lines intersect, there is exactly one plane that contains them.*

In order to prove that a certain set of points T is contained in exactly one plane, we must prove two statements:

(a) There is at least one plane that contains T.
(b) There is not more than one plane that contains T. This statement is equivalent to saying that if each of two planes contains T, then the planes are the same:

$$(T \subseteq M_1 \wedge T \subseteq M_2) \longrightarrow M_1 = M_2.$$

For Theorem 3-12 statements (a) and (b) are:

3-12(1) *If point P is not in line L, then there is a plane that contains L and P.*

3-12(2) *If two planes M_1 and M_2 each contain L and P when $P \notin L$, then $M_1 = M_2$.*

Thus Theorem 3-12 $\longleftrightarrow$ [3-12(1) $\wedge$ 3-12(2)].

PROOF

We now present a ledger proof of 3-12(1).
Given: Point P is not in line L ($P \notin L$).
Conclusion: There is a plane M that contains $L \cup P$ ($L \cup P \subseteq M$).

1. Line L contains two distinct points A and B ($A \neq B \wedge A, B \in L$).	1. Postulate 3.
2. $P \neq A, P \neq B$.	2. Given, Statement 1 ($A, B \in L$), and ST_2.
3. $L = \overleftrightarrow{AB}$.	3. Statement 1 and Corollary 1 to Postulate 2.
4. $P \notin \overleftrightarrow{AB}$.	4. Given, Statement 3, and the substitution property of equality.

5. P, A, B are noncollinear.	5. Statement 1 ($A \neq B$), Statement 4, and Theorem 3-2(4).
6. There is a plane M that contains P, A, and B ($P, A, B \in M$).	6. Statements 1 ($A \neq B$), 2, 5, and Postulate 5.
7. M contains L ($L \subseteq M$).	7. Statements 1 and 6 and Postulate 9.
8. $L \cup P \subseteq M$.	8. Statements 6 ($P \in M$) and 7 and ST_{18}.

PROOF

Consider the following essay proof of 3-12(2): If planes M_1 and M_2 each contain L and P and $P \notin L$, then $M_1 = M_2$.

If M_1 is a plane that contains line L and point P, then M_1 contains points A, B, and P, where A and B are distinct points in L and P is not in L. Thus plane M_1 contains three noncollinear points A, B, and P. Since M_2 also contains line L and point P, M_2 must also contain the three noncollinear points A, B, and P. Therefore $M_1 = M_2$, because Postulate 5 states that if A, B, and P are noncollinear points, there is exactly one plane that contains them. The proof of Theorem 3-12 is complete.

Theorem 3-13 $\longleftrightarrow$ [3-13(1) $\wedge$ 3-13(2)], where 3-13(1) is the statement: There is at least one plane that contains two distinct intersecting lines, and 3-13(2) is the statement: If two planes M_1 and M_2 each contain $L_1 \cup L_2$, where L_1 and L_2 are distinct intersecting lines, then $M_1 = M_2$.

***PROOF OF* 3-13(1)**

Consider the following flow-diagram proof of 3-13(1). We must prove:

$$\left.\begin{array}{l} L_1 \neq L_2 \\ L_1 \cap L_2 \neq \emptyset \end{array}\right\} \longrightarrow \text{There is a plane } M \text{ that contains } L_1 \cup L_2.$$

PROOF

$$\underline{L_1 \cap L_2 \neq \emptyset} \xrightarrow{(1)} \text{There is a point } P \text{ such that } \left\{\begin{array}{l} P \in L_1 \\ P \in L_2 \end{array}\right.$$

$$\left.\begin{array}{l} P \in L_1 \\ P \in L_2 \\ \underline{L_1 \neq L_2} \\ \text{(2) There is a point } Q \in L_2 \text{ such that } P \neq Q \end{array}\right\} \xrightarrow{(3)} Q \notin L_1 \xrightarrow{(4)}$$

$$\rightarrow \text{There is a plane } M \text{ such that } \left\{\begin{array}{l} Q \in M \\ \underline{\underline{L_1 \subseteq M}} \end{array}\right.$$

$$\left.\begin{array}{l} L_1 \subseteq M \\ P \in L_1 \end{array}\right\} \xrightarrow{(5)} P \in M$$

$$\left.\begin{array}{l} Q \in M \\ P \in M \\ P \neq Q \end{array}\right\} \xrightarrow{(6)} \overleftrightarrow{PQ} \subseteq M \qquad \left.\begin{array}{l} P \neq Q \\ P \in L_2 \\ Q \in L_2 \end{array}\right\} \xrightarrow{(7)} \overleftrightarrow{PQ} = L_2$$

$$\left.\begin{array}{l} \overleftrightarrow{PQ} \subseteq M \\ \overleftrightarrow{PQ} = L_2 \end{array}\right\} \xrightarrow{(8)}$$

$$\rightarrow \underline{\underline{L_2 \subseteq M.}}$$

From the double-underlined statements, we have $L_1 \cup L_2 \subseteq M$. The reader may supply the reasons.

PROOF OF 3-13(2)

If planes M_1 and M_2 each contain $L_1 \cup L_2$, where L_1 and L_2 are distinct intersecting lines, then $M_1 = M_2$.

If L_1 and L_2 are distinct lines intersecting in point P, then there is a point Q, where $P \neq Q$, such that $Q \in L_2$ and $Q \notin L_1$. Why? If planes M_1 and M_2 each contain $L_1 \cup L_2$, then each plane contains L_1 and Q. Why? Since $Q \notin L_1$, we can use Theorem 3-12 to conclude that $M_1 = M_2$.

In the exercises you are asked to prove the following

COROLLARY TO THEOREM 3-13 *If two lines intersect, they are coplanar.*

Exercises

1. In bracket form 3-12(2) can be written

$$\left.\begin{array}{l} L \subseteq M_1 \cap M_2 \\ P \in M_1 \cap M_2 \\ P \notin L \end{array}\right\} \longrightarrow M_1 = M_2.$$

   Write each of the three contrapositives of this implication. Write a verbal statement and draw a figure corresponding to each of the contrapositives.

2. In bracket form 3-13(2) can be written

$$\left.\begin{array}{l} L_1 \cap L_2 \neq \emptyset \\ L_1 \neq L_2 \\ L_1 \cup L_2 \subseteq M_1 \cap M_2 \end{array}\right\} \longrightarrow M_1 = M_2.$$

   (a) Write the first contrapositive of this implication. Write a verbal statement of the contrapositive. Can you draw a figure to illustrate it? Explain why the hypothesis is a false statement. Is this contrapositive a true or a false implication? Explain.
   (b) Write the second and third contrapositives of the bracket form of 3-13(2). For each contrapositive, write a verbal statement and draw a figure corresponding to it.

3. Write an indirect proof of Theorem 3-1: If two distinct lines intersect, then their intersection contains exactly one point.

4. Write an indirect proof of Theorem 3-9: If a line intersects a plane not containing it, the intersection is a single point.

5. Prove:

$$\left.\begin{array}{l} \text{Points } A, B, C \text{ are collinear} \\ A, B \in \text{plane } M \\ C \notin M \end{array}\right\} \longrightarrow A = B.$$

L 6. Prove: A, B, C are noncollinear points $\longrightarrow A, B, C$ are distinct.

L 7. Prove: A, B, C, D are noncoplanar points $\longrightarrow A, B, C$ are noncollinear.

L 8. Prove: A, B, C, D are noncoplanar points $\longrightarrow A, B, C, D$ are distinct.

9. Prove:

$$\left.\begin{array}{l} A, B, C, D \text{ are noncoplanar} \\ X \in \overleftrightarrow{AD} \\ X \neq A \end{array}\right\} \longrightarrow A, B, C, X \text{ are noncoplanar.}$$

10. Prove the corollary to Theorem 3-13: If two lines intersect, they are coplanar.

11. Prove: If M_1 and M_2 are distinct planes that intersect in $\overleftrightarrow{AB}$ and T is a point in neither plane and X is a point in $\overleftrightarrow{TA}$ distinct from A and T, then X is in neither plane.

L 12. Prove:

$$\left.\begin{array}{l} L_1 \cap M = P \\ L_2 \subseteq M \\ P \notin L_2 \end{array}\right\} \longrightarrow L_1 \cap L_2 = \emptyset.$$

Axiomatic Systems

Using nine postulates and six definitions, we have thus far proved thirteen theorems and ten corollaries. This set of postulates, definitions, theorems, and corollaries is a part of an *axiomatic system* according to the following definition, in which corollaries and lemmas are considered as theorems.

DEFINITION *An* axiomatic system *consists of a collection of postulates (axioms), definitions, and essential undefined terms together with all the theorems that can be derived from them. An axiomatic system is said to be based on its set of postulates.*

Now it is absolutely essential that an axiomatic system be free of contradictions. If, for example, two of our postulates contradict each other, our system contains a statement of the form $p \wedge \sim p$. The inclusion of such a statement in the hypothesis of any implication gives a true implication. For example, $[a \wedge (p \wedge \sim p)] \longrightarrow q$ is true for *any statements* a and q. In other words, if a contradiction is inadvertently included in our axiomatic system, we can "prove" any statement. We must insist that our axiomatic system be "consistent" according to the following definition.

DEFINITION *An axiomatic system is* consistent *if its set of postulates does not imply two statements that are contradictory.*

Having recognized consistency as an essential property for axiomatic systems, we must now face a very difficult question: How can we determine whether or not a given axiomatic system is consistent? We might consider proving all the theorems and then checking the list of theorems and postulates to see if it is free of contradictions. This, however, is not feasible, because we can never be sure that we have proved *all* possible theorems. There is always the possibility that a contradiction might be found if our list were extended. In other words, our question has no definitive answer.

There is, however, a pragmatic test for consistency, which has long been used by mathematicians. This test requires us to find a way to assign meaning to each of the undefined terms in our system so that all the postulates in the system are true. If such an interpretation exists, our set of postulates is said to be *satisfiable* and the interpretation is called a *model* for our set of axioms. If there exists a model for a set of axioms, we say that the set is consistent. It follows that the axiomatic system based on this set of axioms is consistent, because if we accept our set of axioms as true, then every statement derived from them is true. Thus the system is free of contradictory statements and hence is consistent.

Consider the following set of axioms S_1, in which "lambdas" and "deltas" are undefined objects and each lambda is a set of deltas.

1. There is exactly one lambda containing any two distinct deltas.
2. Every lambda contains at least three distinct deltas.
3. There is at least one lambda.
4. Not all deltas are in the same lambda.

We ask if this system is consistent. It is, if we can construct a model for our set of four axioms. Let "deltas" be people named $A, B, C, \ldots$ and let "lambdas" be committees called $C_1, C_2, C_3, \ldots$. Consider the following set of committees:

$$C_1 = \{A, E, G\}, \quad C_2 = \{A, B, C\}, \quad C_3 = \{C, F, G\},$$
$$C_4 = \{B, E, F\}, \quad C_5 = \{C, D, E\}, \quad C_6 = \{B, D, G\},$$
$$C_7 = \{A, D, F\}.$$

An examination of this set of committees will reveal that our four axioms are satisfied—that is, each is true. Thus we have a model for our axiomatic system that involves seven lambdas (committees) and seven deltas (people). We conclude that the axiomatic system that consists of S_1 and the theorems that can be deduced from S_1 is consistent.

If we form a new set of axioms by annexing a fifth statement, s, to the four axioms in S_1, we are confronted by another crucial question: Is our new statement, s, *decidable* in terms of our original set of axioms? We define "decidable" as follows:

DEFINITION *A statement is* decidable *in terms of a set of axioms if (a) it is implied by the axioms or (b) a contradiction of the statement is implied by the axioms. If a statement is implied by the axioms, we say it is* decidable true *and that it can be* logically derived *from our axioms. If a contradiction of the statement is implied, we say it is* decidable false.

If s is decidable true, then it is really a theorem in our original axiomatic system. If we include it as an axiom, no harm is done, but then we know that we do not have a minimal set of axioms for our system. For example, our list of set postulates is not a minimal set for the axiomatic system based on them. We are not concerned about this because, fortunately, an axiomatic development does not require that the axiom set be minimal. Sometimes, when a theorem is hard to prove, pedagogical considerations may justify our stating it as a postulate. However, we must realize that postulating statements we cannot prove on the grounds that they are "true" is a dangerous procedure that should be used sparingly. It is not our purpose to construct a long list of experimentally verified geometric facts, but rather to study the process whereby theorems are proved from the axioms at hand.

If our new statement, s, is decidable false, then it is inconsistent with our original set of axioms. Any set of axioms that includes S_1 and s is inconsistent.

If our new statement, s, is not decidable in terms of S_1, we may decide to form a new set of axioms by (a) appending s to S_1 or (b) appending a statement, a, such that $(S_1 \wedge a) \longrightarrow s$. For example, the statement "A line contains at least five points" is not decidable on the basis of the postulates now available. After the adoption of Postulate 11 in the next chapter, this statement becomes decidable true.

If $S = S_1 \wedge s$ or $S = S_1 \wedge a$, where $(S_1 \wedge a) \longrightarrow s$, then any model for axiom set S will generally be different from the model that served for S_1. This is true because the new model must satisfy a condition that was not considered when the first model was constructed.

We state a definition and three possible fifth postulates for S_1.

We say that two lambdas are *parallel* if they have no common elements.

5(a) *Any two lambdas have at least two deltas in common.*

5(b) *Each delta is contained in at least three lambdas.*

5(c) *Each delta not in a given lambda is contained in exactly one lambda that is parallel to the given lambda.*

Let $S(a) = S_1 \wedge 5(a)$, $S(b) = S_1 \wedge 5(b)$, and $S(c) = S_1 \wedge 5(c)$.

The set S(a) is inconsistent because statement 5(a) contradicts a statement implied by S_1. In Exercise 3 you are asked to prove that $S_1 \longrightarrow$ Two lambdas have at most one delta in common.

The set S(b) is consistent because the statement 5(b) is implied by S_1 (see Exercise 5).

The set S(c) is consistent because it has the following model involving twelve lambdas (committees) and nine deltas (people).

$$\begin{array}{lll} C_1 = \{A, B, C\}, & C_2 = \{D, E, F\}, & C_3 = \{G, H, I\}, \\ C_4 = \{A, D, I\}, & C_5 = \{C, F, G\}, & C_6 = \{C, E, I\}, \\ C_7 = \{C, D, H\}, & C_8 = \{A, E, G\}, & C_9 = \{A, F, H\}, \\ C_{10} = \{B, F, I\}, & C_{11} = \{B, E, H\}, & C_{12} = \{B, D, G\}. \end{array}$$

We observe that, like the famous fifth postulate of Euclidean geometry, statement 5(c) is undecidable in any axiomatic system based on S_1.

Exercises

1. Consider S_1.
   (a) How do we know that deltas exist?
   (b) What would be the effect of deleting statement (4)?
2. Prove:
   (a) $S_1 \longrightarrow$ There exist three distinct deltas.
   (b) $S_1 \longrightarrow$ There exist three distinct lambdas.
3. Prove: $S_1 \longrightarrow$ Two distinct lambdas have not more than one delta in common.
4. Prove: $S_1 \longrightarrow$ Not all lambdas contain the same delta.

5. Prove: $S_1 \longrightarrow$ Each delta is contained in at least three lambdas.
6. Prove: $S_1 \longrightarrow$ There exist at least seven lambdas and seven deltas.
   [*Note:* The existence of our model does not prove this. Why?]
7. Using S_1, can you prove that any two lambdas must intersect?
8. Construct another model for S_1.
9. Prove: $S(c) \longrightarrow$ There exist at least two lambdas parallel to a given lambda.
10. Prove: $S(c) \longrightarrow$ If a lambda intersects one of two parallel lambdas in a single delta, then it intersects the other.

References

Adler, *Modern Geometry*

Golos, *Foundations of Euclidean and Non-Euclidean Geometry*

Kay, *College Geometry*

Moise, *Elementary Geometry from an Advanced Standpoint*

Ringenberg, *College Geometry*

Chapter 4

Coordinate Systems for a Line

In Chapter 3 we presented nine postulates about points, lines, and planes. From Postulate 3 we know that a line is a set of points and contains *at least* two points. Our experience with lines suggests that there should be infinitely many points in a line, but we cannot prove this from the postulates available at this time. Moreover, we have some strong intuitive ideas about "distance" and "betweenness" that we should try to express as statements about real numbers, so that we can apply our knowledge of real numbers to the study of geometry. A summary of the properties of real numbers and a list of theorems concerning real numbers can be found in Appendix A. We shall make use of this material as the chapter progresses.

Distance

Our previous experience with measuring distance suggests four important ideas.

(1) The measure of the distance between two distinct points is a positive number.
(2) The distance between two points P and Q is zero if and only if $P = Q$.
(3) The distance between P and Q is equal to the distance between Q and P.
(4) The number we assign as the measure of the distance between two distinct points depends on the unit of measure we are using.

The following postulate and definition are consistent with our intuitive notions about distance.

POSTULATE 10 (THE DISTANCE POSTULATE) *Given any pair of distinct points, there exists a correspondence between the set of pairs of distinct points in space and the set of positive real numbers that assigns to each pair of distinct points a unique positive number such that the number assigned to the given pair is one.*

DEFINITION *Any pair of points that is assigned the number one by the correspondence whose existence is asserted by Postulate* 10 *is called a* unit pair *of the correspondence.*

DEFINITION *The number that corresponds by Postulate* 10 *to a pair of distinct points is called the* measure of the distance *between the points relative to any unit pair of the correspondence.*

Notation: In order to emphasize the idea that the measure of the distance between two distinct points is positive, we use a symbol suggested by the absolute-value notation. If P and Q are points, the measure of the distance between P and Q will be denoted by either of the symbols $|PQ|$ or $|QP|$.

DEFINITION *The measure of the distance between any point and itself relative to any unit pair is the number zero.*

If $\{M, N\}$ is a unit pair, the measure of the distance between A and B relative to $\{M, N\}$ may be the number 4. We express this idea by writing $|AB|$ (relative to $\{M, N\}) = 4$. Frequently, instead of using the phrase "the measure of the

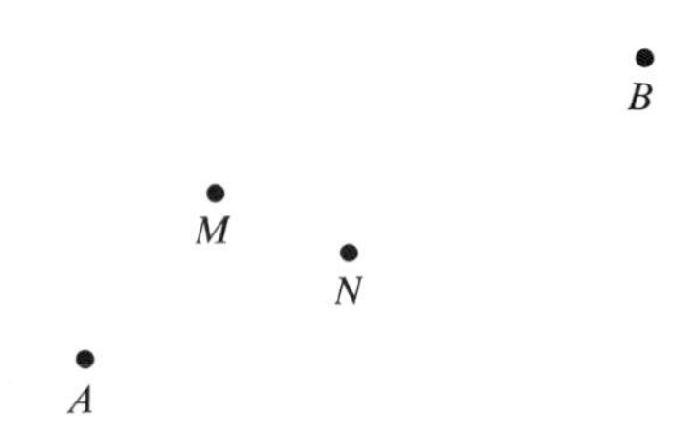

distance between two points" we shall simply say "the distance between two points." The symbol $|AB|$ will be used to denote the distance between A and B when we are sure the unit pair is known. In all cases, the symbol $|AB|$ denotes a nonnegative number.

Our experience with the number line in algebra suggests that we can obtain the measure of the distance between two points in a line by finding the absolute value of the difference between their corresponding numbers. Thus for points E and T in the line shown we have

$$|ET| = |4 - (-1)| = |-1 - 4| = 5.$$

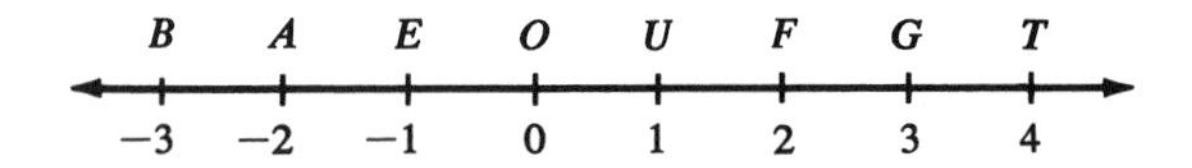

We now define a coordinate system for a line relative to a given unit pair.

DEFINITION *Let A and A' be any pair of distinct points and let L be any line.* A coordinate system for L relative to the unit pair $\{A, A'\}$ *is a one-to-one correspondence between the set of points in L and the real numbers such that the absolute value of the difference between the numbers corresponding to any two points in L is the measure of the distance between these two points relative to the unit pair $\{A, A'\}$.*

Thus if the coordinate system assigns point P the number p and point Q the number q, then $|PQ|$ (relative to $\{A, A'\}$) $= |p - q|$. Note that

(i) $$|AA'| \text{ (relative to } \{A, A'\}) = 1. \qquad \text{Why?}$$

Since a one-to-one correspondence is a set of pairs, we see that a coordinate system is a set of pairs in which each pair consists of a real number and a point in some line L. We can, therefore, speak of a point or a number being *in* a coordinate system. If the pair $(P, 5)$ is one of the pairs in a coordinate system, we say that the point P and the number 5 correspond to each other. We may also say that the coordinate system assigns P the number 5 and assigns 5 to point P.

Observe that our definition of a coordinate system does not tell us *how* the required one-to-one correspondence between the real numbers and the points in line L is to be established or even that such a correspondence exists. However, if we know that a certain one-to-one correspondence between L and R qualifies as a coordinate system relative to a certain unit pair, we are entitled to conclude that the process of finding "the absolute value of the difference between the numbers corresponding to any two points in L" produces a number that, according to Postulate 10 and the three definitions that follow it, is the unique non-negative measure of the distance between these two points relative to the given unit pair.

We can see one implication of this fact by considering two different one-to-one correspondences between L and R. Let O and P be two distinct points and let $L = \overleftrightarrow{OP}$. Our first correspondence is suggested by the numbers written below

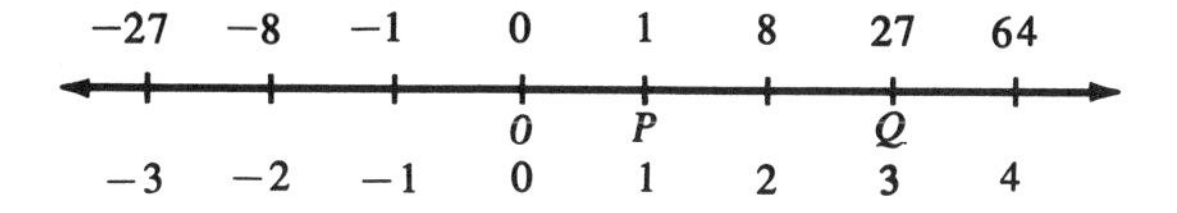

line L in the diagram. We obtain our second one-to-one correspondence between L and R by assigning to each point in L the cube of the number assigned to it by our first correspondence, as indicated by the numbers written above L in the diagram. Now the assertion that *both* of these correspondences are coordinate systems for L relative to unit pair $\{O, P\}$ is untenable, because if the first correspondence is such a coordinate system, we have $|PQ|$ (*relative to* $\{O, P\}$) $=$ $|3 - 1| = 2$, and if the second is such a coordinate system, we have $|PQ|$ (relative to $\{O, P\}$) $= |27 - 1| = 26$. Thus we have different numbers assigned as $|PQ|$ (relative to $\{O, P\}$)—in violation of Postulate 10. We must conclude that if the first correspondence is a coordinate system for L relative to unit pair $\{O, P\}$, the second is not, and conversely.

DEFINITION *The* origin *and* unit point *of a coordinate system for a line are the points in the line that correspond to the numbers* 0 *and* 1, *respectively. The number that the coordinate system assigns to a point in the line is called the* coordinate *of that point, and the point that the coordinate system assigns to a given number is called the* graph *of that number.*

Now that we have defined a coordinate system for a line, we proceed to assert that such a thing exists.

POSTULATE 11 (THE RULER POSTULATE) *If $\{A, A'\}$ is a given unit pair and P and Q are distinct points in line L, then there exists a unique coordinate system for L relative to $\{A, A'\}$ such that the origin of the coordinate system is P and the coordinate of Q is positive.*

If we take $\{P, Q\}$ as the unit pair, we have the following important corollary.

COROLLARY TO THE RULER POSTULATE *If P and Q are distinct points in a line, then there exists a unique coordinate system for the line that has P as its origin and Q as its unit point.*

We note that Postulate 11 suggests there are many coordinate systems for a given line. We shall denote a coordinate system by using the letter C, or the letter C with a prime or a subscript.

Sometimes we shall find it necessary to refer to the number that the coordinate system C assigns to a given point as the *C-coordinate of that point.* The point S that C assigns to the number s is called the graph of s in the coordinate system C. The graph of a set of numbers in the coordinate system C is, of course, the set of points whose C-coordinates are those numbers.

Notation: Given a coordinate system, we shall use the symbol $X(x)$ to represent "the point X whose C-coordinate is x" or "point X has the coordinate x," depending on the context.

The following statements are deduced from the definition of a coordinate system for a line and Postulate 11.

(a) If A, B, and C are collinear points, then there is a coordinate system that assigns coordinates to these three points.
(b) If there is a coordinate system that assigns coordinates to the three points A, B, and C, then these points are collinear.

Thus we have the "if-and-only-if" statement:

(ii) Points A, B, and C are collinear $\longleftrightarrow$ There is a coordinate system that assigns coordinates to points A, B, and C.

The following set definitions and set postulate are useful in proving Theorem 4-1 and its corollaries.

DEFINITIONS *A set S is a* finite set *if it is the empty set or if there is an integer k such that the elements of S can be put into one-to-one correspondence with $\{1, 2, 3, \ldots, k\}$.*

$$[S \text{ is finite} \longleftrightarrow n(S) = 0 \vee n(S) = k \text{ where } k \in C]$$

A set is an infinite set *if it is not a finite set.*

SP_{13}. Each of two sets is finite if and only if their union is finite.

$$[(S \text{ is finite} \wedge T \text{ is finite}) \longleftrightarrow S \cup T \text{ is finite.}]$$

THEOREM 4-1 *There are infinitely many points in (a) a given line, (b) a given plane, and (c) space.*

COROLLARY 1 TO THEOREM 4-1 *There are infinitely many points in a plane that are not in a given line in the plane.*

COROLLARY 2 TO THEOREM 4-1 *There are infinitely many points in space that are not in a given plane.*

Exercises

1. Explain why $|AA'|$ (relative to $\{A, A'\}$) $= 1$.
2. Prove: Points A, B and C are collinear $\longleftrightarrow$ There is a coordinate system that assigns coordinates to points A, B, and C.
3. Prove Theorem 4-1.
4. Points P, X, and Q are in a given coordinate system and $P(5)$, $X(x)$, and $Q(13)$. Find x if
   (a) $|XP| = 20$. (b) $|XP| = 4 \vee |XQ| = 4$.
   (c) $|XP| = 4 \wedge |XQ| = 4$.
5. $P(-1)$, $Q(5)$, and $X(x)$ are in a coordinate system. Find x if
   (a) $|PX| = 3|XQ|$. (b) $|PX| = \frac{1}{5}|XQ| \wedge |XQ| > |PQ|$.
6. $P(p)$ and $Q(q)$ are distinct points in a coordinate system that also contains $X(x)$. Prove: $|PX| = |XQ| \longleftrightarrow x = (p + q)/2$.

L 7. Prove the following statement for sets S and T:

$$\left.\begin{matrix} S \cup T \text{ is infinite} \\ S \text{ is finite} \end{matrix}\right\} \longrightarrow T \text{ is infinite.}$$

8. Use Exercise 7 to prove Corollary 1 to Theorem 4-1.
9. Use Exercise 7 to prove Corollary 2 to Theorem 4-1.

L 10. Given that (1) C is a coordinate system for L relative to $\{A, A'\}$, (2) x is the coordinate that C assigns to point X, (3) D is a correspondence between L and R that assigns $-x$ to $X(x)$. Prove:
   (a) D is a one-to-one correspondence between L and R, the real numbers.
   (b) D is a coordinate system for L relative to $\{A, A'\}$.

L 11. Change (3) in Exercise 10 to "D is a correspondence between L and R that pairs $x + a$ with $X(x)$" and prove statements (a) and (b).

Betweenness

We have strong intuitive ideas about the meaning of the statement that a certain point B is between points A and C. We consider that A, B, and C are distinct collinear points arranged in one of the following ways:

However, in our discussions we are not going to rely on diagrams such as these. In defining "betweenness for points" we must state a definition that is "diagram-free." Such a definition is easily stated in terms of the numbers a, b, and c that are the coordinates of points A, B, and C, respectively, as assigned by a coordinate system.

DEFINITION OF BETWEENNESS FOR POINTS *Point B is between points A and $C \longleftrightarrow$ There is a coordinate system that assigns A, B, and C coordinates a, b, and c so that b is between a and c.*

We shall use the symbol $A\text{-}B\text{-}C$ to represent the sentence "Point B is between points A and C." The symbol $a\text{-}b\text{-}c$ will represent the sentence "The number b is between the numbers a and c" (see Appendix A). Using this notation, we can state the definition above somewhat more compactly as follows:

> $A\text{-}B\text{-}C \longleftrightarrow$ There is a coordinate system that assigns points A, B, and C the coordinates a, b, and c, respectively, and $a\text{-}b\text{-}c$.

At this time it is convenient for us to assume the truth of the following statement that is proved later:

> If the coordinate of B is between the coordinates of A and C in one coordinate system for $\overleftrightarrow{AC}$, the same is true for every coordinate system for $\overleftrightarrow{AC}$.

The proof of this statement (see Corollary 2 to Theorem 7-4) depends on Postulate 12 and may be read at this time. In view of our assumption, we have the following statement:

> (iii) If a coordinate system assigns a, b, and c to points A, B, and C, respectively, then $A\text{-}B\text{-}C \longleftrightarrow a\text{-}b\text{-}c$.

The definition of betweenness has the following corollary, whose proof is required in the exercises.

COROLLARY TO THE DEFINITION OF BETWEENNESS *Given distinct points A and B, there exists* (1) *a point X such that $A\text{-}X\text{-}B$,* (2) *a point Y such that $A\text{-}B\text{-}Y$,* (3) *a point Z such that $Z\text{-}A\text{-}B$.*

We know that each of the following statements is true for the reasons indicated:

1. There is a coordinate system that assigns the points A, B, and C the coordinates a, b, and $c \longleftrightarrow A$, B, and C are collinear (definition of coordinate system).
2. $a\text{-}b\text{-}c \longleftrightarrow |b - a| + |c - b| = |c - a| \wedge a, b, c$ are distinct (RT_{37}—see Appendix A).

If a, b, and c are the coordinates assigned to points A, B, and C by a coordinate system, then by Postulate 11 and our definition of a coordinate system we know that

3. $|AB| = |b - a|$, $|BC| = |c - b|$ and $|AC| = |c - a|$.
4. a, b, and c are distinct if and only if A, B, and C are distinct.

Thus, beginning with our definition, we have the following sequence of equivalent statements.

$A\text{-}B\text{-}C \overset{(1)}{\longleftrightarrow}$ For any coordinate system that assigns points A, B, and C the coordinates a, b, and c, respectively, we have $a\text{-}b\text{-}c \overset{(2)}{\leftrightarrow}$

$\leftrightarrow$ For any coordinate system that assigns points A, B, and C the coordinates a, b, and c, respectively, we have $|b - a| + |c - b| = |c - a|$ and a, b, and c are distinct $\overset{(3)}{\leftrightarrow}$

$\leftrightarrow$ A, B, and C are collinear and $|AB| + |BC| = |AC|$ and A, B, and C are distinct.

Reasons: (1) Statement (iii).
(2) Statement 2 above.
(3) Statement 1, 3, and 4 above.

Observe that each of the reasons applies "both ways"—that is, for the left-to-right implication and for the right-to-left implication. Thus we have the following theorem.

THEOREM 4-2 *Point B is between points A and C if and only if A, B, and C are distinct collinear points and $|AB| + |BC| = |AC|$.*

COROLLARY TO THEOREM 4-2

$$A\text{-}B\text{-}C \longrightarrow \begin{cases} |AC| > |BC| \\ |AC| > |AB|. \end{cases}$$

If A, B, and C are distinct points in a line whose coordinates are a, b, and c, respectively, we know that a, b, and c are distinct real numbers. If a, b, and c are distinct real numbers, we know that exactly one of them is between the other two (see Appendix A). The following statement is a direct consequence of these two statements and our definition of betweenness for points:

> If A, B, and C are distinct points in a line, exactly one of them is between the other two.

Moreover, we know that if exactly one of the points A, B, and C is between the other two, then A, B, and C are distinct points in a line. Thus we have the following theorem.

THEOREM 4-3 *Points A, B, and C are distinct points in a line if and only if exactly one of these points is between the other two.*

We use the symbol $a \veebar b$ to indicate that *exactly one* of the statements a, b is true. Thus the statement $a \veebar b$ has truth values as shown in the truth table below. Since the case where a and b are both true makes $a \vee b$ true, but is *excluded* from the cases that make $a \veebar b$ true, the symbol $\veebar$ is said to denote the "exclusive or."

a	b	$a \veebar b$
T	T	F
T	F	T
F	T	T
F	F	F

Using the symbol $\veebar$ as defined above, we can state Theorem 4-3 as follows:

THEOREM 4-3 *A, B, and C are distinct points in a line* $\longleftrightarrow (A\text{-}B\text{-}C \veebar B\text{-}A\text{-}C \veebar A\text{-}C\text{-}B)$.

Observe that $A\text{-}B\text{-}C \longrightarrow [\sim(B\text{-}A\text{-}C) \wedge \sim(A\text{-}C\text{-}B)]$. Why? In general

$$(a \veebar b \veebar c) \wedge a \longrightarrow (\sim b \wedge \sim c)$$

because *exactly one* of the statements a, b, c is true.

COROLLARY TO THEOREM 4-3

$$(B \in \overleftrightarrow{AC} \wedge |AB| < |AC| \wedge |BC| < |AC|) \longrightarrow A\text{-}B\text{-}C.$$

DEFINITION *The points $A, B, C, D, E, \ldots$ are* collinear in that order *if and only if $A\text{-}B\text{-}C$, $B\text{-}C\text{-}D$, $C\text{-}D\text{-}E, \ldots$.*

Subsets of a Line

In this section we use the idea of betweenness for points to define some important subsets of a line.

DEFINITIONS *If A and B are distinct points,* (1) *the* segment AB *(denoted by the symbol $\overline{AB}$) is the union of A and B and the set of points between A and B,* (2) *the* open segment AB *(denoted by the symbol $\overset{\circ\!-\!\circ}{AB}$) is the set of points between A and B. $\overset{\circ\!-\!\circ}{AB}$ is also called the* interior *of $\overline{AB}$. Any point in the interior of a segment is called an* interior point *of the segment or an* internal point *of the segment. The points A and B are called the* endpoints *of either $\overline{AB}$ or $\overset{\circ\!-\!\circ}{AB}$.*

From these definitions it follows that $\overline{AB} = \{A\} \cup \overset{\circ\!-\!\circ}{AB} \cup \{B\}$. Observe that a segment includes its endpoints, while an open segment does not.

DEFINITION *The union of the interior of a segment with exactly one of the endpoints of the segment is called a* semiopen segment.

If A and B are distinct points, we use the symbol $\overset{-\!\circ}{AB}$ to denote the semiopen segment that is the union of A and the set of points between A and B and we use the symbol $\overset{\circ\!-}{AB}$ to denote the union of B with the set of points between A and B. Thus $\overset{-\!\circ}{AB} = \{A\} \cup \overset{\circ\!-\!\circ}{AB}$ and $\overset{\circ\!-}{AB} = \overset{\circ\!-\!\circ}{AB} \cup \{B\}$.

DEFINITION *The ray AB (denoted by the symbol $\overrightarrow{AB}$) is the union of the segment AB and the set of points X such that A-B-X.* [$\overrightarrow{AB} = \overline{AB} \cup \{X \mid A\text{-}B\text{-}X\}$.]

By this definition we also know that $\overrightarrow{BA} = \overline{BA} \cup \{X \mid B\text{-}A\text{-}X\}$.

DEFINITIONS *The interior of $\overrightarrow{AB}$ is the set of points in $\overrightarrow{AB}$ except A. The interior of $\overrightarrow{AB}$ is called a* half line *and is denoted by the symbol $\mathring{\overrightarrow{AB}}$.* [$\mathring{\overrightarrow{AB}} = \overrightarrow{AB} \sim \{A\}$.] *The point A is called the* endpoint *of either $\overrightarrow{AB}$ or $\mathring{\overrightarrow{AB}}$.*

Note that a ray includes its endpoint while a half line does not.

The ray AB is different from the ray BA. A is the endpoint of $\overrightarrow{AB}$, but B is the endpoint of $\overrightarrow{BA}$. We see that $\overrightarrow{AB} \cup \overrightarrow{BA} = \overleftrightarrow{AB}$ and $\overrightarrow{AB} \cap \overrightarrow{BA} = \overline{AB}$.

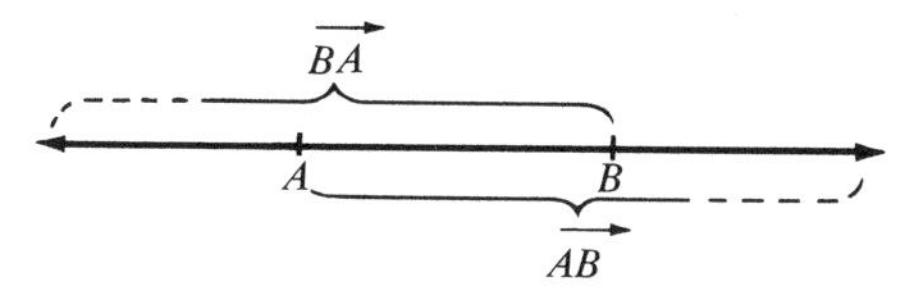

If A and B are distinct points, we know by Postulate 11 that there is a coordinate system for $\overleftrightarrow{AB}$. This coordinate system assigns A and B coordinates, which we can call a and b, respectively, where $a \neq b$. Our definition of betweenness for points tells us that

$$A(a)\text{-}X(x)\text{-}C(c) \longleftrightarrow a\text{-}x\text{-}c.$$

With these facts in mind the above definitions can be restated in terms of coordinates as shown in the following theorem.

THEOREM 4-4 *If A and B are distinct points whose coordinates as assigned by a coordinate system C are a and b, respectively, then*

(1) $\overline{AB} = \{X(x) \mid x = a \vee a\text{-}x\text{-}b \vee x = b\}$,
(2) $\mathring{\overline{AB}} = \{X(x) \mid a\text{-}x\text{-}b\}$,
(3) $\overrightarrow{AB} = \{X(x) \mid x = a \vee a\text{-}x\text{-}b \vee x = b\} \cup \{X(x) \mid a\text{-}b\text{-}x\}$,
(4) $\mathring{\overrightarrow{AB}} = \{X(x) \mid a\text{-}x\text{-}b \vee x = b\} \cup \{X(x) \mid a\text{-}b\text{-}x\}$.

If we know that $a < b$ or that $a > b$, the sets mentioned in Theorem 4-4 can be described more simply as indicated by the first two corollaries to Theorem 4-4.

COROLLARY 1 TO THEOREM 4-4 *If in Theorem 4-4 we know that $a < b$, then*

(1) $\overline{AB} = \{X(x) \mid a \leqq x \leqq b\}$,
(2) $\mathring{\overline{AB}} = \{X(x) \mid a < x < b\}$,
(3) $\overrightarrow{AB} = \{X(x) \mid a \leqq x\}$,
(4) $\mathring{\overrightarrow{AB}} = \{X(x) \mid a < x\}$.

COROLLARY 2 TO THEOREM 4-4 *If in Theorem 4-4 we know that $a > b$, then*

(1) $\overline{AB} = \{X(x) \mid b \leqq x \leqq a\}$,
(2) $\mathring{\overline{AB}} = \{X(x) \mid b < x < a\}$,

(3) $\overrightarrow{AB} = \{X(x) \mid x \leqq a\}$,
(4) $\overset{\circ\!\!\rightarrow}{AB} = \{X(x) \mid x < a\}$.

COROLLARY 3 TO THEOREM 4-4 $\quad (B \in \overrightarrow{AC} \wedge 0 < |AB| < |AC|) \longrightarrow B \in \overset{\circ\!\!-\!\!\circ}{AC}$.

COROLLARY 4 TO THEOREM 4-4

$$C \in \overset{\circ\!\!\rightarrow}{AB} \longrightarrow \begin{cases} \overrightarrow{AB} = \overrightarrow{AC} \\ \overset{\circ\!\!\rightarrow}{AB} = \overset{\circ\!\!\rightarrow}{AC}. \end{cases}$$

If A-B-C, the rays BA and BC appear to extend in opposite direction, as do the half lines BA and BC. This suggests the following definition for opposite rays and opposite half lines.

DEFINITIONS $\quad A$-B-$C \longleftrightarrow \overrightarrow{BA}$ *and* $\overrightarrow{BC}$ *are* opposite rays. A-B-$C \longleftrightarrow \overset{\circ\!\!\rightarrow}{BA}$ *and* $\overset{\circ\!\!\rightarrow}{BC}$ *are* opposite half lines.

COROLLARY 5 TO THEOREM 4-4 *If two rays are collinear and have the same endpoint, they are either opposite rays or the same ray.*

In the following argument a, b, and c are the coordinates of points A, B, and C, respectively, as assigned by a coordinate system for $\overleftrightarrow{AB}$.

$$\left.\begin{array}{l}\overrightarrow{BA} \text{ and } \overrightarrow{BC} \text{ are} \\ \text{opposite rays}\end{array}\right\} \xrightarrow{(1)} A\text{-}B\text{-}C \xrightarrow{(2)} a\text{-}b\text{-}c \overset{(3)}{-\!\!\otimes\!\!<} \begin{array}{l} a < b < c \xrightarrow{(4)} \begin{cases} \overrightarrow{BC} = \{X(x) \mid x \geqq b\} \\ \overrightarrow{BA} = \{X(x) \mid x \leqq b\} \end{cases} \\ a > b > c \xrightarrow{(4)} \begin{cases} \overrightarrow{BC} = \{X(x) \mid x \leqq b\} \\ \overrightarrow{BA} = \{X(x) \mid x \geqq b\}. \end{cases} \end{array}$$

Reasons:

(1) Definition of opposite rays.
(2) Definition of betweenness of points.
(3) Definition of betweenness for real numbers.
(4) Corollaries 1 and 2 to Theorem 4-4.

Note: We use the symbol

$$x -\!\!\otimes\!\!< \begin{array}{l} y \\ z \end{array}$$

in step (3) to indicate that $x \longrightarrow (y \vee z)$.]
Since

$$\left[a \longrightarrow b \longrightarrow c -\!\!\otimes\!\!< \begin{array}{l} d \longrightarrow x \\ e \longrightarrow y \end{array}\right] \longrightarrow \left(a -\!\!\otimes\!\!< \begin{array}{l} x \\ y \end{array}\right)$$

is a tautology, we have proved the following theorem.

THEOREM 4-5 *If a coordinate system is given for a line that contains two opposite rays, then one ray is the set of points having coordinates greater than or equal to the coordinate of the common endpoint, and the other ray is the set of points having coordinates less than or equal to the coordinate of the common endpoint.*

The proofs of the following corollaries to Theorem 4-5 are considered in the exercises.

COROLLARY 1 TO THEOREM 4-5 *$\overrightarrow{BA}$ and $\overrightarrow{BC}$ are opposite rays if and only if $\overrightarrow{BA} \cup \overrightarrow{BC} = \overleftrightarrow{AB}$.*

COROLLARY 2 TO THEOREM 4-5 *$\overrightarrow{BA}$ and $\overrightarrow{BC}$ are opposite rays if and only if $\overrightarrow{BA}$ and $\overrightarrow{BC}$ are collinear and $\overrightarrow{BA} \cap \overrightarrow{BC} = \{B\}$.*

THEOREM 4-6 (POINT-PLOTTING THEOREM) *If p is a positive number and $\{Q, Q'\}$ is a given unit pair, then there is exactly one point P in $\overrightarrow{\mathring{A}B}$ such that $|AP| = p$.*

An essay proof of Theorem 4-6 is requested in the exercises.

Exercises

1. A, B, and C are distinct collinear points. In each case determine whether A-B-C, A-C-B, B-A-C, or betweenness is undetermined.
   (a) $|AB| + |BC| = |AC|$. (b) $\overrightarrow{AC} \cup \overrightarrow{AB} = \overleftrightarrow{AB}$.
   (c) $\overrightarrow{BC} \cap \overrightarrow{AC} = \overrightarrow{AC}$. (d) $\overrightarrow{AC} \cap \overrightarrow{AB} = \{A\}$.
   (e) $\overline{BA} \cap \overline{BC} = \overline{BC}$. (f) $\overline{C\mathring{A}} \cup \overline{C\mathring{B}} = \overline{\mathring{A}\mathring{B}}$.
2. If A-B-C-D-E-F, determine whether each of the following statements is true or false.
   (a) $\overline{BD} \cap \overline{CF} = \overline{CD}$. (b) $\overrightarrow{BC} = \overrightarrow{BD}$.
   (c) $\overline{BF} \cup \overrightarrow{FA} = \overrightarrow{FA}$. (d) $\overline{CD} \subset \overrightarrow{AB}$.
   (e) F and D are in opposite rays that are contained in $\overleftrightarrow{AC}$ and have common endpoint E.
   (f) $\overline{DC} \cup \overrightarrow{CA} = \overrightarrow{DA}$. (g) $\overrightarrow{CB} \cap \overrightarrow{ED} = \overrightarrow{CB}$.
   (h) $\overrightarrow{FD} \subset \overrightarrow{CE}$. (i) $\overrightarrow{CB} \cap \overrightarrow{DE} = \emptyset$.
   (j) $\overrightarrow{DE} \subset \overline{CF}$.

L 3. Prove:
   (a) $\overline{\mathring{A}\mathring{B}} \subset \overline{A\mathring{B}} \subset \overline{AB} \subset \overrightarrow{AB} \subset \overleftrightarrow{AB}$. (b) $\overrightarrow{\mathring{A}B} \subset \overrightarrow{AB} \subset \overleftrightarrow{AB}$.
   (c) $\overline{\mathring{A}\mathring{B}} \subset \overline{\mathring{A}B} \subset \overrightarrow{\mathring{A}B}$.

L 4. Prove:
   (a) $\left.\begin{matrix} A\text{-}B\text{-}C \\ B\text{-}C\text{-}D \end{matrix}\right\} \longrightarrow (A\text{-}C\text{-}D \wedge A\text{-}B\text{-}D)$.
   (b) $\left.\begin{matrix} A\text{-}B\text{-}C \\ A\text{-}C\text{-}D \end{matrix}\right\} \longrightarrow (A\text{-}B\text{-}D \wedge B\text{-}C\text{-}D)$.

L 5. Prove:
   (a) $\overrightarrow{\mathring{A}B} \cup \{A\} = \overrightarrow{AB}$. (b) $\overrightarrow{BA} \cap \overrightarrow{\mathring{A}B} = \overline{\mathring{A}B}$.

L 6. Prove:

$$X \in \overline{\mathring{A}\mathring{B}} \longleftrightarrow A\text{-}X\text{-}B \longrightarrow \begin{cases} \overleftrightarrow{AX} = \overleftrightarrow{XB} = \overleftrightarrow{AB} \\ \overrightarrow{AX} = \overrightarrow{AB} \\ \overrightarrow{\mathring{A}X} = \overrightarrow{\mathring{A}B} \\ \overline{XA} \subset \overline{AB} \\ \overline{XB} \subset \overline{AB}. \end{cases}$$

7. Prove the corollary to the definition of betweenness. (Refer to the real-number properties in Appendix A.)
8. (a) Prove the corollary to Theorem 4-2.
   (b) Prove the corollary to Theorem 4-3.
9. Use Exercise 11, page 34, to prove that

$$\left.\begin{matrix}(p \wedge w) \longrightarrow t \\ (p \wedge y) \longrightarrow t\end{matrix}\right\} \longrightarrow \left[\left.\begin{matrix}p \\ w \vee y\end{matrix}\right\} \longrightarrow t\right]$$

is a tautology.

L 10. Use the tautology in Exercise 8, page 34, to prove each of the following:
   (a) $a < b \longrightarrow (a\text{-}x\text{-}b \longleftrightarrow a < x < b)$.
   (b) $a < b \longrightarrow [(x = a \vee a\text{-}x\text{-}b \vee x = b) \longleftrightarrow a \leqq x \leqq b]$. (You may also use the tautology in Exercise 9 above.)
   (c) $a < b \longrightarrow (b < x \longleftrightarrow a\text{-}b\text{-}x)$.
   (d) $a < b \longrightarrow [a \leqq x \longleftrightarrow (a \leqq x \leqq b \vee b < x)]$.
11. Use Exercise 10 to prove Corollary 1 to Theorem 4-4.
12. Prove Corollary 3 to Theorem 4-4.
13. Prove Corollary 4 to Theorem 4-4.
14. Prove Corollary 5 to Theorem 4-4.
15. Prove Corollary 1 to Theorem 4-5.
16. Prove Corollary 2 to Theorem 4-5.
17. Write an essay proof of Theorem 4-6.

L 18. Prove: If $0 < |A'C'| < |AC|$, then there exists a point $X \in \overset{\circ}{AC}$ such that $|AX| = |A'C'|$.

L 19. If $P \neq Q$, $P(p)$, $Q(q)$, and $X(x)$ in some coordinate system, prove

   (a) $\overrightarrow{PQ} = \left\{X(x) \,\middle|\, \dfrac{x-p}{q-p} \geqq 0\right\}$. (b) $\overset{\circ}{\overrightarrow{PQ}} = \left\{X(x) \,\middle|\, \dfrac{x-p}{q-p} > 0\right\}$.

   (c) $\overline{PQ} = \left\{X(x) \,\middle|\, 0 \leqq \dfrac{x-p}{q-p} \leqq 1\right\}$. (d) $\overset{\circ\circ}{PQ} = \left\{X(x) \,\middle|\, 0 < \dfrac{x-p}{q-p} < 1\right\}$.

20. If $P \neq Q$, $P(p)$, $Q(q)$, and $X(x)$ in some coordinate system, prove

$$\overset{\circ\circ}{PQ} = \{X(x) \mid (x-p)(x-q) < 0\}.$$

Relationships Between Different Coordinate Systems for a Line

Suppose that we have two coordinate systems C and C' for line L as shown below.

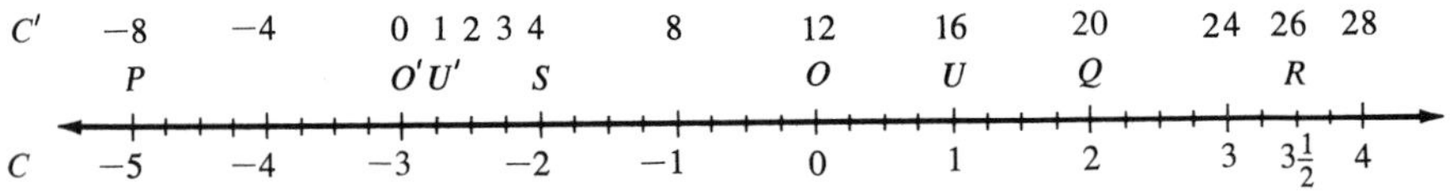

C'	−8	−4	0	1	2	3	4	8	12	16	20	24	26	28
	P		O'	U'			S		O	U	Q		R	
C	−5	−4	−3				−2	−1	0	1	2	3	$3\frac{1}{2}$	4

If X is any point whose C-coordinate is x and whose C'-coordinate is x', we would like a formula that would enable us to find x' if x is given and vice versa. From our diagram we obtain the following pairs of numbers:

$|PQ|$ (relative to $\{O', U'\}$) $= 28$, $|RS|$ (relative to $\{O', U'\}$) $= 22$,
$|PQ|$ (relative to $\{O, U\}$) $= 7$, $|RS|$ (relative to $\{O, U\}$) $= 5\frac{1}{2}$.

We see that the ratio of the first number to the second in each pair is equal to 4. We observe that $|OU|$ (relative to $\{O', U'\}$) $= 4$. Thus we have

$$\frac{|PQ| \text{ (relative to } \{O', U'\})}{|PQ| \text{ (relative to } \{O, U\})} = \frac{|RS| \text{ (relative to } \{O', U'\})}{|RS| \text{ (relative to } \{O, U\})}$$
$$= |OU| \text{ (relative to } \{O', U'\}).$$

We accept this principle as stated in the following postulate.

POSTULATE 12 *If unit pairs $\{O, U\}$ and $\{O', U'\}$ are given, then for every pair of distinct points P and Q and for every pair of distinct points R and S it is true that*

$$\frac{|PQ| \text{ (relative to } \{O', U'\})}{|PQ| \text{ (relative to } \{O, U\})} = \frac{|RS| \text{ (relative to } \{O', U'\})}{|RS| \text{ (relative to } \{O, U\})}$$
$$= |OU| \text{ (relative to } \{O', U'\}).$$

We can use Postulate 12 to obtain a formula for x' in terms of x as indicated by the following theorem.

THEOREM 4-7 (THE TWO-COORDINATE-SYSTEMS THEOREM) *If C and C′ are two coordinate systems for $\overleftrightarrow{AB}$ that assign the point A coordinates a and a′ respectively, the point B coordinates b and b′ respectively, and the point X coordinates x and x′ respectively, then there exist two numbers e and f with $e \neq 0$ such that $x' = ex + f$.*

OUTLINE OF PROOF

Case I. $X \neq A$ and $X \neq B$.
Using Postulate 12, we have

$$\frac{|x - a|}{|x' - a'|} = \frac{|b - a|}{|b' - a'|}.$$

Since $|y| = |a| \longrightarrow y = a \vee y = -a$, we have

$$\frac{x - a}{x' - a'} = \frac{b - a}{b' - a'} \vee \frac{x - a}{x' - a'} = -\frac{b - a}{b' - a'}.$$

Solving each of these equations for x', we have

$$x' = \frac{b' - a'}{b - a}x + \frac{a'b - ab'}{b - a} \vee x' = -\frac{b' - a'}{b - a}x + \frac{a'b + ab' - 2aa'}{b - a}.$$

To facilitate discussion, we shall let r and s, respectively, represent the last two equations.

Referring once again to Postulate 12, we can also write

$$\frac{|x - b|}{|x' - b'|} = \frac{|b - a|}{|b' - a'|}.$$

Proceeding precisely as we did before, we have

$$x' = \frac{b' - a'}{b - a}x + \frac{a'b - ab'}{b - a} \vee x' = -\frac{b' - a'}{b - a}x + \frac{2bb' - a'b - ab'}{b - a}.$$

We note that the first statement in the disjunction is statement r, and we shall denote the second statement by t.

At this point we have the statement $(r \vee s) \wedge (r \vee t)$. In the exercises you are asked to prove that the statement $s \wedge t$ is false. You will also be asked to show that $[(r \vee s) \wedge (r \vee t) \wedge \sim(s \wedge t)] \longrightarrow r$ is a tautology. Using these facts, we can show that

$$x' = \frac{b' - a'}{b - a} x + \frac{a'b - ab'}{b - a}.$$

Thus if

$$e = \frac{b' - a'}{b - a} \quad \text{and} \quad f = \frac{a'b - ab'}{b - a},$$

we have $x' = ex + f$. Furthermore, $e, f \in R$ and $e \neq 0$. Why?

Case II. The proof of Case II $(X = A \vee X = B)$ is requested in the exercises.

EXAMPLE

Find x' in terms of x for the situation depicted below.

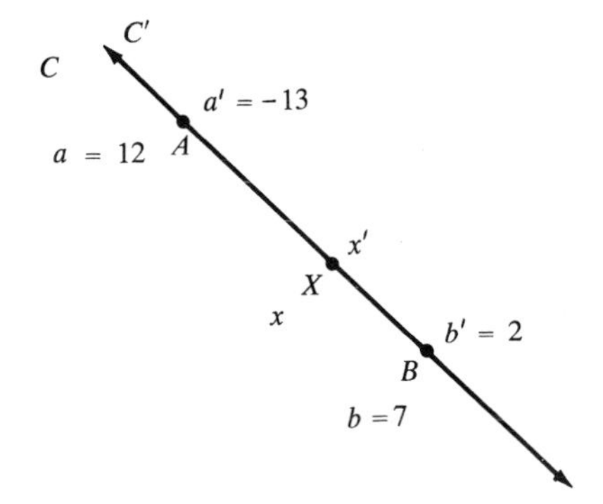

SOLUTION

From Theorem 4-7 we know that $x' = ex + f$. We must find e and f. We see that if $X = A$, then $x = 12$ and $x' = -13$. If $X = B$, then $x = 7$ and $x' = 2$. Thus we have the two equations

$$\text{(i)} \; -13 = 12e + f, \qquad \text{(ii)} \; 2 = 7e + f.$$

Solving these equations for e and f, we have $x' = -3x + 23$.

Now we consider an important special case of Theorem 4-7 as indicated in the diagram below.

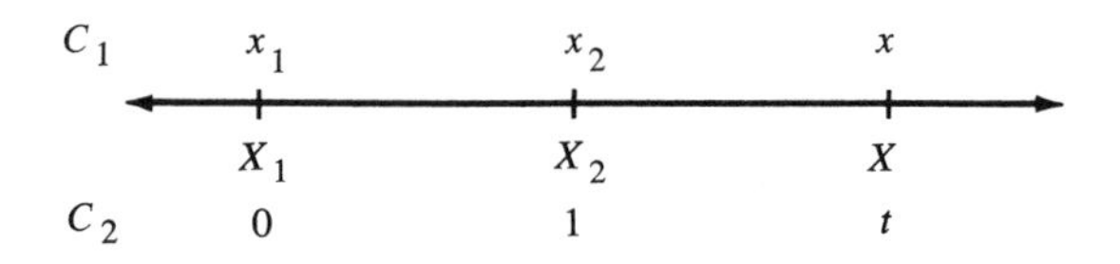

Suppose that the origin and unit point for the coordinate system C_2 have coordinates x_1 and x_2, respectively, in coordinate system C_1. Let t represent the C_2-coordinate of any point X and let x represent its C_1-coordinate. We wish to express x in terms of t and the known numbers x_1 and x_2.

By Theorem 4-7, we have $x = et + f$, where e and f are to be determined. If $X = X_1$, we have $t = 0$ and $x = x_1$. If $X = X_2$, we know that $t = 1$ and $x = x_2$. Accordingly we have the two equations

$$\text{(i) } x_1 = e\cdot 0 + f, \qquad \text{(ii) } x_2 = e\cdot 1 + f.$$

From (i) we have $f = x_1$. Substituting this value in (ii), we have $x_2 = e + x_1$. Hence $e = x_2 - x_1$, and the desired equation is

$$x = x_1 + (x_2 - x_1)t.$$

We state this important result in the following corollary.

COROLLARY 1 TO THEOREM 4-7 (THE TWO-POINT THEOREM) *In any coordinate system for line L let x_1 and x_2 be the respective coordinates of distinct points X_1 and X_2 in L. Then the formula*

$$x = x_1 + t(x_2 - x_1)$$

expresses the coordinate x of any point in L in terms of the coordinate t of the same point relative to the coordinate system with origin X_1 and unit point X_2.

If C is a coordinate system for $\overleftrightarrow{X_1X_2}$, we can use the phrase X_1-X-X_2 for C to mean that C assigns points X_1, X, and X_2 coordinates in such a way that the coordinate of X is between the coordinates of X_1 and X_2. Thus if x_1, x, and x_2 are the coordinates assigned by C to points X_1, X, and X_2, respectively, we have

$$X_1\text{-}X\text{-}X_2 \text{ for } C \longleftrightarrow x_1\text{-}x\text{-}x_2.$$

Let C' be any other coordinate system for $\overleftrightarrow{X_1X_2}$ and let x_1', x', and x_2' be the coordinates assigned by C' to points X_1, X, and X_2, respectively. Then we have

$$X_1\text{-}X\text{-}X_2 \text{ for } C' \longleftrightarrow x_1'\text{-}x'\text{-}x_2'.$$

Theorem 4-7 enables us to prove the statement made earlier that

$$X_1\text{-}X\text{-}X_2 \text{ for } C \longleftrightarrow X_1\text{-}X\text{-}X_2 \text{ for } C'.$$

Since C and C' are two coordinate systems for the same line, Theorem 4-7 applies, and we have $x' = ex + f$. We consider the case where $e > 0$ and $x_1 < x_2$. For this case we have the following flow-diagram proof.

$$X_1\text{-}X\text{-}X_2 \text{ for } C \xrightarrow{(1)} x_1\text{-}x\text{-}x_2 \xrightarrow{(2)} ex_1 < ex < ex_2 \overset{(3)}{\dashv}$$
$$\leftrightarrow ex_1 + f < ex + f < ex_2 + f \overset{(4)}{\leftrightarrow}$$
$$\leftrightarrow x_1' < x' < x_2' \xrightarrow{(5)} X_1\text{-}X\text{-}X_2 \text{ for } C'.$$

Therefore

$$(6)\ X_1\text{-}X\text{-}X_2 \text{ for } C \longrightarrow X_1\text{-}X\text{-}X_2 \text{ for } C'.$$

Reasons:

(1) Definition of X_1-X-X_2 for C.
(2) O_5 (multiplication property of order) and Exercise 10(a), page 84.
(3) O_3 (addition property of order).
(4) Theorem 4-7.
(5) Definition of X_1-X-X_2 for C'.
(6) Transitive property of implication.

This proof is easily modified for the remaining three cases—namely,

$e > 0$ and $x_1 > x_2$, $e < 0$ and $x_1 < x_2$, and $e < 0$ and $x_1 > x_2$.

Thus we have proved the following corollary.

COROLLARY 2 TO THEOREM 4-7 *If X is between X_1 and X_2 for one coordinate system for X_1X_2, then X is between X_1 and X_2 for every coordinate system for $\overleftrightarrow{X_1X_2}$.*

We sometimes use the statement "betweenness for points is independent of the coordinate system" to convey the idea expressed in Corollary 2.

Observe that our definition of betweenness for points tells us that if X is between X_1 and X_2, there is at least one coordinate system for $\overleftrightarrow{X_1X_2}$ for which the coordinate of X is between the coordinates of X_1 and X_2. From this and Corollary 2 we have

COROLLARY 3 TO THEOREM 4-7 X_1-X-$X_2 \longleftrightarrow$ *Every coordinate system for $\overleftrightarrow{X_1X_2}$ assigns X_1, X, and X_2 coordinates in such a way that the coordinate of X is between the coordinates of X_1 and X_2.*

We are now ready to prove the following useful theorem.

THEOREM 4-8 *If P and Q are distinct points in L, and p and q are different real numbers, then there exists a unique coordinate system for L that assigns to P the coordinate p and to Q the coordinate q.*

In order to prove Theorem 4-8 we must prove two statements:

(1) *Existence:* There exists at least one coordinate system for L that assigns P the coordinate p and Q the coordinate q.
(2) *Uniqueness:* There is not more than one coordinate system for L that assigns P the coordinate p and Q the coordinate q.

Analysis of the existence proof: First we observe that our statement of existence can be expressed as follows:

$$\left.\begin{array}{l} P \neq Q \\ p, q \in R \wedge p \neq q \end{array}\right\} \longrightarrow \left\{\begin{array}{l} \text{There is a coordinate system for } \overleftrightarrow{PQ} \\ \text{such that } P(p) \text{ and } Q(q). \end{array}\right.$$

Since $P \neq Q$, there is a coordinate system C with origin P and unit point Q.

Suppose now that there is a coordinate system C' that assigns P the coordinate p and Q the coordinate q. If x and x' are respectively the C-coordinate and the C'-coordinate of point X, then by Corollary 1. to Theorem 4-7 we have $x' = (q - p)x + p$. We seek the C-coordinates of the origin and the unit point of the coordinate system C'. If $x' = 0$, we have $x = p/(p - q)$; if $x' = 1$, we have $x = (p - 1)/(p - q)$. Now $p/(p - q)$ and $(p - 1)/(p - q)$ are unequal real numbers and hence are assigned distinct points A and B in $\overleftrightarrow{PQ}$ by C. By the corollary to the ruler postulate, we know that there exists a coordinate system C_1 that has A for its origin and B for its unit point. We can establish our existence statement by proving that C_1 assigns P the coordinate p and Q the coordinate q.

PROOF OF EXISTENCE

Statement	Reason
1. There is a coordinate system C that assigns P the coordinate 0 and Q the coordinate 1.	1. $P \neq Q$ and corollary to the ruler postulate.
2. $p/(p - q)$ and $(p - 1)/(p - q)$ are distinct real numbers.	2. Why?
3. C assigns $p/(p - q)$ and $(p - 1)/(p - q)$ to distinct points A and B.	3. Why?
4. There is a coordinate system C_1 having A for its origin and B for unit point.	4. Why?
5. If x and x_1 are respectively the C-coordinate and C_1-coordinate of X, then $x = \left(\frac{p-1}{p-q} - \frac{p}{p-q}\right)x_1 + \frac{p}{p-q}$, $x = \frac{1}{q-p}x_1 + \frac{p}{p-q}$.	5. Corollary 1 to Theorem 4-7.
6. $x = 0 \longrightarrow x_1 = p$, $x = 1 \longrightarrow x_1 = q$. Therefore C_1 assigns P the coordinate p, C_1 assigns Q the coordinate q. Hence, there exists a coordinate system such that $P(p)$ and $Q(q)$.	6. Why?

To prove the uniqueness part of Theorem 4-8, we shall show that under the conditions given ($P \neq Q \wedge p, q \in R \wedge p \neq q$), if C'' is a coordinate system that assigns P and Q the coordinates p and q, respectively, then $C_1 = C''$. In order to prove this, we may rely on the corollary to the ruler postulate, which states that two coordinate systems for the same line are equal if they have the same origin and the same unit point. You are asked to supply the details in the exercises. This will complete our outline of the proof of Theorem 4-8.

If there exist two constants e and f with $e \neq 0$ such that the variables x and x' satisfy the equation $x' = ex + f$, we say that x' is linearly dependent on x. It follows that x is linearly dependent on x'. When each of two variables is linearly dependent on the other, we say that they are linearly related. Thus we have

$$x' \text{ is linearly dependent on } x \longleftrightarrow x \text{ is linearly dependent on } x' \leftrightarrow$$
$$\leftrightarrow x \text{ and } x' \text{ are linearly related.}$$

Theorem 4-7 says that the coordinates assigned by two coordinate systems for the same line are linearly related.

(1) C is a coordinate system for L
(2) C' is a coordinate system for L
(3) x is the C-coordinate of X
(4) x' is the C'-coordinate of X
$\longrightarrow$ x and x' are linearly related.

Let us consider the situation suggested by the second converse. Suppose that C is a coordinate system for L and that D is a correspondence between L and R that assigns the number $ax + b$ to $X(x)$, where $a \neq 0$. Is D a coordinate system for L? It is easy to prove that D is a one-to-one correspondence between L and R. In the exercises you are asked to prove:

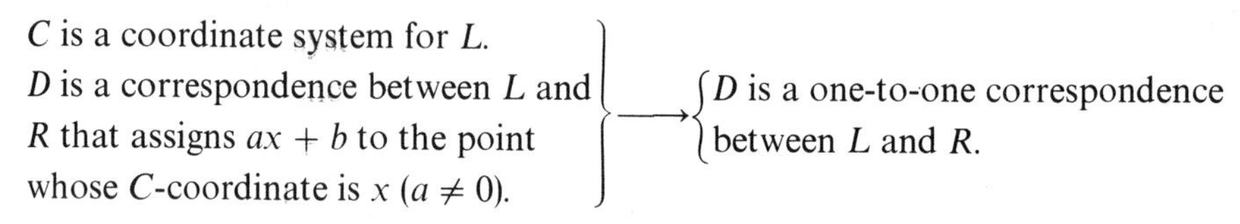
C is a coordinate system for L.
D is a correspondence between L and R that assigns $ax + b$ to the point whose C-coordinate is x ($a \neq 0$).
$\longrightarrow$ D is a one-to-one correspondence between L and R.

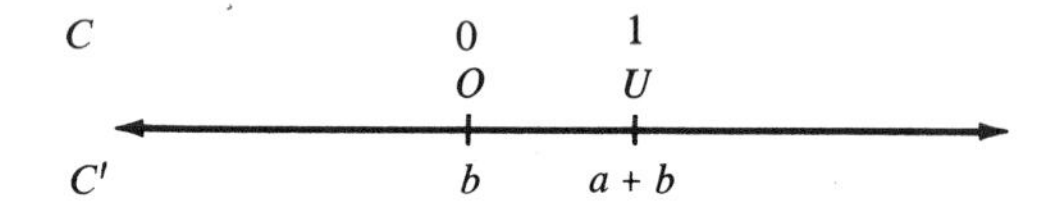

Let O and U be the origin and the unit point for C. By Theorem 4-8 there exists a coordinate system C' for L that assigns O the coordinate b and U the coordinate $a + b$. By Corollary 1 to Theorem 4-7, we have $x' = ax + b$, where x is the C-coordinate of a given point and x' is its C'-coordinate. But $ax + b$ is the number assigned to $X(x)$ by the correspondence D. Thus, given a point in L, D and C' pair the same number with this point. Moreover, given a real number s, D and C' pair the same point in L to this number, namely the point whose C-coordinate is $(s - b)/a$. Therefore, the two one-to-one correspondences C' and D are equal because they consist of exactly the same pairs. (See the definition of one-to-one correspondence, page 3.) Thus we have

C' is a coordinate system for L
$C' = D$
$\longrightarrow$ D is a coordinate system for L.

We have proved the following theorem:

THEOREM **4-9**

C is a coordinate system for L.
D is a correspondence between L and R that assigns $ax + b$ to the point whose C-coordinate is x ($a \neq 0$).
$\longrightarrow$ *D is a coordinate system for L.*

Some observations about distance, betweenness, and coordinate systems are appropriate at this point. First, we observe that distance has not been defined, although we have asserted that devices exist (coordinate systems) that enable us to measure distance relative to a specified unit pair. The situation is analogous to saying that although we have not defined electricity, we know that there are devices (ammeters) that measure its flow.

Again we note that our definition of a coordinate system tells us nothing about *how* the one-to-one correspondence between R and L is to be effected. We have a strong tendency to visualize an "infinite ruler" having a "uniform scale" such as the one shown below.

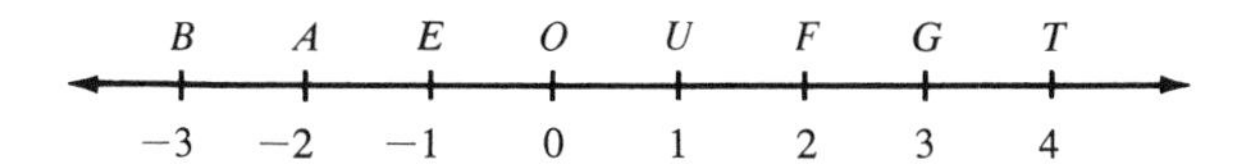

However, insofar as our postulates and definitions are concerned, the picture could look like this:

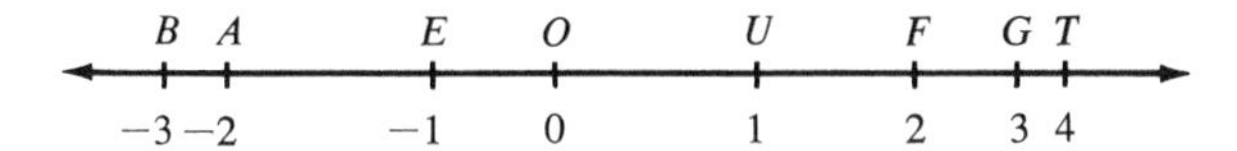

It might be argued that this picture is wrong, because the distance between E and U is greater than the distance between F and T and yet $|EU| = 2$ and $|FT| = 2$. The answer is that the distance between E and U may *appear* to be greater than the distance between F and T, but it is not.

Again, we might ask if a coordinate system could look like this:

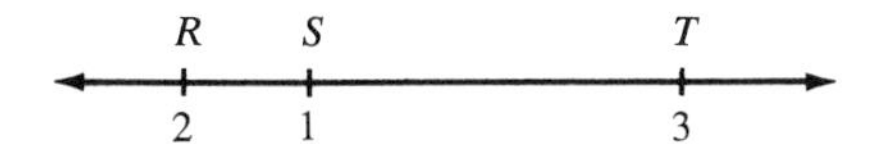

In replying, we must remember that our formal development tells us nothing about how to proceed when we draw pictures. Betweenness for points is defined in terms of betweenness of the real numbers that are the coordinates of these points in some coordinate system. If the coordinate system C assigns 1 to S, 2 to R, and 3 to T, then according to our definition of betweenness, R is between S and T regardless of appearances. The ruler postulate does not insure the existence of an "infinite ruler" with a "uniform scale"; therefore, we are not entitled to infer anything from the properties of such a ruler. Our conclusions must be based entirely on the definitions and postulates we have stated.

The Midpoint of a Segment

DEFINITION *A point M is a midpoint of $\overline{AB}$ if A-M-B and $|AM| = |MB|$.*

THEOREM 4-10 *Every segment $\overline{PQ}$ has exactly one midpoint.*

In proving Theorem 4-10, we must prove two things for segment PQ.

Existence: There exists at least one point that is a midpoint.
Uniqueness: There is at most one such midpoint.

PROOF

Let p and q be real numbers such that $p < q$. By Theorem 4-8, there is a coordinate system C for $\overleftrightarrow{PQ}$ such that $P(p)$ and $Q(q)$. $(p + q)/2 \in R$. Why? Therefore, there is a point $M \in \overleftrightarrow{PQ}$ such that $M[(p + q)/2]$. For the existence part of the theorem we shall show that M is a midpoint of $\overline{PQ}$.

$$\left.\begin{array}{r} P(p),\ Q(q),\ M\left(\dfrac{p+q}{2}\right) \\ p<q \longrightarrow \left\{\begin{array}{l}2p<p+q\\ p+q<2q\end{array}\right\} \longrightarrow p<\dfrac{p+q}{2}<q\end{array}\right\} \longrightarrow P\text{-}M\text{-}Q$$

$$P(p),\ Q(q),\ M\left(\frac{p+q}{2}\right) \longrightarrow \left\{\begin{array}{l} |PM| = \left|\dfrac{p+q}{2}-p\right| \longrightarrow |PM| = \left|\dfrac{q-p}{2}\right| \\ |MQ| = \left|q-\dfrac{p+q}{2}\right| \longrightarrow |MQ| = \left|\dfrac{q-p}{2}\right| \end{array}\right\} \rightarrow$$

$$\rightarrow \left.\begin{array}{r}|PM|=|MQ|\\ P\text{-}M\text{-}Q\end{array}\right\}\longrightarrow M \text{ is a midpoint.}$$

For uniqueness we shall show that if $T(t)$ is a midpoint of $\overline{PQ}$, then $T = M$.

$$T \text{ is a midpoint} \longrightarrow \left\{\begin{array}{l} \left.\begin{array}{r} T(t),\ P(p),\ Q(q) \\ |TP| = |TQ| \end{array}\right\} \longrightarrow |t-p| = |q-t| \\ \left.\begin{array}{r} P\text{-}T\text{-}Q \longrightarrow p\text{-}t\text{-}q \\ p<q \end{array}\right\} \xrightarrow{*} p<t<q \longrightarrow \left\{\begin{array}{l} t-p>0 \\ q-t>0 \end{array}\right\} \end{array}\right\} \rightarrow$$

$$\rightarrow t-p = q-t \longrightarrow \left.\begin{array}{r} \left.\begin{array}{r} t = \dfrac{p+q}{2} \\ T(t) \end{array}\right\} \longrightarrow T\left(\dfrac{p+q}{2}\right) \\ M\left(\dfrac{p+q}{2}\right) \end{array}\right\} \longrightarrow T = M.$$

You may supply the reasons for this proof.

The proof of the following corollary to Theorem 4-10 is contained in the proof of the theorem.

COROLLARY TO THEOREM 4-10 *If a coordinate system for $\overleftrightarrow{PQ}$ assigns P the coordinate p and Q the coordinate q, then it assigns $(p + q)/2$ as the coordinate of the midpoint of $\overline{PQ}$.*

---

* See L-Exercise 10(a), page 84.

DEFINITION *The midpoint of a segment is said to* bisect *the segment. More generally, if M is the midpoint of $\overline{AB}$ and S is a set of points such that $S \cap \overline{AB} = M$, then the set S bisects $\overline{AB}$.*

Suppose that $Y \in \overset{\circ\!-\!\circ}{AB}$ and $|AY| = 2|YB|$ and that $A(3)$, $Y(y)$, and $B(15)$. We can find y as follows:

$$\left.\begin{array}{l} Y \in \overset{\circ\!-\!\circ}{AB} \\ A(3) \wedge B(15) \end{array}\right\} \longrightarrow 3 < y < 15 \longrightarrow \left\{\begin{array}{l} y - 3 > 0 \longrightarrow y - 3 = |y - 3| \\ 15 - y > 0 \longrightarrow 15 - y = |15 - y| \end{array}\right. \left.\begin{array}{l} \\ \\ |AY| = 2|YB| \longrightarrow |y - 3| = 2|15 - y| \end{array}\right\} \dashv$$

$$\leftrightarrow y - 3 = 2(15 - y) \longrightarrow y = 11.$$

If $Y \in \overset{\circ\!-\!\circ}{AB}$ and either of the numbers $|AY|$ or $|YB|$ is equal to $\frac{1}{3}|AB|$, the point Y is called a *point of trisection* of $\overline{AB}$. Clearly, each segment has two points of trisection. In the case of the segment AB above, the other point of trisection is $P(7)$. In the exercises you will be asked to find a general formula for obtaining the coordinates of the points of trisection for any segment AB.

Congruent Segments

In this section we shall suppose that a unit pair has been chosen and is fixed. For this reason we shall omit the phrase "relative to a given unit pair." The results we obtain are valid for any unit pair we choose.

We begin by defining the familiar word "length."

DEFINITION *The distance between two distinct points is called the* length *of the segment joining the two points.*

Thus if S and T are distinct points, the number $|ST|$ is the length of the segment ST.

Let AB and CD be segments. A segment is a set of points, Consequently, the statement $\overline{AB} = \overline{CD}$ means that $\overline{AB}$ and $\overline{CD}$ are names for the same set of points, as shown in the diagram.

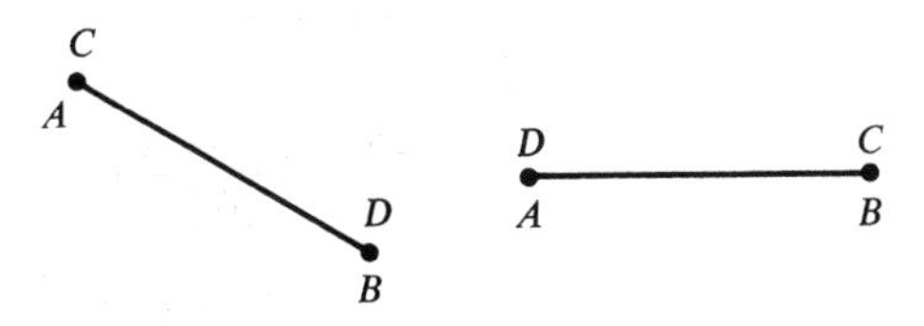

On the other hand, the statement $|AB| = |CD|$ means that the number $|AB|$ is the same as the number $|CD|$. In other words, the segments AB and CD have the same length.

Later, in Chapter 7, we develop in considerable detail the idea of *congruence*. Informally, one geometric figure is congruent to another if they have the "same size and shape." If we apply this idea to segments, we are led to the following definition.

DEFINITION *Segments having the same length are* congruent.

Notation: The statement "the segment AB is congruent to the segment CD" is denoted by $\overline{AB} \cong \overline{CD}$, where the symbol $\cong$ represents the phrase "is congruent to."

Using this notation, we can express our definition compactly as follows:

$$\overline{AB} \cong \overline{CD} \longleftrightarrow |AB| = |CD|.$$

DEFINITION *A relation R on set S is the truth set of a statement s about ordered pairs of elements in S. If (a, b) makes s true, then we write $a \, Ⓢ \, b$ and say that $(a, b) \in R$. Thus*

$$(a, b \in S \wedge a \, Ⓢ \, b) \longleftrightarrow (a, b) \in R.$$

The statement s is said to define a relation on S.

EXAMPLE 1

Let S be the set $\{1, 2, 3, 4\}$ and let s be the statement "x is greater than y," where $x, y \in S$. We note that certain ordered pairs of elements of S make the statement s true while others make s false when x and y are replaced by the first and second numbers of the ordered pair, respectively. The relation R defined on set S by statement s consists of the ordered pairs of elements of S that make the statement true. Thus

$$R = \{(2, 1), (3, 1), (3, 2), (4, 1), (4, 2), (4, 3)\}.$$

EXAMPLE 2

Let s be the statement "x has the same final digit as y," where $x, y \in W$, the set of whole numbers. We see that $3 \, Ⓢ \, 23$, $75 \, Ⓢ \, 35$, $18 \, Ⓢ \, 18$, and so on, while it is false that $7 \, Ⓢ \, 16$, $35 \, Ⓢ \, 17$, and so on. The relation defined on the set of whole numbers is the set of ordered pairs of whole numbers that make s true. Thus $R = \{(3, 23), (75, 35), (18, 18), \ldots\}$.

DEFINITION *Let R be a relation on set S that is defined by statement s. If $a, b, c \in S$ and*

(*i*) $(a, a) \in R$ [*reflexive property*],
(*ii*) $(a, b) \in R \longrightarrow (b, a) \in R$ [*symmetric property*],
(*iii*) $[(a, b) \in R \wedge (b, c) \in R] \longrightarrow (a, c) \in R$ [*transitive property*],

then statement s is said to define an equivalence relation *on set S. We also say that R is an equivalence relation.*

We can verify that the relation defined in Example 2 is an equivalence relation because it possesses the reflexive, symmetric, and transitive properties. However, the relation defined in Example 1 is not an equivalence relation. We can show that this relation does not possess the reflexive and symmetric properties.

Let S be the set of segments and let s be the statement "a is congruent to b," where $a, b \in S$. The following theorem asserts that s defines an equivalence relation for set S. The proof of Theorem 4-11 is requested in the exercises.

THEOREM 4-11 *Congruence between segments defines an equivalence relation on the set of segments.*

We note that Theorem 4-11 is sometimes expressed by the statement "congruence between segments is an equivalence relation." In the future we shall state theorems about equivalence relations in this more compact form.

Exercises

1. Given the data indicated in the diagrams, express x_2 in terms of x_1 in each case.

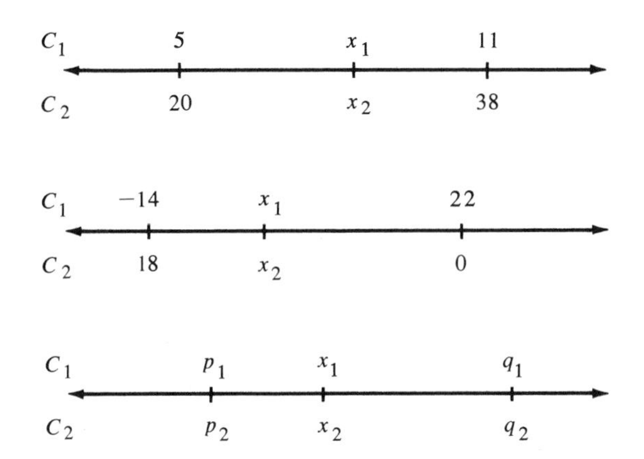

2. Prove: If $|PQ|$ (relative to $\{E, F\}$) $= 1$, then $|AB|$ (relative to $\{P, Q\}$) $= |AB|$ (relative to $\{E, F\}$).
3. A given coordinate system assigns points A, B, and C the coordinates -4, 8, and 12, respectively.
   (a) In the coordinate system with origin A and unit point C, find the coordinate of B using the formula from the two-point theorem. In this coordinate system does B lie in $\overline{AC}$? In $\overrightarrow{AC}$? In the ray opposite $\overrightarrow{AC}$?
   (b) In the coordinate system with origin A and unit point B, find the coordinate of C. In this coordinate system does C lie in $\overline{AB}$? In $\overrightarrow{AB}$? In the ray opposite $\overrightarrow{AB}$?
4. Coordinate system C assigns A and B the coordinates -4 and 8, respectively, and the coordinate system C' assigns them the coordinates 0 and q, respectively. P is any point in $\overleftrightarrow{AB}$ whose C-coordinate is x and whose C'-coordinate is k. Express x in terms of k. Using this information, complete the following table. (Assume $q > 0$.)

	k	x	Point in $\overleftrightarrow{AB}$ or subset of $\overleftrightarrow{AB}$
(a)	0	-4	A
(b)			B
(c)	$\frac{1}{2}$		
(d)	$\frac{1}{3}$		
(e)			$P \in \overrightarrow{AB}$ such that $\lvert AP \rvert = 3\lvert AB \rvert$
(f)		$-4 \leqq x \leqq 8$	
(g)	$k \geqq 0$		
(h)			$\overset{\circ\!-\!\circ}{AB}$
(i)			Interior of the ray opposite $\overrightarrow{AB}$
(j)	4		
(k)		-48	

5. Given $A(5)$, $B(-23)$, and $T(t)$ with $T \in \overset{\circ\!-\!\circ}{AB}$, find t if $\lvert TB \rvert = 3\lvert AT \rvert$.
6. If in Exercise 5 we remove the restriction that $T \in \overset{\circ\!-\!\circ}{AB}$, what other solution do we obtain?
7. Given $A(a)$ and $B(b)$, with $S(s)$ and $T(t)$ the points of trisection of $\overline{AB}$ and A-S-T-B, find s and t in terms of a and b.
8. If A-T-B, $A(a)$, $B(b)$, $T(t)$, and k is a number between 0 and 1, use the two-point theorem to express t in terms of a, k, and b.
9. Let R, S, and T be points with C-coordinates 7, -3, and -10.
   (a) Does there exist a coordinate system for $\overleftrightarrow{RT}$ that assigns R, S, and T the numbers -14, 16, and 37, respectively?
   (b) Does there exist a coordinate system for $\overleftrightarrow{RT}$ that assigns R, S, and T the numbers 17, 0, -17, respectively?
10. If D, E, and F are points and $E \neq F$ and if $\{A, A'\}$ and $\{B, B'\}$ are unit pairs and if

$$\frac{\lvert DE \rvert \text{ (relative to } \{A, A'\})}{\lvert EF \rvert \text{ (relative to } \{A, A'\})} = 7, \qquad \text{find} \qquad \frac{\lvert EF \rvert \text{ (relative to } \{B, B'\})}{\lvert DE \rvert \text{ (relative to } \{B, B'\})}.$$

11. Write a flow-diagram proof for the following statement.

$$\left.\begin{array}{l}\text{Points } B \text{ and } C \text{ are between points } A \text{ and } D \\ \overline{AB} \cong \overline{CD}\end{array}\right\} \longrightarrow \overline{AC} \cong \overline{BD}.$$

12. Prove: A-B-C-$D \longrightarrow \lvert AB \rvert \cdot \lvert CD \rvert + \lvert BC \rvert \cdot \lvert AD \rvert = \lvert AC \rvert \cdot \lvert BD \rvert$.
13. Prove:

$$\left.\begin{array}{l}M \text{ is the midpoint of } \overline{AB} \\ M \text{ is the midpoint of } \overline{CD} \\ A, B, C, D \text{ are distinct collinear points}\end{array}\right\} \longrightarrow \overline{AC} \cong \overline{BD}.$$

14. Prove $\sim(s \wedge t)$, where s and t are the statements identified in Theorem 4-7.
15. Prove that $[(r \vee s) \wedge (r \vee t) \wedge \sim(s \wedge t)] \longrightarrow r$ is a tautology.
16. Prove Case II of Theorem 4-7.
17. Prove Corollary 2 to Theorem 4-7 by applying Corollary 1.

18. Prove that linear dependence is an equivalence relation.
19. Prove the uniqueness part of Theorem 4-8.
20. Prove Theorem 4-11.

References

Birkhoff and Beatley, *Basic Geometry*
Kay, *College Geometry*
Ringenberg, *College Geometry*
Zippin, *Uses of Infinity*

Chapter 5

Convexity and Separation

In this chapter we consider two important geometric concepts: convexity and separation. The results obtained here will have many applications to succeeding chapters.

Convex Sets

In drawing (a) below, the shading represents a "circular region." We see that if P and Q are any distinct points of this region, then the segment PQ is contained in the region. However, in drawing (b), if M and N are two points of the region, $\overline{MN}$ may be contained in the region, but clearly in some cases (as in the one pictured) some points of $\overline{MN}$ may not be points of the region.

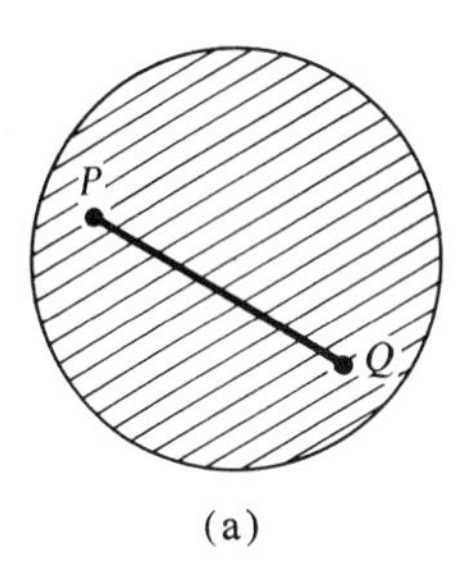

(a)

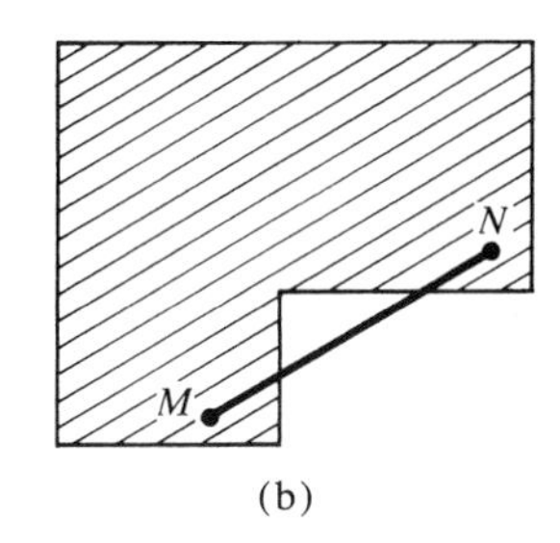

(b)

We say that the set of points pictured in drawing (a) is a convex set because a segment joining any two of its points lies entirely within the set. The set of points shown in drawing (b) is not convex. We define a convex set as follows:

DEFINITION *A set is* convex *if for every two distinct points in the set, the segment joining the two points is contained in the set.*

COROLLARY TO THE DEFINITION OF CONVEX SET *Every set that consists of no more than one point is convex.*

Our first two theorems are easily proved.

THEOREM 5-1 *If A and B are distinct points, then each of the following is a convex set:* (1) $\overleftrightarrow{AB}$, (2) $\overrightarrow{AB}$, (3) $\overset{\circ\!\!\longrightarrow}{AB}$, (4) $\overline{AB}$, (5) $\overset{\circ\!\!-\!\!-}{AB}$ *or* $\overset{-\!\!-\!\!\circ}{AB}$, (6) $\overset{\circ\!\!-\!\!\circ}{AB}$.

PROOF OF (1)

Let $\overleftrightarrow{AB}$ be the line determined by distinct points A and B (as the symbol $\overleftrightarrow{AB}$ indicates) and let P and Q be any two distinct points in $\overleftrightarrow{AB}$. Then by our definition of segment, $\overline{PQ}$ is a subset of $\overleftrightarrow{AB}$. By our definition of convex set it follows that $\overleftrightarrow{AB}$ is a convex set.

The proofs of parts (2), (3), (4), (5), and (6) are left as exercises.

The drawing below suggests that the union of two convex sets may not be a convex set.

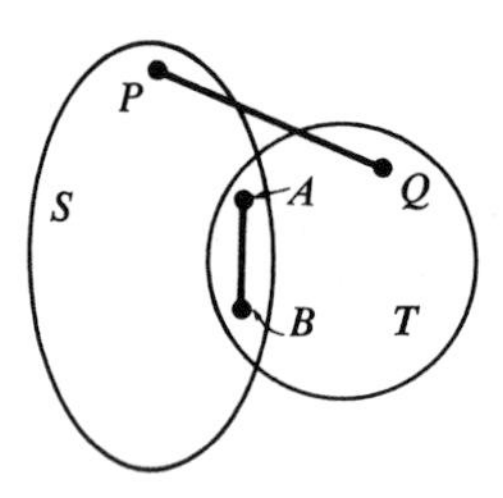

Consider the intersection of convex sets S and T. If $n(S \cap T) = 0$ or $n(S \cap T) = 1$, then $S \cap T$ is convex by the corollary to the definition of convex set. If $n(S \cap T) > 1$, then let A and B be any two distinct points in $S \cap T$. In order to prove that $S \cap T$ is convex according to our definition, we must show that $\overline{AB} \subseteq S \cap T$.

PROOF

$$A \text{ and } B \text{ are distinct points in } S \cap T \xrightarrow{(1)} \left.\begin{cases} A \in S \wedge B \in S \xrightarrow{(2)} \overline{AB} \subseteq S \\ A \in T \wedge B \in T \xrightarrow{(2)} \overline{AB} \subseteq T \end{cases}\right\} \xrightarrow{(3)} \overline{AB} \subseteq S \cap T.$$

Reasons:

(1) and (3) Definition of the intersection of two sets.
(2) Definition of convex set.

We have proved the following theorem:

THEOREM 5-2 *If two sets are convex, then their intersection is convex.*

Separation

Suppose we have a set K that is the union of two nonempty, nonintersecting sets A and B and suppose that set S, which does not intersect K, is so situated that the interior of any segment XY that joins a point X in A to a point Y in B must intersect S. [See figures (a) and (b) below.] Consider all sets of points

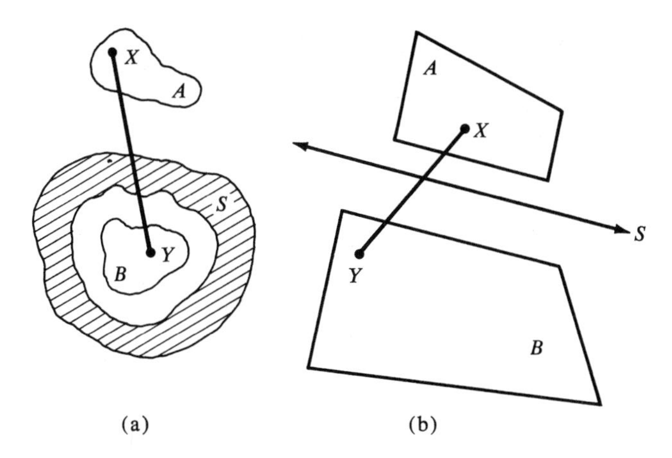

(a) (b)

as being coplanar. Under these circumstances we say that S *separates* K into two sets A and B, or that S separates A and B. Thus we have the following definition:

DEFINITION *Set S separates set K into two sets A and B if and only if all of the following statements are true:*

(1) $S \cap K = \emptyset$.
(2) $K = A \cup B$ *and* $A \cap B = \emptyset$.
(3) $A \neq \emptyset \wedge B \neq \emptyset$.
(4) $[(X \in A) \wedge (Y \in B)] \longrightarrow \overleftrightarrow{XY} \cap S \neq \emptyset$.

Under these circumstances, we also say that S *separates* sets A and B whose union is K.

If square S is in plane M, it separates $M \sim S$ into two sets I and E, where I is the "interior" of the square and E is the "exterior" of the square. If we let S represent space, then a cube C separates $S \sim C$ into two sets I, the "interior" of the cube, and E, the "exterior" of the cube. In this case $S = I \cup C \cup E$.

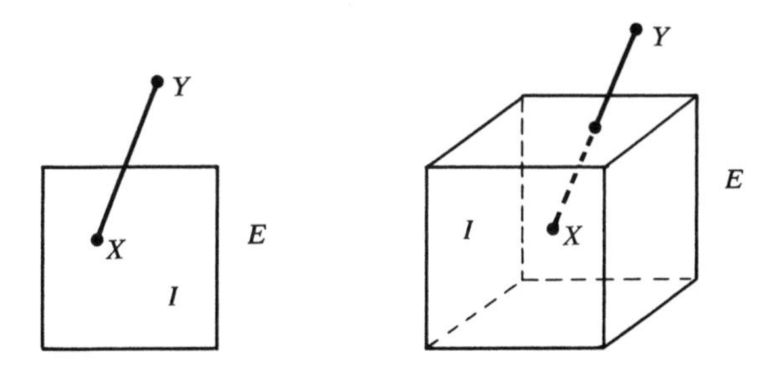

Exercises

1. Make a drawing to illustrate each of the following situations.
   (a) Convex sets A and B such that $A \cup B$ is convex.

(b) Convex sets A and B such that $A \cup B$ is not convex.
(c) A is a convex set and B is not a convex set such that $A \cup B$ is convex.
(d) A is a convex set and B is not a convex set such that $A \cup B$ is not convex.
(e) A is a convex set and B is not a convex set such that $A \cap B$ is not convex.
(f) A is a convex set and B is not a convex set such that $A \cap B$ is a convex set.

2. (a) Explain why space is a convex set.
   (b) Explain why a plane is a convex set.
3. Determine which of the following statements are true.
   (a) If set F separates A and B, then A and B are nonempty.
   (b) If set S separates convex sets E and F, then E and F are disjoint sets.
   (c) If set W separates convex sets A and B, then W is a convex set.
   (d) If set J separates sets A and B, then $A \cap B = \emptyset$.
   (e) If $X \in A$ and $Y \in B$ and $\overleftrightarrow{XY} \cap S = \emptyset$, then set S does not separate sets A and B.
   (f) If the intersection of two sets is not empty, then there is no set that separates them.
   (g) If $S \cap A \neq \emptyset$, then S cannot separate A from any set.
   (h) If set D separates sets X and Y and also separates sets X and Z, then set D separates sets X and $Y \cap Z$.
4. Prove parts (2), (3), (4), (5), and (6) of Theorem 5-1.
5. Decide which of the following implications are true. For those that are true, write a proof to verify your conclusion. If a statement is false, draw a figure to illustrate that it is false.
   (a) $\left.\begin{array}{l} S \text{ is convex} \\ S \cup T \text{ is convex} \end{array}\right\} \longrightarrow T \text{ is convex.}$
   (b) $\left.\begin{array}{l} S \text{ is convex} \\ S \cap T \text{ is not convex} \end{array}\right\} \longrightarrow T \text{ is not convex.}$
   (c) $\left.\begin{array}{l} B \cup C \text{ is convex} \\ A \cup B \text{ is convex} \end{array}\right\} \longrightarrow B \text{ is convex.}$
   (d) $\left.\begin{array}{l} A \subset B \\ A \text{ is convex} \end{array}\right\} \longrightarrow B \sim A \text{ is not convex.}$
6. Let $W = \{E, F, G, H\}$, $A = \{R, S, T\}$ and $B = \{X, Y\}$. Assuming the incidence conditions shown in the diagram, does the set W separate A and B?

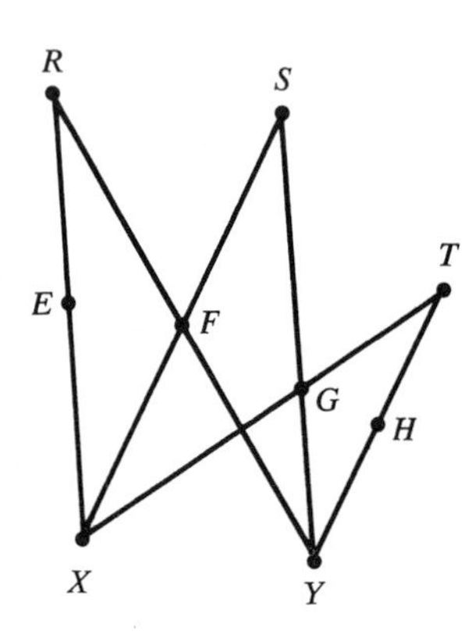

7. Prove the corollary to the definition of convexity.
8. Prove: If point A is in line L, then A separates $L \sim \{A\}$ into two half lines.

Pasch's Postulate

In many proofs presented in school geometry, it is assumed that certain lines intersect because they appear to intersect in a drawing. Thus if X and Y are points in "opposite sides" of L, we state with confidence that $\overline{XY}$ intersects L.

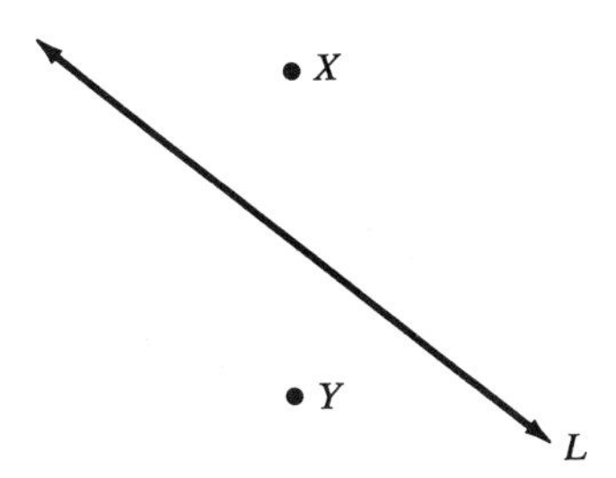

In the late nineteenth century, the German mathematician Pasch observed that, obvious as such statements are, there is really no way to prove them using only the standard postulates stated by Euclid. Believing that a statement of this kind should be provable by appeal to a carefully stated assumption, Pasch stated a postulate, which we now introduce as the basis for our discussion of separation in a plane.

POSTULATE* 13 *(PASCH'S POSTULATE) *Suppose that L is a line in the plane determined by three noncollinear points and that L does not contain any of the three points. If L intersects one of the segments determined by these points, then it will intersect exactly one of the other segments.*

Postulate 13 can be expressed symbolically as follows:

$$\left.\begin{array}{l} A, B, C \text{ are noncollinear} \\ L \subseteq \text{pl } ABC \\ L \cap \{A, B, C\} = \emptyset \\ L \cap \overline{AB} \neq \emptyset \end{array}\right\} \longrightarrow (L \cap \overline{BC} \neq \emptyset \veebar L \cap \overline{AC} \neq \emptyset).$$

This symbolic statement has the form $h \longrightarrow (b \veebar c)$, where h is the conjunction of the four statements in the hypothesis. We can show (see Exercise 2) that

$$[h \longrightarrow (b \veebar c)] \longrightarrow [(h \wedge \sim b) \longrightarrow c].$$

Thus the following statement is implied by Pasch's postulate.

$$\text{P}_1: \quad \left.\begin{array}{l} A, B, C \text{ are noncollinear} \\ L \subseteq \text{pl } ABC \\ L \cap \{A, B, C\} = \emptyset \\ L \cap \overline{AB} \neq \emptyset \\ L \cap \overline{BC} = \emptyset \end{array}\right\} \longrightarrow L \cap \overline{AC} \neq \emptyset.$$

If we write the fourth contrapositive of Pasch's postulate, we have

$$\text{P}_2: \quad \left.\begin{array}{l} A, B, C \text{ are noncollinear} \\ L \subseteq \text{pl}\, ABC \\ L \cap \{A, B, C\} = \emptyset \\ \sim(L \cap \overline{BC} = \emptyset \veebar L \cap \overline{AC} \neq \emptyset) \end{array}\right\} \longrightarrow L \cap \overline{AB} = \emptyset.$$

It can be shown (see Exercise 2) that

$$(1) \quad \sim(a \veebar b) \longleftrightarrow [(a \wedge b) \vee (\sim a \wedge \sim b)]$$

is a tautology. We have, then,

$$\begin{aligned} &\sim(L \cap \overline{BC} \neq \emptyset \veebar L \cap \overline{AC} \neq \emptyset) \leftrightarrow \\ &\qquad \leftrightarrow [\underbrace{(L \cap \overline{BC} = \emptyset \wedge L \cap \overline{AC} = \emptyset)}_{r} \vee \underbrace{(L \cap \overline{BC} \neq \emptyset \wedge L \cap \overline{AC} \neq \emptyset)}_{s}]. \end{aligned}$$

Thus the fourth statement in the hypothesis of P_2 is replacable by the disjunction $r \vee s$, where r and s are the indicated statements in the right-hand bracket. If we let p represent the conjunction of the first three statements in the hypothesis of P_2, we see that P_2 has the form $[p \wedge (r \vee s)] \longrightarrow t$. In Exercise 2 you are asked to prove that

$$(2) \quad ([p \wedge (r \vee s)] \longrightarrow t) \longleftrightarrow ([(p \wedge r) \longrightarrow t] \wedge [(p \wedge s) \longrightarrow t])$$

is a tautology. Applying tautologies (1) and (2) to P_2, we obtain two more useful statements implied by Pasch's postulate.

$$\text{P}_3: \quad \left.\begin{array}{l} A, B, C \text{ are noncollinear} \\ L \subseteq \text{pl}\, ABC \\ L \cap \{A, B, C\} = \emptyset \\ L \cap \overline{BC} = \emptyset \\ L \cap \overline{AC} = \emptyset \end{array}\right\} \longrightarrow \overline{AB} \cap L = \emptyset.$$

$$\text{P}_4: \quad \left.\begin{array}{l} A, B, C \text{ are noncollinear} \\ L \subseteq \text{pl}\, ABC \\ L \cap \{A, B, C\} = \emptyset \\ L \cap \overline{BC} \neq \emptyset \\ L \cap \overline{AC} \neq \emptyset \end{array}\right\} \longrightarrow \overline{AB} \cap L = \emptyset.$$

Since each of the statements P_1, P_2, P_3, and P_4 is obtainable from Postulate 13 by simple logical transformations, we shall refer to each of them as Postulate 13 of Pasch's postulate.

COROLLARY TO PASCH'S POSTULATE

$$\left.\begin{array}{l} A, B, C \text{ are distinct points} \\ \{A, B, C\} \text{ and line } L \text{ are coplanar} \\ L \cap \{A, B, C\} = \emptyset \end{array}\right\} \longrightarrow \left\{\begin{array}{l} L \text{ intersects an even number} \\ \text{of the segments } AB, BC, AC. \end{array}\right.$$

If A, B, and C are noncollinear points, the proof of the corollary follows immediately from Pasch's postulate as stated in P_1, P_3, and P_4. The proof for the case in which A, B, and C are distinct collinear points is left as an exercise.

We now define a set of points called a half plane. We shall prove that a half plane is convex and show that if line L is in plane M, then L separates $M \sim L$ into two half planes.

Notation: If $A \notin L$, we shall denote the plane determined by A and L by the symbol $\underline{L/A}$.

DEFINITIONS *If* $A \notin L$, $L/A = \{A\} \cup \{X \mid X \in \underline{L/A} \wedge \overline{AX} \cap L = \emptyset\}$. L/A *is called the half plane determined by* L *and* A. *The line* L *is the* edge *of the half plane. The union of* L *and* L/A *is a* closed half plane *and denoted by* $\overline{L/A}$. L/A *is also called the* A*-side of* L.

COROLLARY TO THE DEFINITION OF HALF PLANE *If* $A \notin L$, *then*

$$L/A \cap L = \emptyset \qquad \textit{and} \qquad L/A \subseteq \underline{L/A}.$$

THEOREM 5-3 *If* $A \notin L$, *then* L/A *is a convex set.*

We must show:

$$\left.\begin{array}{l} X \neq Y \\ X, Y \in L/A \end{array}\right\} \longrightarrow \overline{XY} \subseteq L/A.$$

PROOF

Case I: $X = A$ or $Y = A$. Suppose $X = A$; then

$$\left.\begin{array}{l} \left.\begin{array}{l} \left.\begin{array}{l} \left.\begin{array}{r} X = A \\ X \neq Y \end{array}\right\} \longrightarrow Y \neq A \\ \qquad\qquad\qquad Y \in L/A \end{array}\right\} \longrightarrow \overline{AY} \cap L = \emptyset \\ \qquad\qquad\qquad\qquad\qquad\qquad X = A \end{array}\right\} \longrightarrow \overline{XY} \cap L = \emptyset \\ \text{Let } P \text{ be any point in } \overset{\circ\!-\!\circ}{XY}, \text{ then } P \in \overset{\circ\!-\!\circ}{XY} \longrightarrow \overline{PX} \subseteq \overline{XY} \end{array}\right\} \dashv$$

$$\left.\begin{array}{r} \mapsto \overline{PX} \cap L = \emptyset \\ X = A \end{array}\right\} \longrightarrow \overline{PA} \cap L = \emptyset.$$

It is easily shown that $P \in \underline{L/A}$. From the definition of half plane, we have $P \in L/A$. Thus since $P \in \overset{\circ\!-\!\circ}{XY} \longrightarrow P \in L/A$, we know

$$\left.\begin{array}{l} \overset{\circ\!-\!\circ}{XY} \subseteq L/A \\ X, Y \in L/A \end{array}\right\} \longrightarrow \overline{XY} \subseteq L/A.$$

If $Y = A$, a similar argument shows that $\overline{XY} \subseteq L/A$.

Case II: $X \neq A$ and $Y \neq A$. Since $X \neq Y$, we have three distinct points A, X, and Y in L/A. This statement implies that $A, X, Y \in \underline{L/A}$ and $L \cap \{A, X, Y\} = \emptyset$. From the corollary to Pasch's postulate, we know that L intersects an even number of the segments AX, AY, and XY. But

$$\left.\begin{array}{l} X \neq A \\ X \in L/A \end{array}\right\} \longrightarrow \overline{AX} \cap L = \emptyset \quad \text{and} \quad \left.\begin{array}{l} Y \neq A \\ Y \in L/A \end{array}\right\} \longrightarrow \overline{AY} \cap L = \emptyset.$$

Thus we conclude that $\overline{XY} \cap L = \emptyset$. Let $P \in \overset{\circ\!-\!\circ}{XY}$. If $P = A$, then $P \in L/A$. Suppose $A \neq P$. We can show that A, P, and X are distinct points in $\underline{L/A}$. Since $\overline{XY} \cap L = \emptyset$ and $P \in \overset{\circ\!-\!\circ}{XY}$, we know that $P \notin L$. Thus $L \cap \{A, P, X\} = \emptyset$. We have previously shown that $\overline{AX} \cap L = \emptyset$. Also,

$$\left.\begin{array}{l} \overline{XY} \cap L = \emptyset \\ P \in \overset{\circ\!-\!\circ}{XY} \longrightarrow \overline{PX} \subseteq \overline{XY} \end{array}\right\} \longrightarrow \overline{PX} \cap L = \emptyset.$$

Using the corollary to Pasch's postulate, we conclude that $\overline{AP} \cap L = \emptyset$. But $P \in \underline{L/A}$ and, therefore, $P \in L/A$. We have

$$\left.\begin{array}{l} \overset{\circ\!-\!\circ}{XY} \subseteq L/A \\ X, Y \in L/A \end{array}\right\} \longrightarrow \overline{XY} \subseteq L/A$$

and the proof is complete.

We now proceed with the discussion by showing that if $A \notin L$, the set $S = \underline{L/A} \sim \overline{L/A}$ is also a half plane.

THEOREM 5-4 *$A \notin L \longrightarrow S = \underline{L/A} \sim \overline{L/A}$ is a half plane.*

DISCUSSION OF PROOF

We show first that $S \neq \emptyset$. Let T be any point in L; then $A \neq T$ and there is a point Z such that A-T-Z. Why? We shall prove that $Z \in S$ by showing $Z \in \underline{L/A}$ and $Z \notin L \wedge Z \notin L/A$. It is clear that $Z \in \underline{L/A}$ because $Z \in \overleftrightarrow{AT}$ and $\overleftrightarrow{AT} \subseteq \underline{L/A}$. Why? Furthermore, $\overleftrightarrow{AT}$ and L are distinct lines whose intersection is T. Since $Z \in \overleftrightarrow{AT}$ and $Z \neq T$, we know $Z \notin L$. To prove that $Z \notin L/A$, we have only to show that $\overline{AZ} \cap L \neq \emptyset$. Why? But A-T-Z implies that $T \in \overset{\circ\!-\!\circ}{AZ}$ and, therefore, $\overline{AZ} \cap L \neq \emptyset$.

Now we have a point $Z \in S$. But $Z \notin L$; thus there is a half plane $L/Z = \{Z\} \cup \{X \mid X \in \underline{L/Z} \wedge \overline{XZ} \cap L = \emptyset\}$. The proof of our theorem will depend upon our showing $S = L/Z$. This can be accomplished by establishing that $S \subseteq L/Z \wedge L/Z \subseteq S$. A discussion of the first statement is included. The remainder of the proof will be left to the exercises.

Let R be a point in S. If $R = Z$, then $R \in L/Z$ and our proof is complete. If $R \neq Z$, then we must show $R \in \underline{L/Z}$ and $\overline{RZ} \cap L = \emptyset$. It can be easily verified that $\underline{L/Z} = \underline{L/A}$, since each of these planes contains the point Z and line L, and $Z \notin L$. Since $R \in S$, we know $R \in \underline{L/A}$ and thus $R \in \underline{L/Z}$. We have to prove $\overline{RZ} \cap L = \emptyset$. The following argument shows that $\overline{RA} \cap L \neq \emptyset$:

$$\begin{array}{l} R \notin L/A \longrightarrow R \notin \{X \mid X \in \underline{L/A} \wedge \overline{XA} \cap L = \emptyset\} \dashv \\ \quad \leftrightarrow \left.\begin{array}{r} R \notin \underline{L/A} \vee \overline{RA} \cap L \neq \emptyset \\ R \in \underline{L/A} \end{array}\right\} \longrightarrow \overline{RA} \cap L \neq \emptyset. \end{array}$$

Also, $\overline{AZ} \cap L \neq \emptyset$.

Now we have L coplanar with $\{R, Z, A\}$ and $L \cap \{R, Z, A\} = \emptyset$. Why? Moreover, $\overline{AZ} \cap L \neq \emptyset$ and $\overline{RA} \cap L \neq \emptyset$. It follows by the corollary to Pasch's postulate that $\overline{RZ} \cap L = \emptyset$. We have also noted that $R \in \underline{L/Z}$. The underlined statements imply that $R \in L/Z$. Thus we have shown that $R \in S \longrightarrow R \in L/Z$. Therefore, $S \subseteq L/Z$.

The result of Theorem 5-4 is that we can claim that when $A \notin L$, the set $\underline{L/A} \setminus \overline{L/A}$ is a half plane. We frequently speak of this half plane as being the half plane opposite L/A and denote it by the symbol $L/{\sim}A$.

DEFINITION *If $A \notin L$, the set $L/{\sim}A = \underline{L/A} \setminus \overline{L/A}$ is the half plane opposite L/A.*

It can be proved that L/A is opposite $L/{\sim}A$ (see Exercise 11). Hence we can say that the half planes L/A and $L/{\sim}A$ are opposite each other.

THEOREM 5-5 (PLANE SEPARATION THEOREM) *If L is a line contained in plane M, then L separates $M \setminus L$ into two half planes.*

DISCUSSION OF PROOF

Let A be a point in plane M that is not in L. We have two half planes L/A and $L/{\sim}A$. Why? $M = \underline{L/A}$. Why? We shall show that L separates $\underline{L/A} \setminus L$ into L/A and $L/{\sim}A$. According to the definition of separation we must show

(i) $L \cap (\underline{L/A} \setminus L) = \emptyset$,
(ii) $\underline{L/A} \setminus L = L/A \cup L/{\sim}A$ and $L/A \cap L/{\sim}A = \emptyset$,
(iii) $L/A \neq \emptyset$ and $L/{\sim}A \neq \emptyset$,
(iv) $[X \in L/A \wedge Y \in L/{\sim}A] \longrightarrow \overleftrightarrow{XY} \cap L \neq \emptyset$.

We shall outline a proof for statement (iv). The other parts will be considered in the exercises. Since $Y \in L/{\sim}A$, we know that $Y \in \underline{L/A}$, $Y \notin L \wedge Y \notin L/A$. Therefore, $\overline{AY} \cap L \neq \emptyset$. Since $X \in L/A$ and $Y \in L/{\sim}A$, have $X \neq Y$. Also $A \neq Y$. Why? If $X = A$, then $\overline{AY} \cap L \neq \emptyset$ implies that $\overline{XY} \cap L \neq \emptyset$. This statement, together with the fact that $X \notin L$ and $Y \notin L$, implies that $\overleftrightarrow{XY} \cap L \neq \emptyset$.

If $X \neq A$, then A, X, and Y are distinct points coplanar with line L, which does not contain any of them. Moreover, we have $\overline{AY} \cap L \neq \emptyset$, and, since $X \in L/A$, we know that $\overline{AX} \cap L = \emptyset$. It follows from the corollary to Pasch's postulate that $\overline{XY} \cap L \neq \emptyset$. Since $X, Y \notin L$, we must have $\overleftrightarrow{XY} \cap L \neq \emptyset$.

COROLLARY TO THEOREM 5-5 $A \notin L \longrightarrow \underline{L/A} = L/A \cup L \cup L/{\sim}A$.

The following lemma will prove useful in many situations.

PLANE SEPARATION LEMMA

(*a*) $A \in L/B \longleftrightarrow B \in L/A \longleftrightarrow L/A = L/B$.
(*b*) $L/A = L/B \longleftrightarrow L/{\sim}A = L/{\sim}B$.
(*c*) $A \in L/{\sim}B \longleftrightarrow B \in L/{\sim}A \longleftrightarrow L/B = L/{\sim}A \longleftrightarrow L/A = L/{\sim}B$.

(d) $(A \in L/C \wedge B \in L/C) \longrightarrow A \in L/B$.
(e) $(A \in L/C \wedge B \in L/{\sim}C) \longrightarrow A \in L/{\sim}B$.

Observe that part (a) of this lemma justifies our naming a half plane by naming its edge and any point in it.

Exercises

1. Write verbal statements corresponding to Pasch's postulate as stated in P_1, P_3, and P_4.
2. Prove that each of the following is a tautology.
   (a) ${\sim}(a \veebar b) \longleftrightarrow [(a \wedge b) \vee ({\sim}a \wedge {\sim}b)]$.
   (b) $[h \longrightarrow (b \veebar c)] \longrightarrow [(h \wedge {\sim}b) \longrightarrow c]$.
   (c) $([p \wedge (r \vee s)] \longrightarrow t) \longleftrightarrow ([(p \wedge r) \longrightarrow t] \wedge [(p \wedge s) \longrightarrow t])$.
3. Prove the corollary to the definition of half planes.

L 4. (a) Prove the following set theorem where A and B are subsets of universe U.

$$B \cap (A \smallsetminus B) = \emptyset.$$

   (b) Prove statement (i) in Theorem 5-5.
   (c) Prove the second part of statement (ii) in Theorem 5-5.

L 5. (a) Prove the following set theorem, where A, B, and C are subsets of universe U.

$$\left.\begin{array}{l} B \subseteq A \\ B \cap C = \emptyset \end{array}\right\} \longrightarrow B \cup [A \smallsetminus (B \cup C)] = A \smallsetminus C.$$

   (b) Prove the first part of statement (ii) in Theorem 5-5.
   (c) Prove: $B \subseteq A \longrightarrow (A \smallsetminus B) \cup B = A$.

6. (a) Prove the corollary to Theorem 5-5.
   (b) Prove: $A \notin L \longrightarrow [L/A \cup \overline{L/{\sim}A} = \underline{L/A} \wedge \overline{L/A} \cap L/{\sim}A = \emptyset]$.
7. (a) Prove: $A \in L/B \longrightarrow \underline{L/A} = \underline{L/B}$.
   (b) Supply reasons to the following proof of the statement $A \in L/B \longrightarrow B \in L/A$.

$$\underline{A \in L/B} \xrightarrow{(1)} \otimes \begin{cases} A \in \{B\} \xrightarrow{(2)} B \in \{A\} \xrightarrow{(3)} B \in L/A \\ A \in \{X \mid X \in \underline{L/B} \wedge \overline{XB} \cap L = \emptyset\} \xrightarrow{(4)} \overline{AB} \cap L = \emptyset \end{cases}$$

$$\left.\begin{array}{l} \overline{AB} \cap L = \emptyset \\ \left.\begin{array}{l} (5)\ B \in \underline{L/B} \\ (6)\ L/B = \underline{L/A} \end{array}\right\} \xrightarrow{(7)} B \in \underline{L/A} \end{array}\right\} \overset{(8)}{\dashv}$$

$$\vdash B \in \{X \mid X \in \underline{L/A} \wedge \overline{XA} \cap L = \emptyset\} \xrightarrow{(9)} B \in L/A.$$

   (c) Prove $A \in L/B \longrightarrow L/A \subseteq L/B$.
   (d) Complete the proof of part (a) of the plane separation lemma.
8. Prove parts (b) through (e) of the plane separation lemma.
9. Prove the corollary to Pasch's postulate for the case in which A, B, and C are distinct collinear points.
10. Complete the proof of Theorem 5-4 by proving

$$\left.\begin{array}{l} A \notin L \\ A\text{-}T\text{-}Z \\ T \in L \\ R \in L/Z \end{array}\right\} \longrightarrow R \in S, \qquad \text{where } S = \underline{L/A} \smallsetminus \overline{L/A}.$$

11. If L/A is a half plane, we have defined the opposite half plane (denoted by $L/{\sim}A$) to be $\underline{L/A} \smallsetminus \overline{L/A}$. Similarly, the half plane opposite $L/{\sim}A$ is defined to be $\underline{L/A} \smallsetminus \overline{L/{\sim}A}$. Prove: $\underline{L/A} \smallsetminus \overline{L/{\sim}A} = L/A$.

More About Separation

If $\{A\}$ is the intersection of $\overrightarrow{AB}$ and line L and $\overrightarrow{AB}$ contains a point in L/P, it appears that the interior of $\overrightarrow{AB}$ should lie in L/P. We use our definition of a half plane to prove that this is true.

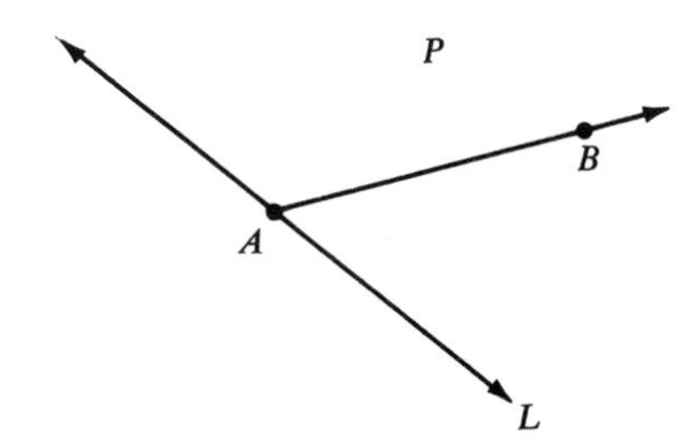

THEOREM 5-6 *If the intersection of a ray with the edge of a half plane is the endpoint of the ray and the ray contains a point in the half plane, then the interior of the ray lies in the half plane.*

We must prove:

$$\left.\begin{array}{r}\overrightarrow{AB} \cap L = \{A\} \\ \overrightarrow{AB} \cap L/P \neq \emptyset\end{array}\right\} \longrightarrow \overset{\circ}{\overrightarrow{AB}} \subseteq L/P.$$

The following proof is provided.

Part I: $\overrightarrow{AB} \cap L = \{A\} \longrightarrow \overset{\circ}{\overrightarrow{AB}} \subseteq \underline{L/B}$.

$$\left.\begin{array}{r}\left.\begin{array}{r}A \notin \overset{\circ}{\overrightarrow{AB}} \\ B \in \overset{\circ}{\overrightarrow{AB}}\end{array}\right\} \longrightarrow A \neq B \\ \underline{\overrightarrow{AB} \cap L = \{A\}} \\ B \in \overrightarrow{AB}\end{array}\right\} \longrightarrow B \notin L \begin{array}{l}\nearrow L \text{ and } B \text{ determine } \underline{L/B} \\ \searrow L/B = \{B\} \cup \{Y \mid Y \in \underline{L/B} \wedge \overline{BY} \cap L = \emptyset\}.\end{array}$$

$$\left.\begin{array}{r}\underline{\overrightarrow{AB} \cap L = \{A\}} \longrightarrow A \in L \\ L \subseteq \underline{L/B}\end{array}\right\} \longrightarrow \left.\begin{array}{r}B \in \underline{L/B} \\ A \in \underline{L/B} \\ A \neq B\end{array}\right\} \longrightarrow \left.\begin{array}{r}\overleftrightarrow{AB} \subseteq \underline{L/B} \\ \overset{\circ}{\overrightarrow{AB}} \subseteq AB\end{array}\right\} \longrightarrow \overset{\circ}{\overrightarrow{AB}} \subseteq \underline{L/B}.$$

Part II: $\overrightarrow{AB} \cap L = \{A\} \longrightarrow \overset{\circ}{\overrightarrow{AB}} \cap L = \emptyset$.

$$\left.\begin{array}{r}\overset{\circ}{\overrightarrow{AB}} \cap \{A\} = \emptyset \\ \underline{\{A\} = \overrightarrow{AB} \cap L}\end{array}\right\} \longrightarrow \left.\begin{array}{r}\overset{\circ}{\overrightarrow{AB}} \cap (\overrightarrow{AB} \cap L) = \emptyset \longrightarrow (\overset{\circ}{\overrightarrow{AB}} \cap \overrightarrow{AB}) \cap L = \emptyset \\ \overset{\circ}{\overrightarrow{AB}} \subseteq \overrightarrow{AB} \longrightarrow \overset{\circ}{\overrightarrow{AB}} \cap \overrightarrow{AB} = \overset{\circ}{\overrightarrow{AB}}\end{array}\right\} \dashv$$

$$\nrightarrow \overset{\circ}{\overrightarrow{AB}} \cap L = \emptyset.$$

Part III: $\overrightarrow{AB} \cap L = \{A\} \longrightarrow \overset{\circ}{\overrightarrow{AB}} \subseteq L/B$.

$$\text{Let } X \in \overset{\circ}{\overrightarrow{AB}} \multimap \begin{array}{l}\left.\begin{array}{r}B \in L/B \\ X = B\end{array}\right\} \longrightarrow X \in L/B \\ \left.\begin{array}{r}X \neq B \\ X \in \overset{\circ}{\overrightarrow{AB}} \\ B \in \overset{\circ}{\overrightarrow{AB}}\end{array}\right\} \longrightarrow \left.\begin{array}{l}\left.\begin{array}{r}\overline{XB} \subseteq \overset{\circ}{\overrightarrow{AB}} \\ \overset{\circ}{\overrightarrow{AB}} \cap L = \emptyset\end{array}\right\} \longrightarrow \overline{XB} \cap L = \emptyset \\ \left.\begin{array}{r}X \in \overset{\circ}{\overrightarrow{AB}} \\ \overset{\circ}{\overrightarrow{AB}} \subseteq \underline{L/B}\end{array}\right\} \longrightarrow X \in \underline{L/B}\end{array}\right\} \longrightarrow X \in L/B.\end{array}$$

Thus we have $X \in \overset{\circ}{\overrightarrow{AB}} \longrightarrow X \in L/B$. Therefore $\overset{\circ}{\overrightarrow{AB}} \subseteq L/B$.

Part IV:

$$\left.\begin{array}{l}\overrightarrow{AB} \cap L = \{A\} \\ \overrightarrow{AB} \cap L/P \neq \emptyset\end{array}\right\} \longrightarrow \overset{\circ}{\overrightarrow{AB}} \subseteq L/P.$$

$$\underline{\overrightarrow{AB} \cap L/P \neq \emptyset} \longrightarrow \text{There is a point } W \text{ such that } \left\{\begin{array}{l} W \in \overrightarrow{AB} \\ W \in L/P \end{array}\right.$$

$$\left.\begin{array}{r} W \in L/P \\ L/P \cap L = \emptyset \end{array}\right\} \longrightarrow \left.\begin{array}{r} W \notin L \\ A \in L \end{array}\right\} \dashv$$

$$\left.\begin{array}{r} \nrightarrow A \neq W \\ W \in \overrightarrow{AB} \end{array}\right\} \longrightarrow \left.\begin{array}{r} W \in \overset{\circ}{\overrightarrow{AB}} \\ \overset{\circ}{\overrightarrow{AB}} \subseteq L/B \end{array}\right\} \longrightarrow \left.\begin{array}{l} W \in L/B \longrightarrow L/W = L/B \\ W \in L/P \longrightarrow L/W = L/P \end{array}\right\} \dashv$$

$$\left.\begin{array}{r} \nrightarrow L/P = L/B \\ \overset{\circ}{\overrightarrow{AB}} \subseteq L/B \end{array}\right\} \longrightarrow \overset{\circ}{\overrightarrow{AB}} \subseteq L/P.$$

COROLLARY 1 TO THEOREM 5-6 $\quad \overrightarrow{AB} \cap L = \{A\} \longrightarrow \overset{\circ}{\overrightarrow{AB}} \subseteq L/B.$

COROLLARY 2 TO THEOREM 5-6

$$\left.\begin{array}{l} T \in \overleftrightarrow{XY} \\ A \in \overleftrightarrow{XY}/B \end{array}\right\} \longrightarrow \left\{\begin{array}{l} \overset{\circ}{\overrightarrow{TA}} \subseteq \overleftrightarrow{XY}/B \\ \overrightarrow{TA} \subseteq \overline{\overleftrightarrow{XY}/B}. \end{array}\right.$$

The proofs for the corollaries to Theorem 5-6 and the following theorems and corollaries are considered in the exercises.

THEOREM 5-7 *If lines L and T intersect in a single point X, and A and B are points in L that are in opposite half planes with edge T, then* $\overrightarrow{XA}$ *and* $\overrightarrow{XB}$ *are opposite rays.*

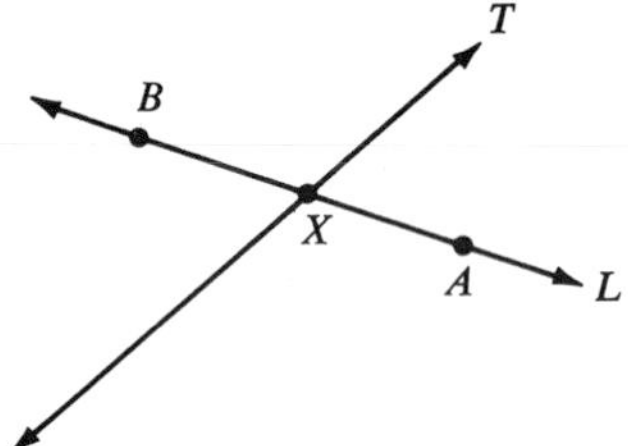

THEOREM 5-8 $L \cap \overset{\circ\!-\!\circ}{AB}$ *is a single point if and only if A and B are in opposite half planes with common edge L.* ($L \times \overset{\circ\!-\!\circ}{AB} \longleftrightarrow A \in L/\sim B$.)

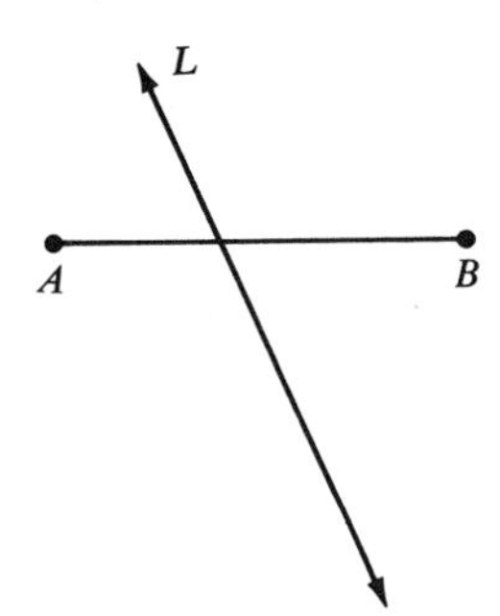

COROLLARY 1 TO THEOREM 5-8 *If $\overrightarrow{TA}$ and $\overrightarrow{TB}$ are opposite rays and $W \notin \overleftrightarrow{AB}$, then $\overset{\circ}{\overrightarrow{TA}}$ and $\overset{\circ}{\overrightarrow{TB}}$ lie in opposite half planes with common edge $\overleftrightarrow{TW}$.*

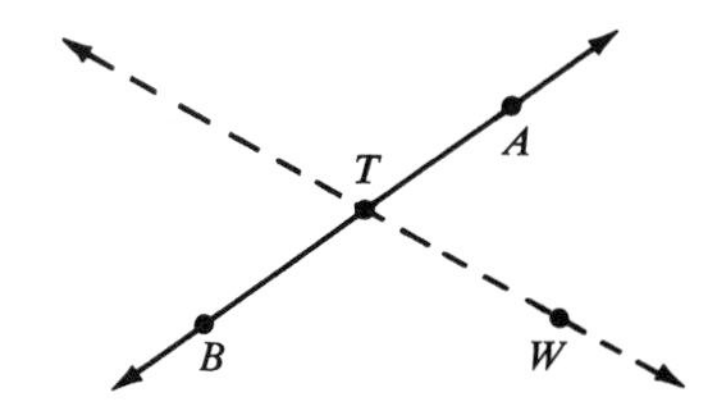

COROLLARY 2 TO THEOREM 5-8 *If line L and segment AB are coplanar but do not intersect, then A and B are in the same half plane with edge L.*

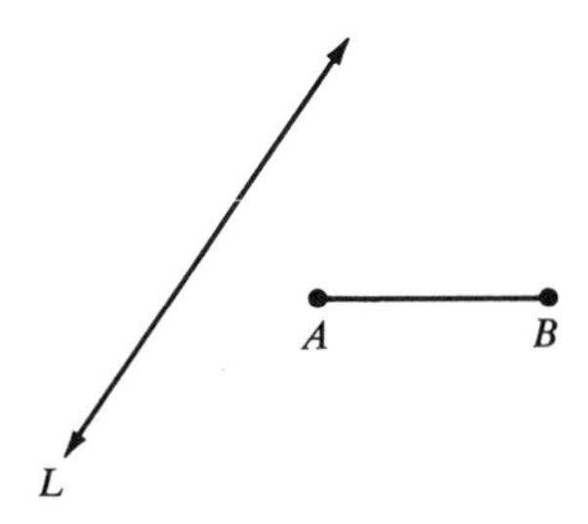

COROLLARY 3 TO THEOREM 5-8

$$X \notin \overleftrightarrow{AB} \longrightarrow \begin{cases} \overleftrightarrow{AB}/X \cap \overleftrightarrow{AX} = \overset{\circ}{\overrightarrow{AX}} \\ \overline{\overleftrightarrow{AB}/X} \cap \overleftrightarrow{AX} = \overrightarrow{AX}. \end{cases}$$

THEOREM 5-9 *A closed half plane is a convex set.*

Exercises

1. Prove Corollary 1 to Theorem 5-6.
2. Supply reasons for the following proof of Corollary 2 to Theorem 5-6: $A \in \overleftrightarrow{XY}/B$ implies that $A \notin \overleftrightarrow{XY}$ (1). Since $A \in \overleftrightarrow{TA}$, it follows that $\overleftrightarrow{TA} \neq \overleftrightarrow{XY}$ (2). $\overleftrightarrow{TA}$ and $\overleftrightarrow{XY}$ each contain point T, hence $\overleftrightarrow{XY} \cap \overleftrightarrow{TA} = \{T\}$ (3). But $\overrightarrow{TA} \subseteq \overleftrightarrow{TA}$ and $T \in \overleftrightarrow{TA}$, therefore $\overset{\circ}{\overrightarrow{TA}} \cap \overleftrightarrow{XY} = \{T\}$ (4). Thus $\overset{\circ}{\overrightarrow{TA}} \subseteq \overleftrightarrow{XY}/A$ (5). We have $T \in \overleftrightarrow{XY}$ and $\overset{\circ}{\overrightarrow{TA}} \subseteq \overleftrightarrow{XY}/A$. These statements imply that $(\{T\} \cup \overset{\circ}{\overrightarrow{TA}}) \subseteq (\overleftrightarrow{XY} \cup \overleftrightarrow{XY}/A)$ (6). We conclude that $\overrightarrow{TA} \subseteq \overline{\overleftrightarrow{XY}/A}$ (7).
3. Supply reasons for the following proof of Theorem 5-7. We must prove

$$\left.\begin{array}{l} L \cap T = \{X\} \\ A, B \in L \\ A \in T/{\sim}B \end{array}\right\} \longrightarrow \overrightarrow{XA} \text{ and } \overrightarrow{XB} \text{ are opposite rays.}$$

$$\left.\begin{array}{l} \underline{A \in T/{\sim}B} \\ (1)\ T/{\sim}B \cap T/B = \emptyset \end{array}\right\} \xrightarrow{(2)} \left.\begin{array}{l} A \notin T/B \\ (3)\ B \in T/B \end{array}\right\} \xrightarrow{(4)} \left.\begin{array}{l} A \neq B \\ \underline{A, B \in L} \end{array}\right\} \overset{(5)}{\dashv}$$

$$\left.\begin{array}{l} L \cap T = \{X\} \\ \hline \nrightarrow \overleftrightarrow{AB} = L \end{array}\right\} \xrightarrow{(6)} \overleftrightarrow{AB} \cap T = \{X\}$$

$$\left.\begin{array}{l} A \in T/\sim B \\ \hline (7)\ B \in T/B \end{array}\right\} \xrightarrow{(8)} \overset{\circ\!-\!\circ}{AB} \cap T \neq \emptyset \quad (9)\ \overset{\circ\!-\!\circ}{AB} \subseteq \overleftrightarrow{AB}$$

$$\xrightarrow{(10)} \overset{\circ\cdots\circ}{AB} \cap T = \{X\} \overset{(11)}{\dashv}$$

$$\longmapsto X \in \overset{\circ\!-\!\circ}{AB} \xrightarrow{(12)} \overrightarrow{XA} \text{ and } \overrightarrow{XB} \text{ are opposite rays.}$$

4. Prove Theorem 5-8.
5. Prove Corollary 1 to Theorem 5-8.
6. Prove Corollary 2 to Theorem 5-8.
7. Prove Corollary 3 to Theorem 5-8.
8. Prove Theorem 5-9.

L 9. Prove:
   (a) $C \notin \overleftrightarrow{AB} \longrightarrow \overset{\longrightarrow\circ}{CA} = \overrightarrow{CA} \cap \overleftrightarrow{AB}/C$.
   (b) $C \notin \overleftrightarrow{AB} \longrightarrow \overset{\circ\!\longrightarrow\!\circ}{CA} = \overset{\circ\!\longrightarrow}{CA} \cap \overleftrightarrow{AB}/C$.

L 10. Prove:

$$\left.\begin{array}{r} A \notin \overleftrightarrow{BC} \\ X \in \overset{\circ\!-\!\circ}{AB} \end{array}\right\} \longrightarrow X \in \overleftrightarrow{BC}/A.$$

11. Prove the following implication:

$$\left.\begin{array}{l} T \in L \\ X \in L' \\ T\text{-}X\text{-}A \\ L' \subseteq L/A \end{array}\right\} \longrightarrow L'/A \subseteq L/A.$$

Space Separation

Having seen that a point A in line L separates $L \sim \{A\}$ into two convex sets called half lines (Exercise 8, page 102) and that a line L in plane M separates $M \sim L$ into two convex sets called half planes, we consider a sequence of definitions and theorems designed to prove that a plane M separates $S \sim M$ into two convex sets called half spaces. Proofs for these theorems are assigned as exercises.

DEFINITIONS *If point A is not in plane M, then $M/A = \{A\} \cup \{X \mid \overline{AX} \cap M = \emptyset\}$. M/A is called the* half space *determined by M and A. Plane M is called the* face *of half space M/A. $M \cup M/A$ is called the* closed half space *determined by M and A and is denoted by $\overline{M/A}$. M/A is also called the A side of M.*

COROLLARY TO THE DEFINITION OF HALF SPACE *If $A \notin M$, then*

(*i*) $M \cap M/A = \emptyset$,
(*ii*) $M/A \subset \overline{M/A} \subset S$.

THEOREM 5-10

$$\left.\begin{array}{l} L \subseteq M \\ A \notin M \end{array}\right\} \longrightarrow L/A \subseteq M/A.$$

THEOREM 5-11 *If $A \notin M$, then M/A is a convex set.*

Observe that part (ii) of the corollary to the definition implies that $S \smallsetminus \overline{M/A} \neq \emptyset$. In fact, we can prove that it is a half space.

THEOREM 5-12 *If $A \notin M$, then $S \smallsetminus \overline{M/A}$ is a half space.*

DEFINITION *If $A \notin M$, then $M/\sim A = S \smallsetminus \overline{M/A}$ is the half space opposite M/A.*

THEOREM 5-13 (*SPACE SEPARATION THEOREM*) *If $A \notin M$, then M separates $S \smallsetminus M$ into two half spaces M/A and $M/\sim A$.*

Observe that in order to prove Theorem 5-13, we must appeal to the definition of separation. Thus we must prove

(i) $M \cap (S \smallsetminus M) = \emptyset$,
(ii) $S \smallsetminus M = M/A \cup M/\sim A$ and $M/A \cap M/\sim A = \emptyset$,
(iii) $M/A \neq \emptyset$ and $M/\sim A \neq \emptyset$,
(iv) $(X \in M/A \wedge Y \in M/\sim A) \longrightarrow \overline{\mathring{X}\mathring{Y}} \cap M \neq \emptyset$.

SPACE SEPARATION LEMMA

(*a*) $A \in M/B \longleftrightarrow B \in M/A \longleftrightarrow M/A = M/B$.
(*b*) $M/A = M/B \longleftrightarrow M/\sim A = M/\sim B$.
(*c*) $A \in M/\sim B \longleftrightarrow B \in M/\sim A \longleftrightarrow M/A = M/\sim B \longleftrightarrow M/B = M/\sim A$.
(*d*) $(A \in M/C \wedge B \in M/C) \longrightarrow A \in M/B$.
(*e*) $(A \in M/C \wedge B \in M/\sim C) \longrightarrow A \in M/\sim B$.

Exercises

1. Prove the corollary to the definition of half space.
2. Prove: $A \notin M \longrightarrow S \smallsetminus \overline{M/\sim A} = M/A$.
3. Prove Theorem 5-10.
4. Prove Theorem 5-11.
5. Prove Theorem 5-12.
6. Prove Theorem 5-13.
7. Prove the space separation lemma.
8. Prove: $A \notin M \longleftrightarrow (M/A = M'/A \longleftrightarrow M = M')$.
9. Prove: $M' \subseteq M/A \longrightarrow M \parallel M'$.

L 10. Prove: $\overrightarrow{AX} \cap M = \{A\} \longrightarrow \overrightarrow{\mathring{A}X} \subseteq M/X$.

11. Prove that a closed half space is a convex set.

References

Golos, *Foundations of Euclidean and Non-Euclidean Geometry*
Kay, *College Geometry*
Moise, *Elementary Geometry from an Advanced Standpoint*
Prenowitz and Jordan, *Basic Concepts of Geometry*
Ringenberg, *College Geometry*

Chapter 6
Angles and Angle Measurement

There are many ways in which the word "angle" may be defined in mathematics. It may be considered as an ordered pair of concurrent rays, as a rotation of a ray around its endpoint, or as the union of two rays with all the rays "between" them. In trigonometry, an angle is regarded as a rotation from the initial side to the terminal side. The measure of the rotation is positive or negative according as the rotation is counterclockwise or clockwise. In such an interpretation, "zero angles" and "straight angles" are included in the angle concept.

In our study of geometry, we wish to regard an angle as a geometric figure. This means that we begin our description of an angle by saying that an angle is a set of points. We observe that the union of two rays having a common endpoint is a set of points that corresponds to our mental picture of an angle. According to this description, each of the drawings shown below represents an angle, because each can be described as the union of two rays VA and VB. However, we do not wish to accept the figures shown in (c) and (d) as representing angles. In our study, it is necessary for each angle to have a nonempty, well-defined *interior*, as described in a later section. This means that so-called "zero angles" [figure (d)], which have empty interiors, and "straight angles" [figure (c)], whose interior cannot be defined, must be ruled out.

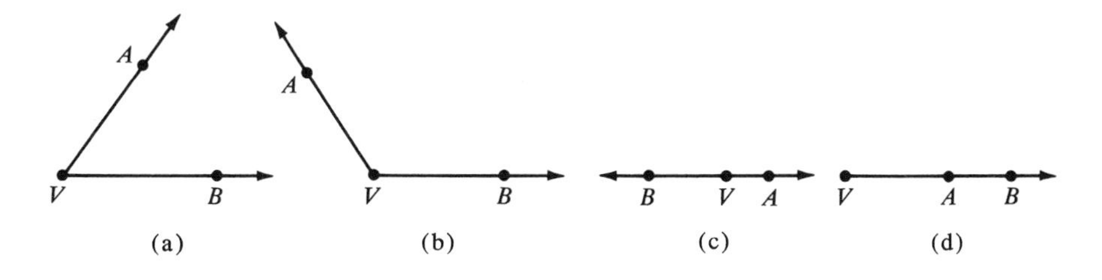

With these considerations in mind, we adopt the following definition.

DEFINITION *An* angle *is the union of two rays that have a common endpoint but do not lie in the same line.*

Notation: $A\underline{B}C = \overrightarrow{BA} \cup \overrightarrow{BC}$.

The following symbolic statements of the definition are useful.

$$A\underline{B}C \text{ is an angle} \longleftrightarrow \overleftrightarrow{BA} \neq \overleftrightarrow{BC}.$$
$$A\underline{B}C \text{ is an angle} \longleftrightarrow A \notin \overleftrightarrow{BC}.$$

We know that if $A\underline{B}C$ is an angle, then $\overleftrightarrow{AB}$ and $\overleftrightarrow{BC}$ are distinct lines. Since $\overrightarrow{BA}$ and $\overrightarrow{BC}$ have a common point B and are subsets of distinct lines, it follows from the corollary to Theorem 3-1 that this point is their intersection. We have proved the first part of the following corollary to the definition of angle.

COROLLARY 1 TO THE DEFINITION OF ANGLE *If $A\underline{B}C$ is an angle, then*

(i) $\overrightarrow{BA} \cap \overrightarrow{BC} = \{B\}$,
(ii) $\overrightarrow{BA} \cap \overleftrightarrow{BC} = \{B\} = \overleftrightarrow{AB} \cap \overrightarrow{BC}$,
(iii) $\overleftrightarrow{AB} \cap \overleftrightarrow{BC} = \{B\}$,
(iv) $A \neq B \wedge B \neq C \wedge A \neq C \wedge C \notin \overleftrightarrow{AB}$.

The proof of the following corollary is requested in the exercises.

COROLLARY 2 TO THE DEFINITION OF ANGLE

$$\left.\begin{array}{l} A \in \overleftrightarrow{XY}/B \\ \overrightarrow{XB} \neq \overrightarrow{XA} \end{array}\right\} \longrightarrow A\underline{X}B \text{ is an angle.}$$

DEFINITIONS *If the union of two rays is an angle, each ray is a* side *of the angle, and the common endpoint is called the* vertex *of the angle.*

Notation: We often name the angle formed by the rays $\overrightarrow{AB}$ and $\overrightarrow{AC}$ by the symbol $\angle BAC$ or the symbol $\angle CAB$. When using this notation, it is important to remember that the middle letter always names the vertex of the angle. Sometimes, when there is no possibility of confusion, we shall abbreviate $\angle BAC$ as $\angle A$.

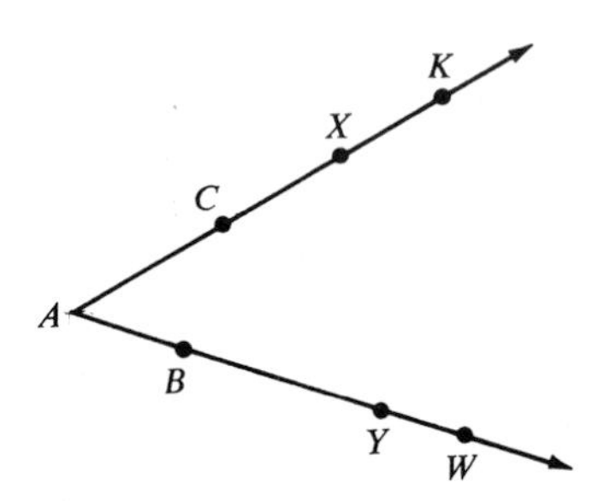

We observe that many "three-letter" symbols can be used to name a given angle. This is true because, as we noted in Chapter 4, a ray can be named by its endpoint and any other point in its interior. In the figure, we have $\overrightarrow{AC} = \overrightarrow{AX} = \overrightarrow{AK}$ and $\overrightarrow{AB} = \overrightarrow{AY} = \overrightarrow{AW}$. Therefore, $\overrightarrow{AC} \cup \overrightarrow{AB} = \overrightarrow{AX} \cup \overrightarrow{AY} = \overrightarrow{AX} \cup \overrightarrow{AB}$, and so on. Accordingly, $\angle BAC = \angle XAY = \angle XAB$, and so on.

Observe that if $A\underline{B}D$ and $A\underline{B}C$ are angles, then

(i) $$\overrightarrow{BD} = \overrightarrow{BC} \longrightarrow \angle ABC = \angle ABD.$$

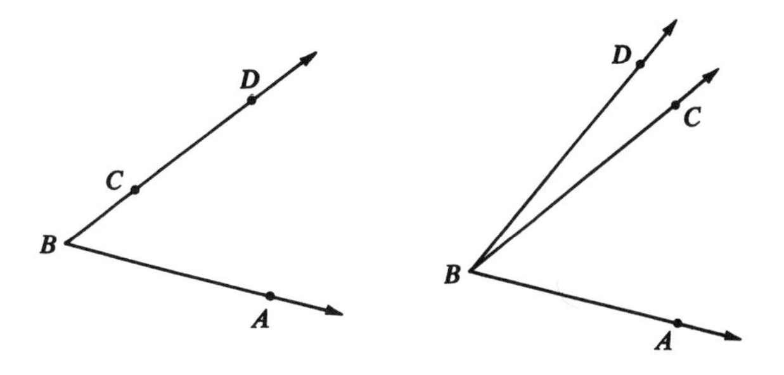

A contrapositive of (i) is

(ii) $$\angle ABC \neq \angle ABD \longrightarrow \overrightarrow{BD} \neq \overrightarrow{BC}.$$

Sometimes we shall name an angle by placing a lower-case letter (often from the Greek alphabet) near the vertex and between its sides. For the angles in the figure below, we have $\angle KOY = \angle\alpha$ and $\angle YOX = \angle\beta$.

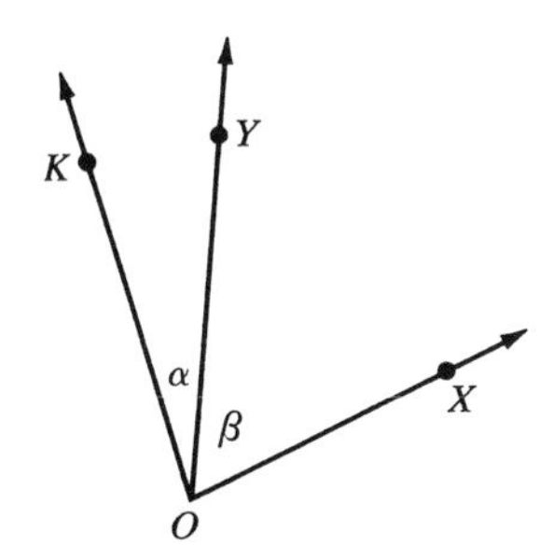

If RST is an angle, then points R, S, and T are noncollinear. By Postulate 5, there is a plane containing R, S, and T. We call this plane RST. $\overleftrightarrow{SR}$ and $\overleftrightarrow{ST}$ lie in pl RST. Therefore $\overrightarrow{SR}$ and $\overrightarrow{ST}$ lie in pl RST. Why? Thus we have proved that there is a plane (namely, pl RST) that contains $\angle RST$.

If M is a plane that contains $\angle RST$, then M contains the three noncollinear points R, S, and T. Why? It follows by Postulate 5 that $M =$ pl RST. Thus we have

$$\angle RST \subseteq M \longrightarrow M = \text{pl } RST.$$

The contrapositive is

$$M \neq \text{pl } RST \longrightarrow \angle RST \nsubseteq M.$$

We have proved that no plane other than pl RST contains $\angle RST$.

The conjunction of the underlined statements leads us to state the following important theorem.

THEOREM 6-1 *Each angle is contained in exactly one plane.*

We sometimes express Theorem 6-1 by saying that every angle *determines* a plane.

Exercises

1. Which of the following statements are true?
   (a) If $R\underline{S}T$ is not an angle, then $\overrightarrow{SR} = \overrightarrow{ST}$.
   (b) If $R\underline{S}T$ is not an angle, then $\overleftrightarrow{SR} = \overleftrightarrow{ST}$.
   (c) If two segments have a common point, four angles are determined.
   (d) An angle is the union of two distinct intersecting rays.
   (e) If $\overleftrightarrow{AB} \cap \overleftrightarrow{CD} = \{P\}$, then $\angle CPA \neq \angle DPB$.
   (f) If A, B, and C are distinct noncollinear points, then $A\underline{B}C$ is an angle.
   (g) Two distinct angles cannot have more than four distinct points in common.
   (h) The intersection of two angles can be a segment.
2. Complete the proof of Corollary 1 to the definition of angle.
3. Write an essay proof of Corollary 2 to the definition of angle.
4. Prove that an angle is not a convex set.

Ray-Coordinate System

In elementary geometry we use circular or semicircular protractors to find the measure of an angle in degrees. Although other units such as radians, mils, right angles, or revolutions are frequently used to express the measure of an angle, we shall restrict ourselves to degree measure, as indicated by our angle-measure postulate.

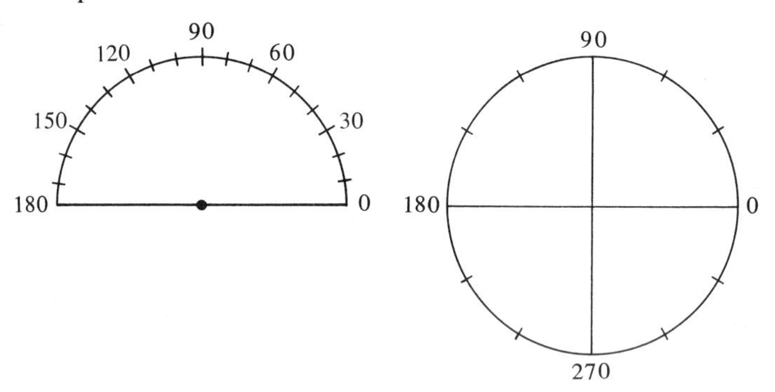

POSTULATE 14 (ANGLE-MEASURE POSTULATE) *There exists a correspondence that associates with each angle in space a unique number between* 0 *and* 180.

Notation: We shall use the symbol $\langle a, b\rangle$ to represent the set consisting of a and b and all the numbers between them. Thus $\langle a, b\rangle = \{x \mid x = a \vee a\text{-}x\text{-}b \vee x = b\} = \langle b, a\rangle$. The symbol $\rangle a, b\langle$ is used to denote all the numbers between a and b. Thus

$$\rangle a, b\langle = \{x \mid a\text{-}x\text{-}b\} = \rangle b, a\langle.$$

In view of Postulate 14 we cay say there exists a correspondence that associates with each angle in space a number in $\rangle 0, 180\langle$.

DEFINITION *The number that corresponds by Postulate 14 to an angle is called the* measure of the angle.

Notation: If $A\underline{B}C$ is an angle, its measure is denoted by $m\angle ABC$. Thus $m\angle ABC$ is a number in $\rangle 0, 180\langle$.

In Chapter 4 we derived the properties of distance by defining a coordinate system for a line and then adopting the ruler postulate. In this chapter we derive the properties of angle measure by defining a ray-coordinate system with respect to a given point in a given plane and then adopting a protractor postulate. Angle measure, like distance measure, is obtained by subtracting coordinates. However, in view of Postulate 14, we must make an adjustment if the difference is greater than 180.

DEFINITION *Let O be a point in plane M. A ray-coordinate system for M relative to O is a one-to-one correspondence between the set of all rays in M with endpoint O and the set of numbers x such that $0 \leqq x < 360$ with the following property: If r and s correspond to distinct rays OR and OS, then*

(1) $m\angle ROS = |r - s|$ *if* $|r - s| < 180$,
(2) $m\angle ROS = 360 - |r - s|$ *if* $|r - s| > 180$,
(3) $\overrightarrow{OR}$ *and* $\overrightarrow{OS}$ *are opposite rays if and only if* $|r - s| = 180$.

DEFINITIONS *The number assigned to a ray in a ray-coordinate system is called the* ray-coordinate *of the ray. The ray whose ray-coordinate is zero is called the* zero ray *of the ray-coordinate system.*

POSTULATE 15 (THE PROTRACTOR POSTULATE) *For any angle AOB there exists a unique ray-coordinate system C for plane AOB relative to O such that $\overrightarrow{OA}$ is the zero ray and such that for any point T in the B-side of $\overleftrightarrow{OA}$, $\overrightarrow{OT}$ corresponds to a number in $\rangle 0, 180 \langle$.*

If AOB is an angle, Postulate 15 tells us that there exists a unique coordinate system C for plane AOB relative to O such that $\overrightarrow{OA}$ is the zero ray and b, the coordinate of $\overrightarrow{OB}$, is in $\rangle 0, 180 \langle$. Let T be any point in pl $AOB \sim \{O\}$ and let t be the coordinate assigned by C to $\overrightarrow{OT}$. Then according to the postulate $T \in \overleftrightarrow{OA}/B \longrightarrow t \in \rangle 0, 180 \langle$.

Under the conditions stated in the postulate, it is also possible to prove the converse of this implication; that is, we can prove $t \in \rangle 0, 180 \langle \longrightarrow T \in \overleftrightarrow{OA}/B$. Thus we have the following lemma, which is useful in proving Theorem 6-2.

LEMMA 1

$$\left.\begin{array}{l} \overrightarrow{OA} \text{ is the zero ray of a coordinate system } C \\ \text{that assigns } \overrightarrow{OB} \text{ a coordinate } b \in \rangle 0, 180 \langle. \\ C \text{ assigns } \overrightarrow{OT} \text{ the coordinate } t. \\ t \in \rangle 0, 180 \langle. \end{array}\right\} \longrightarrow T \in \overleftrightarrow{OA}/B.$$

We shall prove the third contrapositive:

$$\left.\begin{array}{l} \overrightarrow{OA} \text{ is the zero ray of a coordinate system } C \\ \text{that assigns } \overrightarrow{OB} \text{ a coordinate } b \in \rangle 0, 180 \langle. \\ C \text{ assigns } \overrightarrow{OT} \text{ the coordinate } t. \\ T \notin \overleftrightarrow{OA}/B. \end{array}\right\} \longrightarrow t \notin \rangle 0, 180 \langle.$$

Since C is a coordinate system for plane AOB and C assigns $\overrightarrow{OT}$ a coordinate, we know that $T \in \text{pl } AOB$.

$$\left.\begin{array}{l} T \notin \overleftrightarrow{OA}/B \\ T \in \text{pl } AOB \end{array}\right\} \longrightarrow (T \in \overleftrightarrow{OA} \vee T \in \overleftrightarrow{OA}/\sim B).$$

There is a point A' such that A'-O-A. $(T \in \overleftrightarrow{OA} \wedge OT \text{ is a ray}) \longrightarrow T \in \overrightarrow{\mathring{O}A} \vee T \in \overrightarrow{\mathring{O}A'}$. If $T \in \overrightarrow{\mathring{O}A}$, then $\overrightarrow{OT} = \overrightarrow{OA}$ and $t = 0$. If $T \in \overrightarrow{\mathring{O}A'}$, then $t = 180$. In either case $t \notin \rangle 0, 180 \langle$. Suppose $T \in \overleftrightarrow{OA}/\sim B$. There is a point T' such that T-O-T'.

$$\left.\begin{array}{l} T \in \overleftrightarrow{OA}/\sim B \\ T\text{-}O\text{-}T' \end{array}\right\} \longrightarrow T' \in \overleftrightarrow{OA}/B.$$

If $T' \in \overleftrightarrow{OA}/B$ and t' is the coordinate of $\overrightarrow{OT'}$ assigned by C, then by Postulate 15 $t' \in \rangle 0, 180 \langle$. Since $\overrightarrow{OT}$ and $\overrightarrow{OT'}$ are opposite rays, $|t - t'| = 180$.

$$\left.\begin{array}{l} |t - t'| = 180 \\ 0 < t' < 180 \\ t > 0 \end{array}\right\} \longrightarrow t > 180.$$

Thus $T \in \overleftrightarrow{OA}/\sim B \longrightarrow t \notin \rangle 0, 180 \langle$. This completes our outline of the proof of Lemma 1.

THEOREM 6-2 (THE ANGLE-CONSTRUCTION THEOREM) *If $B \notin \overleftrightarrow{OA}$ and $r \in \rangle 0, 180 \langle$, then there is a unique ray OR such that $R \in \overleftrightarrow{OA}/B$ and $m \angle AOR = r$.*

PROOF

If $B \notin \overleftrightarrow{OA}$, then AOB is an angle, and by Postulate 15 there is a unique ray-coordinate system for plane AOB relative to O such that $\overrightarrow{OA}$ corresponds to 0 and every ray OX with $X \in \overleftrightarrow{OA}/B$ corresponds to a number in $\rangle 0, 180 \langle$. Since $r \in \rangle 0, 180 \langle$, there is in this ray-coordinate system a ray OR such that $R \in \overleftrightarrow{OA}/B$ and $\overrightarrow{OR}$ corresponds to r by Lemma 1. Thus $\overrightarrow{OR}$ is a ray with $R \in \overleftrightarrow{OA}/B$ and $m \angle AOR = r$. This completes our proof of the existence of the required ray. It remains to show that it is unique. Suppose $\overrightarrow{OS}$ is any ray such that $S \in \overleftrightarrow{OA}/B$ and $\underline{m \angle AOS = r}$. Let the coordinate of $\overrightarrow{OS}$ be s. By the protractor postulate, $0 < s < 180$, and therefore $\underline{m \angle AOS = s - 0 = s}$. The underlined statements imply that $s = r$. Since a ray-coordinate system is a one-to-one correspondence between rays and numbers, it follows that $\overrightarrow{OS} = \overrightarrow{OR}$. Hence $\overrightarrow{OR}$ is unique.

Betweenness for Rays

Our definition of betweenness for three collinear points was stated in terms of the coordinates assigned to these points by a coordinate system for the line containing them.

DEFINITION *A set of rays is* concurrent *if and only if the rays have the same endpoint.*

We shall describe betweenness for concurrent coplanar rays in terms of a ray-coordinate system. Our intuitive ideas about rays suggest that $\overrightarrow{OB}$ is between $\overrightarrow{OA}$ and $\overrightarrow{OC}$ in the first drawing but not in the last three.

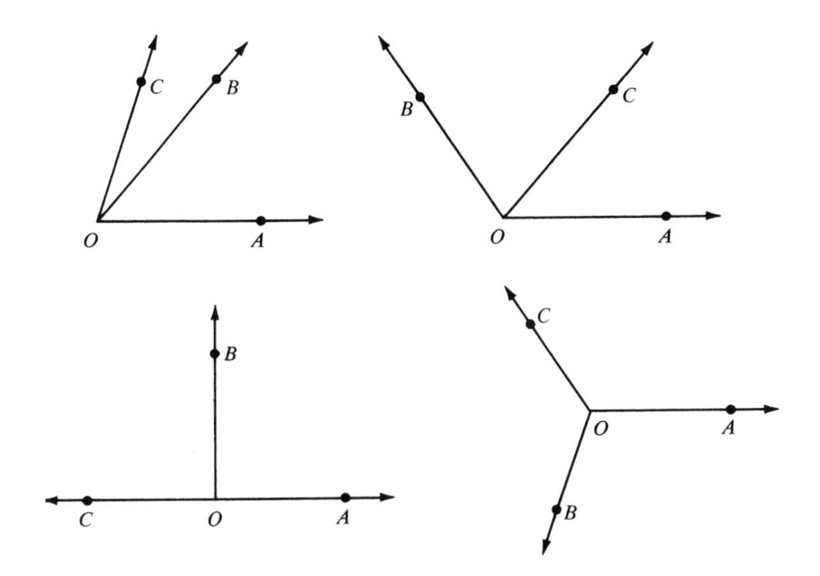

DEFINITION $\overrightarrow{OB}$ *is between* $\overrightarrow{OA}$ *and* $\overrightarrow{OC}$ *if and only if there is a ray-coordinate system that has* $\overrightarrow{OA}$ *as the zero ray and that assigns* $\overrightarrow{OB}$ *and* $\overrightarrow{OC}$ *the numbers b and c, respectively, so that* $0 < b < c < 180$.

The existence of such a ray-coordinate system guarantees that rays $\overrightarrow{OA}$, $\overrightarrow{OB}$, and $\overrightarrow{OC}$ are coplanar and that no two of them fall in the same line. It follows that AOB, BOC, and AOC are angles.

Observe that our definition of betweenness for rays is not symmetric with respect to the roles played by $\overrightarrow{OA}$ and $\overrightarrow{OC}$. This suggests the possibility that $\overrightarrow{OB}$ is between $\overrightarrow{OA}$ and $\overrightarrow{OC}$ but not between $\overrightarrow{OC}$ and $\overrightarrow{OA}$. This possibility is eliminated by the following theorem.

THEOREM 6-3 $\overrightarrow{OB}$ *is between* $\overrightarrow{OA}$ *and* $\overrightarrow{OC} \longrightarrow \overrightarrow{OB}$ *is between* $\overrightarrow{OC}$ *and* $\overrightarrow{OA}$.

PROOF

If $\overrightarrow{OB}$ is between $\overrightarrow{OA}$ and $\overrightarrow{OC}$, then there exists a coordinate system C such that 0, b, and c are assigned to rays OA, OB, and OC, respectively, and $0 < b < c < 180$. It follows that (1) $m\angle AOC = |c - 0| = c$, $m\angle AOB = |b - 0| = b$, and $m\angle BOC = |c - b| = c - b$.

Again, since AOC is an angle, there is a coordinate system C' for plane AOC relative to O such that $\overrightarrow{OC}$ is the zero ray and y, the C' coordinate of $\overrightarrow{OA}$, is between 0 and 180. Accordingly the measure of $\angle COA$ in the C'-coordinate system is given by (2) $m\angle COA = y$. From (1) and (2) it follows that the C'-coordinate of $\overrightarrow{OA}$ is c. Now C' assigns a ray-coordinate to $\overrightarrow{OB}$, which we shall call x. We shall prove that $x = c - b$.

Suppose $x > c$. Then by our definition of a ray-coordinate system $m\angle BOC = x \vee m\angle BOC = 360 - x$. Since $m\angle BOC = c - b$, we have either $x = c - b$ or $360 - x = c - b$. However, $x > c \longrightarrow x \neq c - b$. Therefore

$$x > c \longrightarrow 360 - x = c - b \longrightarrow 360 - c + b = x.$$

Now $m\angle AOB = b$. In C' we have $m\angle AOB = |x - c| = x - c$ or $m\angle AOB = 360 - (x - c)$. Thus $b = x - c$ or $b = 360 - (x - c)$. In the first case, substituting $x = 360 - c + b$, we obtain

$$b = 360 - c + b - c \longrightarrow c = 180.$$

But $c \neq 180$. In the second case, again substituting $x = 360 - c + b$, we have

$$b = 360 - [(360 - c + b) - c] \longrightarrow b = c.$$

But $b \neq c$. Our supposition that $x > c$ has led to contradictions. Therefore $x < c$. But then $m\angle BOC = x$. Recalling that $m\angle BOC = c - b$, we have $x = c - b$. Now we have $0 < c - b < c < 180$. Therefore $\overrightarrow{OB}$ is between $\overrightarrow{OC}$ and $\overrightarrow{OA}$ according to our definition of the phrase "$\overrightarrow{OB}$ is between $\overrightarrow{OC}$ and $\overrightarrow{OA}$."

Observe that Theorem 6-3 gives precise meaning to the statement "$\overrightarrow{OB}$ is between the sides of $\angle AOC$."

The proof of the following theorem is similar to that of Theorem 4-2. In the exercises you will be asked to prove Theorem 6-4 and its corollaries.

THEOREM 6-4 (BETWEENNESS THEOREM FOR ANGLES) *If $\overrightarrow{OB}$ is between $\overrightarrow{OA}$ and $\overrightarrow{OC}$, then $m\angle AOB + m\angle BOC = m\angle AOC$.*

COROLLARY 1 TO THEOREM 6-4 *If $\overrightarrow{OB}$ is between $\overrightarrow{OA}$ and $\overrightarrow{OC}$, then*

(1) $m\angle AOC > m\angle AOB$,
(2) $m\angle AOC > m\angle BOC$.

COROLLARY 2 TO THEOREM 6-4 *If $m\angle AOB < m\angle AOC$ and $B \in \overleftrightarrow{OA}/C$, then $\overrightarrow{OB}$ is between $\overrightarrow{OA}$ and $\overrightarrow{OC}$.*

We shall sometimes speak of $\overrightarrow{OA}$, $\overrightarrow{OB}$, and $\overrightarrow{OC}$ as being *concurrent in that order*. This will mean that $\overrightarrow{OB}$ is between $\overrightarrow{OA}$ and $\overrightarrow{OC}$. We say that four rays, $\overrightarrow{OA}$, $\overrightarrow{OB}$, $\overrightarrow{OC}$, and $\overrightarrow{OD}$, are concurrent in that order if and only if $\overrightarrow{OA}$ and $\overrightarrow{OD}$ are noncollinear and $\overrightarrow{OB}$ is between $\overrightarrow{OA}$ and $\overrightarrow{OC}$, and $\overrightarrow{OC}$ is between $\overrightarrow{OB}$ and $\overrightarrow{OD}$.

DEFINITION *A ray is a* midray of an angle *if it is between the sides of the angle and forms with them two angles of equal measure.*

In the exercises you will be asked to prove the following theorem.

THEOREM 6-5 *Every angle has a unique midray.*

DEFINITION *The midray of an angle* bisects *the angle and is called the* angle bisector.

Exercises

1. Let C be a ray-coordinate system in plane M relative to point O. Let A, A', B, B' be points such that A-O-A' and B-O-B' and let a, a', b, b' be the ray-coordinates assigned by C to $\overrightarrow{OA}$, $\overrightarrow{OA'}$, $\overrightarrow{OB}$, and $\overrightarrow{OB'}$, respectively. Let m represent the measure of $\angle AOB$.
   (a) If $a = 60$ and $b = 140$, find m.
   (b) If $a = 60$ and $b = 210$, find m.
   (c) If $a = 110$ and $b = 65$, find a' and b'.
   (d) If $a = 290$ and $b = 120$, find m.
   (e) If $a = 310$, $m = 110$, and $b < 80$, find b.
   (f) If $a = 20$, $m = 150$, and $b > 200$, find b.
   (g) If $a = 320$ and $b = 210$, find a' and b'.
   (h) If $a = 120$ and $m = 90$, find two possible values for b.
   (i) If $a = 320$ and $m = 80$, find two possible values for b.
   (j) If $a = 110$ and $b = 160$, find $m\angle A'OB'$.
   (k) If $a = 90$, $m = 120$, and $b > a$, what are the possible values of b?
   (l) If $a = 310$, $m = 80$, and $b < a$, what are the possible values of b?
2. Which of the following statements are true?
   (a) If a coordinate system assigns distinct coordinates to $\overrightarrow{OA}$ and $\overrightarrow{OB}$, then $A \notin \overleftrightarrow{OB}$.
   (b) If a coordinate system assigns distinct coordinates to rays OA, OB, and OC, then $O \in \text{pl } ABC$.
   (c) If AOB is an angle, then there is a number in $\rangle 0, 180 \langle$ that corresponds to $\angle AOB$, called the measure of $\angle AOB$.
   (d) Suppose there is a coordinate system C that assigns rays OA and OB the coordinates a and b, respectively; then $m\angle AOB = a - b$.
   (e) If $\bar{H}$ is a closed half plane with edge L and $O \in L$, then there is a one-to-one correspondence between the set of rays in $\bar{H}$ having endpoint O and the numbers in $\rangle 0, 180 \langle$.
   (f) If a coordinate system assigns coordinates 20 and 200 to rays OX and OY, respectively, then $\overleftrightarrow{OX} = \overleftrightarrow{OY}$.
   (g) If a coordinate system assigns coordinate x to ray OX, then C assigns the coordinate $180 + x$ to the ray opposite $\overrightarrow{OX}$.
   (h) If a coordinate system assigns coordinate x to ray OX and coordinate $180 - x$ to ray OY and $m\angle YOX = 90$, then the coordinate of $\overrightarrow{OX}$ is 45.
3. $\overrightarrow{OB}$ is the midray of $\angle AOC$. A ray-coordinate system assigns $\overrightarrow{OA}$ and $\overrightarrow{OC}$ the coordinates a and c, respectively. Find b if
   (a) $a = 30$ and $c = 170$.
   (b) $a = 150$ and $c = 350$.
   (c) $a = 340$ and $c = 40$.
   (d) $a > c$ and $a - c < 180$.
   (e) $a > c$ and $a - c > 180$.
4. Prove Theorem 6-4.
5. Prove Corollary 1 to Theorem 6-4.
6. Prove Corollary 2 to Theorem 6-4.
7. If B is a point in the plane of $\angle AOC$ such that $m\angle AOB + m\angle BOC = m\angle AOC$, prove $\overrightarrow{OB}$ is between $\overrightarrow{OA}$ and $\overrightarrow{OC}$.
8. Prove: If $r \in \rangle 0, 180 \langle$ and $\overrightarrow{OA}$ is a ray in plane M, there are exactly two rays OX and OY in M such that $m\angle AOX = m\angle AOY = r$.

9. Prove: $\left.\begin{array}{l} C, D \in \overleftrightarrow{OA}/B \\ \overrightarrow{OB} \text{ is between } \overrightarrow{OA} \text{ and } \overrightarrow{OC} \\ \overrightarrow{OC} \text{ is between } \overrightarrow{OB} \text{ and } \overrightarrow{OD} \end{array}\right\} \longrightarrow \left\{\begin{array}{l} \overrightarrow{OB} \text{ is between } \overrightarrow{OA} \text{ and } \overrightarrow{OD} \\ \overrightarrow{OC} \text{ is between } \overrightarrow{OA} \text{ and } \overrightarrow{OD}. \end{array}\right.$

10. Prove Theorem 6-5.

Interior of an Angle

In the drawing below, T is a point of side $\overrightarrow{BA}$ of angle ABC. We therefore say that T is in the angle. The point X, however, is said to be in the interior of angle according to the following definition.

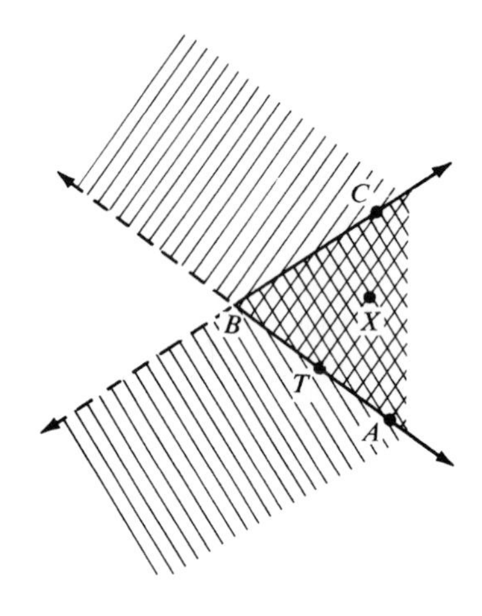

DEFINITION *If I represents the interior of* $\angle ABC$, *then* $I = \overleftrightarrow{BA}/C \cap \overleftrightarrow{BC}/A$.

COROLLARY TO THE DEFINITION $I \neq \emptyset$.

There are two other ways in which the interior of an angle ABC might have been defined.

(1) We might have defined the interior of an angle ABC as R, the set of interior points of all rays between $\overrightarrow{BA}$ and $\overrightarrow{BC}$.

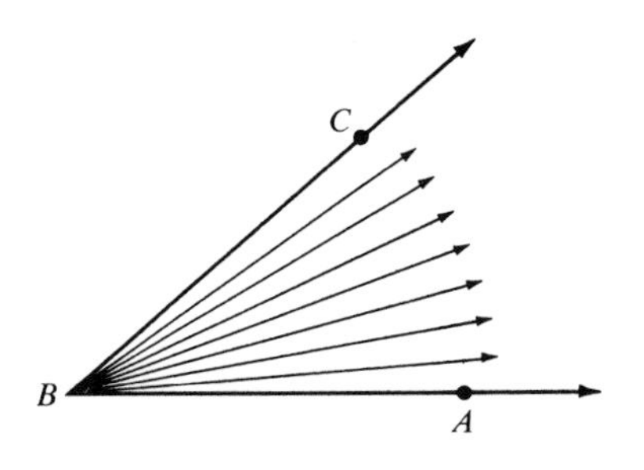

(2) We might have defined the interior of an angle as the set S that consists of the interior points of all segments that join a point in the interior of one side to a point in the interior of the other side.

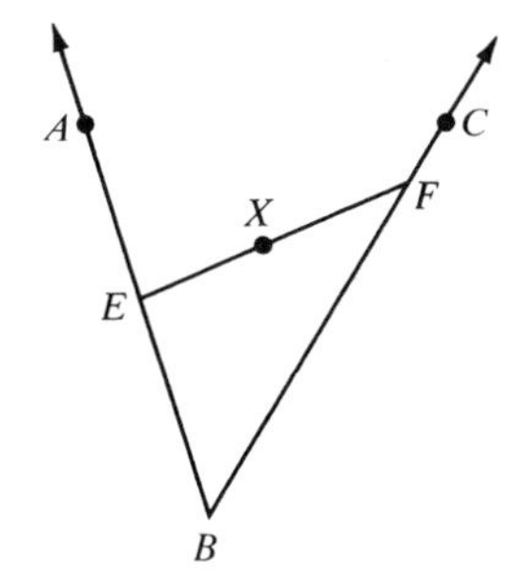

The following theorems state the relationship between sets I, R, and S.

THEOREM 6-6 *The interior of an angle is the set of interior points of all rays between the sides of the angle.* ($I = R$.)

PROOF THAT $R \subseteq I$

Let $\angle AOC$ be a given angle. We shall show $P \in R \longrightarrow P \in I$ as follows: $P \in R \longrightarrow \overrightarrow{OP}$ is between $\overrightarrow{OA}$ and $\overrightarrow{OC} \longrightarrow$ There is a coordinate system C that assigns $\overrightarrow{OA}$, $\overrightarrow{OP}$, and $\overrightarrow{OC}$ the coordinates 0, p, and c, respectively, such that

$$0 < p < c < 180 \longrightarrow 0 < p < 180 \longrightarrow \underline{P \in \overleftrightarrow{OA}/C}.$$

Also, $P \in R \longrightarrow \overrightarrow{OP}$ is between $\overrightarrow{OA}$ and $\overrightarrow{OC} \longrightarrow \overrightarrow{OP}$ is between $\overrightarrow{OC}$ and $\overrightarrow{OA} \longrightarrow$ There is a coordinate system C' that assigns $\overrightarrow{OC}$, $\overrightarrow{OP}$, and $\overrightarrow{OA}$ the numbers 0, p', and c', respectively, so that

$$0 < p' < c' < 180 \longrightarrow 0 < p' < 180 \longrightarrow \underline{P \in \overleftrightarrow{OC}/A}.$$

The underlined statements imply that $P \in I$.

PROOF THAT $I \subseteq R$

Again $\angle AOC$ is the given angle. We shall show $P \in I \longrightarrow P \in R$. The protractor postulate tells us that there exists a coordinate system C in which $\overrightarrow{OA}$ is the zero ray and $\overrightarrow{OC}$ has a coordinate c between 0 and 180. $P \in I \longrightarrow P \in \overleftrightarrow{OA}/C \longrightarrow p$, the ray-coordinate of $\overrightarrow{OP}$, is between 0 and 180. Since $\overrightarrow{OP} \neq \overrightarrow{OC}$, there are two possibilities: (1) $0 < c < p < 180$ or (2) $0 < p < c < 180$. We shall prove that (2) is true by proving that (1) is false.

If $p > c$, we have $m\angle AOC = c$, $m\angle AOP = p$, and $m\angle POC = p - c$. Now we apply the protractor postulate again to establish the existence of a coordinate system C' in which $\overrightarrow{OC}$ is the zero ray and $\overrightarrow{OA}$ corresponds to a number y that is less than 180. Since $m\angle AOC = c$ in the C-coordinate system and

$m\angle COA = y$ in the C'-coordinate system, we have $y = c$. Therefore, the C'-coordinate of $\overrightarrow{OA}$ is c. $P \in I \longrightarrow P \in \overleftrightarrow{OC}/A \longrightarrow x$, the C'-coordinate of $\overrightarrow{OP}$, is between 0 and 180. Therefore, using the C'-coordinate system, we have $m\angle COP = x$. Since $m\angle POC = p - c$, we have $x = p - c$. Therefore, the C'-coordinate of $\overrightarrow{OP}$ is $p - c$. In the C'-coordinate system

$$m\angle AOP = (p - c) - c = p - 2c \quad \text{or} \quad m\angle AOP = c - (p - c) = 2c - p.$$

But we had earlier that $m\angle AOP = p$.

$$p = p - 2c \longrightarrow 2c = 0 \longrightarrow c = 0 \qquad \text{(false)},$$
$$p = 2c - p \longrightarrow 2p = 2c \longrightarrow p = c \qquad \text{(false)}.$$

Our supposition that $p > c$ has led to a contradiction; hence (1) is false. Therefore (2) is true, and we have $0 < p < c < 180$. This implies that $\overrightarrow{OP}$ is between $\overrightarrow{OA}$ and $\overrightarrow{OC}$, and therefore $P \in R$. This completes our proof that $I = R$.

THEOREM 6-7 *The set of interior points of all segments that join a point in the interior of one side of an angle to an interior point of the other side is a subset of the interior of the angle.* ($S \subseteq I$.)

COROLLARY TO THEOREM 6-7

$$\left.\begin{array}{l} B\underline{C}D \text{ is an angle} \\ B\text{-}A\text{-}D \end{array}\right\} \longrightarrow \overrightarrow{CA} \text{ is between } \overrightarrow{CB} \text{ and } \overrightarrow{CD}.$$

Proofs for Theorem 6-7 and its corollary are considered in the exercises.

Reflecting on the statement of Theorem 6-7, we might ask why we do not complete the task of proving that $S = I$ by proving that $I \subseteq S$. The answer is that, using only the postulates we have stated thus far, it is not possible to prove that I is a subset of S. We could announce that $I \subseteq S$ is a postulate, but this would have some surprising and far-reaching results. In fact, it can be shown that $I \subseteq S$ implies a postulate that we have not yet accepted, namely the parallel postulate, which we formally adopt in Chapter 8. We prefer to continue to study theorems whose proofs do not involve the parallel postulate. Such theorems are theorems in *neutral* or *absolute* geometry. Since the parallel postulate distinguishes Euclidean geometry from other geometries, theorems whose proof involve the parallel postulate are called theorems in Euclidean geometry. Our study of Euclidean geometry begins in Chapter 8 with the adoption of Postulate 17.

DEFINITION *The exterior of an angle is the set of points in the plane of the angle that are not in the angle and not in the interior of the angle.* [*Exterior* $\angle ABC = \text{pl } ABC \sim (\angle ABC \cup \textit{interior } \angle ABC)$.]

If $A\underline{B}C$ is an angle, then pl $ABC = I \cup \angle ABC \cup E$, where I is the interior of $\angle ABC$ and E is its exterior. Observe that sets I, $\angle ABC$, and E are disjoint.

Using our definition of the interior of an angle, we can easily prove the following theorem.

THEOREM 6-8 *The interior of an angle is a convex set.*

Next we state a theorem that will be used in Chapter 7 for proving the exterior angle theorem.

THEOREM 6-9

$$\left.\begin{array}{l} C\underline{A}B \text{ is an angle} \\ C\text{-}E\text{-}B \\ A\text{-}E\text{-}F \\ A\text{-}B\text{-}D \end{array}\right\} \longrightarrow \overrightarrow{BF} \text{ is between } \overrightarrow{BC} \text{ and } \overrightarrow{BD}.$$

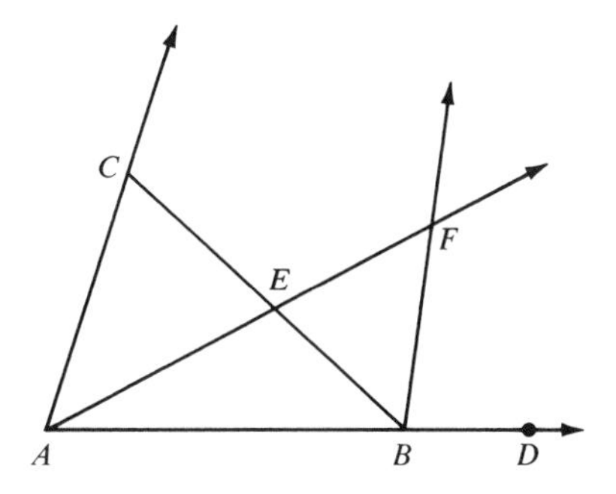

Suggestions for the proof of Theorem 6-9 appear in the exercises.

Exercises

1. Which of the following implications are true? Explain your answers.
   (a) $X, Y \in \text{interior } \angle ABC \longrightarrow \overline{XY} \subseteq \text{interior } \angle ABC$.
   (b) $X, Y \in \angle ABC \longrightarrow \overline{XY} \subseteq \angle ABC$.
   (c) $X, Y \in \text{exterior } \angle ABC \longrightarrow \overline{XY} \subseteq \text{exterior } \angle ABC$.
2. (a) Prove: $A\underline{O}B$ is an angle $\longrightarrow$ interior $\angle AOB \neq \emptyset$.
   (b) Supply reasons for the following proof of Theorem 6-7. (Refer to **L**-Exercise 9, page 112.) We wish to prove

$$A\underline{B}C \text{ is an angle} \longrightarrow \overset{\circ\!-\!\circ}{AC} \subseteq \text{interior } \angle ABC.$$

$$\underline{A\underline{B}C \text{ is an angle}} \xrightarrow{(1)} \left\{\begin{array}{l} C \notin \overleftrightarrow{AB} \xrightarrow{(2)} \overset{\circ\!-\!\circ}{AC} = \overset{\circ\!\rightarrow}{AC} \cap \overleftrightarrow{AB}/C \xrightarrow{(3)} \overset{\circ\!-\!\circ}{AC} \subseteq \overleftrightarrow{AB}/C \\ A \notin \overleftrightarrow{BC} \xrightarrow{(2)} \overset{\circ\!-\!\circ}{AC} = \overset{\circ\!\rightarrow}{CA} \cap \overleftrightarrow{BC}/A \xrightarrow{(3)} \overset{\circ\!-\!\circ}{AC} \subseteq \overleftrightarrow{BC}/A \end{array}\right\} \xrightarrow{(4)}$$

$$\left.\begin{array}{l} \rightarrow \overset{\circ\!-\!\circ}{AC} \subseteq \overleftrightarrow{AB}/C \cap \overleftrightarrow{BC}/A \\ \underline{A\underline{B}C \text{ is an angle}} \xrightarrow{(5)} \text{interior } \angle ABC = \overleftrightarrow{AB}/C \cap \overleftrightarrow{BC}/A \end{array}\right\} \xrightarrow{(6)} \overset{\circ\!-\!\circ}{AC} \subseteq \text{interior } \angle ABC.$$

3. Prove the corollary to Theorem 6-7.
4. Prove Theorem 6-8.

L 5. Supply reasons for the following proof that the union of an angle and its interior is the intersection of two closed half planes.

We prove: $A \notin \overleftrightarrow{BC} \longrightarrow \angle ABC \cup \text{interior } \angle ABC = \overline{\overleftrightarrow{BC}/A} \cap \overline{\overleftrightarrow{AB}/C}$.

$$\begin{aligned}
\overline{\overleftrightarrow{BC}/A} \cap \overline{\overleftrightarrow{AB}/C} &\overset{(1)}{=} (\overleftrightarrow{BC} \cup \overleftrightarrow{BC}/A) \cap (\overleftrightarrow{AB} \cup \overleftrightarrow{AB}/C) \\
&\overset{(2)}{=} [\overleftrightarrow{BC} \cap (\overleftrightarrow{AB} \cup \overleftrightarrow{AB}/C)] \cup [\overleftrightarrow{BC}/A \cap (\overleftrightarrow{AB} \cup \overleftrightarrow{AB}/C)] \\
&\overset{(3)}{=} (\overleftrightarrow{BC} \cap \overleftrightarrow{AB}) \cup (\overleftrightarrow{BC} \cap \overleftrightarrow{AB}/C) \cup (\overleftrightarrow{BC}/A \cap \overleftrightarrow{AB}) \cup (\overleftrightarrow{BC}/A \cap \overleftrightarrow{AB}/C) \\
&\overset{(4)}{=} \{B\} \cup \overset{\circ}{\overrightarrow{BC}} \cup \overset{\circ}{\overrightarrow{BA}} \cup (\overleftrightarrow{BC}/A \cap \overleftrightarrow{AB}/C) \overset{(5)}{=} \overrightarrow{BC} \cup \overrightarrow{BA} \cup (\overleftrightarrow{BC}/A \cap \overleftrightarrow{AB}/C) \\
&\overset{(6)}{=} \angle ABC \cup \text{interior } \angle ABC.
\end{aligned}$$

L 6. Prove: The union of an angle and its interior is a convex set.

7. If $A\underline{O}B$, $B\underline{O}C$, and $C\underline{O}A$ are coplanar angles such that no two of their interiors intersect, then $m\angle AOB + m\angle BOC + m\angle COA = 360$.

8. Prove: If I is the interior of an angle and E is the exterior of the angle, then the angle separates E and I.

9. Prove Theorem 6-9,

$$\left.\begin{array}{l} C\underline{A}B \text{ is an angle} \\ C\text{-}E\text{-}B \\ A\text{-}E\text{-}F \\ A\text{-}B\text{-}D \end{array}\right\} \longrightarrow \overrightarrow{BF} \text{ is between } \overrightarrow{BC} \text{ and } \overrightarrow{BD},$$

by constructing flow proofs for each of the following:

(a) $F \in \overleftrightarrow{BD}/C$. (b) $D \in \overleftrightarrow{CB}/{\sim}A$.

(c) $F \in \overleftrightarrow{CB}/{\sim}A$.

Then use these results to complete the proof.

Linear Pairs of Angles

In the diagram below, $\overrightarrow{OA}$ and $\overrightarrow{OC}$ are opposite rays. This statement is equivalent to saying that $O \in \overset{\circ\!-\!\circ}{AC}$. The point B is not in $\overleftrightarrow{AC}$; hence $\overrightarrow{OA}$, $\overrightarrow{OB}$, and $\overrightarrow{OC}$ are distinct, concurrent rays. Therefore $A\underline{O}B$ and $B\underline{O}C$ are angles. It is convenient to have a name for such pairs of angles as $\angle AOB$ and $\angle BOC$.

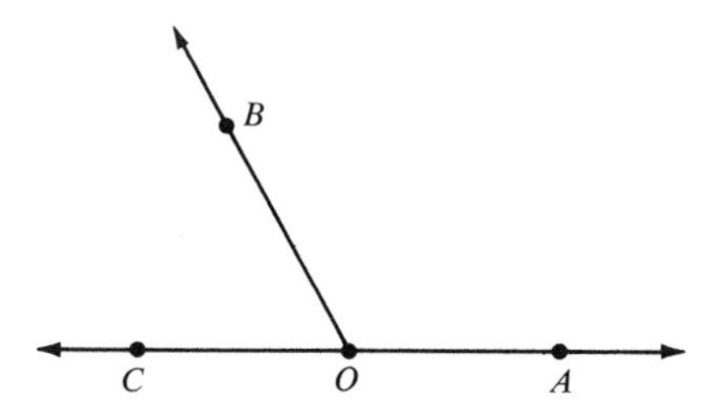

DEFINITION *Two angles form a* linear pair of angles *if they are formed by three distinct concurrent rays of which two are opposite rays.*

Stated symbolically, this definition becomes

$$(O \in \overset{\circ\!-\!\circ}{AC} \wedge B \notin \overleftrightarrow{AC}) \longleftrightarrow A\underline{O}B \text{ and } B\underline{O}C \text{ are a linear pair of angles.}$$

Since $O \in \overset{\circ\!-\!\circ}{AC} \longleftrightarrow A\text{-}O\text{-}C$, we may write

$$(A\text{-}O\text{-}C \wedge B \notin \overleftrightarrow{AC}) \longleftrightarrow A\underline{O}B \text{ and } B\underline{O}C \text{ are a linear pair of angles.}$$

DEFINITION *Two angles are* supplementary *if the sum of their measures is* 180. *In such a case, each angle is called the* supplement *of the other.*

The following theorem serves to confirm your conjecture that the angles of a linear pair of angles are supplementary.

THEOREM 6-10 *The angles of any linear pair are supplementary.*

To prove Theorem 6-10, we must show:

$(\angle AOB$ and $\angle BOC$ are a linear pair$) \longrightarrow (m\angle AOB + m\angle BOC = 180)$.

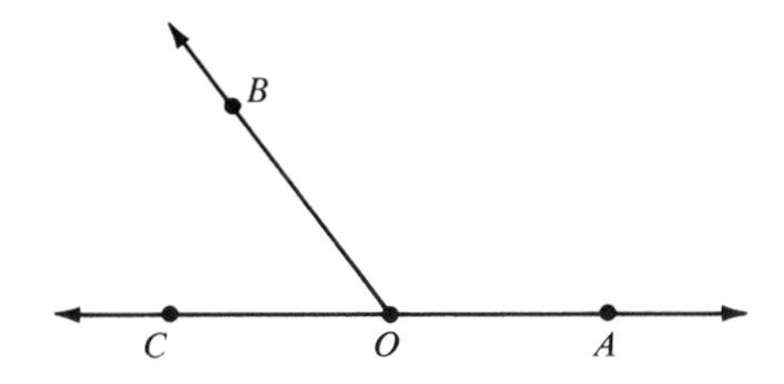

PROOF

1. $O \in \overset{\circ\!\!-\!\!\circ}{AC} \wedge B \notin \overleftrightarrow{AC}$.	1. Hypothesis and definition of a linear pair of angles.
2. $\overrightarrow{OA}$ and $\overrightarrow{OC}$ are opposite rays.	2. Statement 1 and definition of opposite rays.
3. For $\angle AOB$ there exists a unique ray-coordinate system with $\overrightarrow{OA}$ as zero ray such that $\overrightarrow{OB}$ corresponds to a number b such that $0 < b < 180$.	3. Protractor postulate.
4. The ray-coordinate of $\overrightarrow{OC}$ is 180.	4. Statement 2 and definition of ray-coordinate system.
5. $m\angle AOB = \lvert b - 0\rvert = b$. $m\angle BOC = \lvert 180 - b\rvert = 180 - b$.	5. Statement 3 and definition of ray-coordinate system.
6. $m\angle AOB + m\angle BOC = 180$.	6. Addition property of equality.

COROLLARY TO THEOREM 6-10

$$\left.\begin{array}{l} O \in \overset{\circ\!\!-\!\!\circ}{XY} \\ A \in \overleftrightarrow{XY}/B \\ \overrightarrow{OA} \text{ between } \overrightarrow{OX} \text{ and } \overrightarrow{OB} \end{array}\right\} \longrightarrow m\angle XOA + m\angle AOB + m\angle BOY = 180.$$

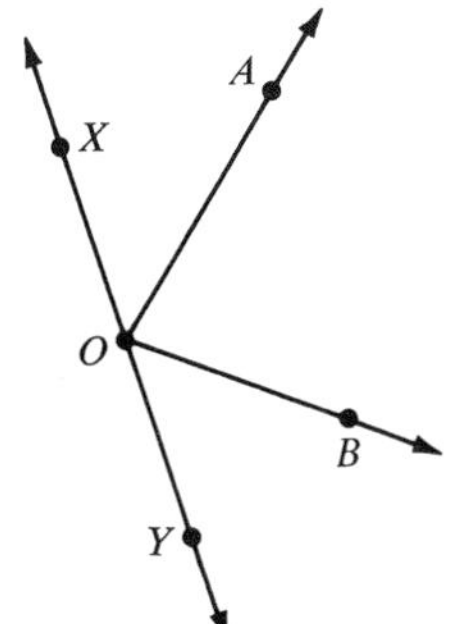

In each of the diagrams below α and β are described as adjacent angles.

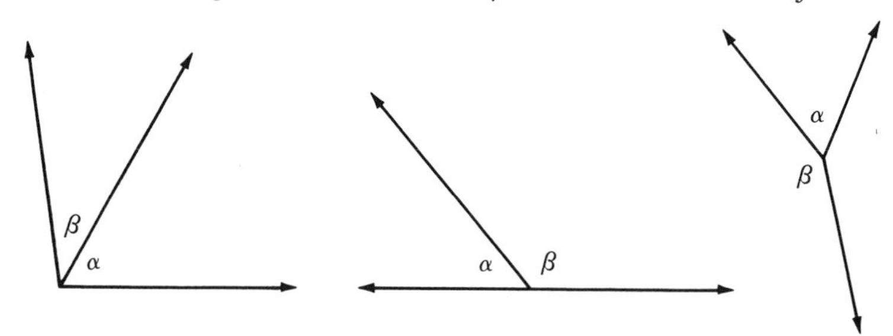

In each case, the two angles have a common side. Moreover, the interiors of the remaining sides of these angles are in opposite half planes with respect to the line that contains their common side. These two ideas suggest the following definition.

DEFINITION *Two angles are* adjacent angles *if they have a common side and the interiors of the remaining sides are in opposite half planes whose common edge is the line containing the common side.*

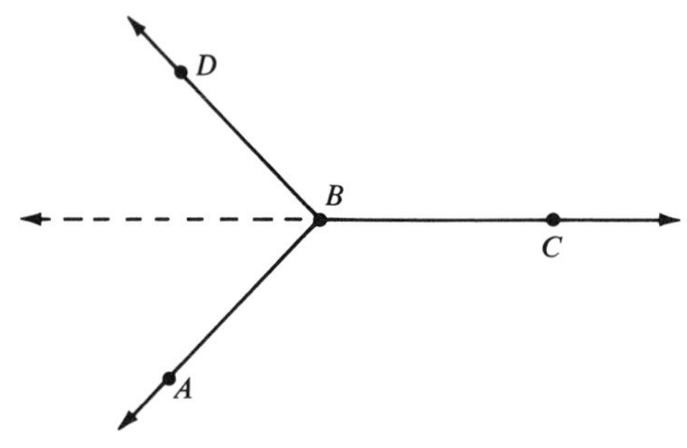

In view of Theorem 5-6, we see that an equivalent form of this definition is the statement:

$$C\underline{B}D \text{ and } C\underline{B}A \text{ are adjacent angles} \longleftrightarrow A \in \overleftrightarrow{BC}/\sim D.$$

THEOREM 6-11 *If two angles are adjacent and supplementary, then they are a linear pair of angles.*

We must prove that

$$\left.\begin{array}{l}\angle AOB \text{ and } \angle BOC \text{ are adjacent angles} \\ m\angle AOB + m\angle BOC = 180\end{array}\right\} \longrightarrow \angle AOB \text{ and } \angle BOC \text{ are a linear pair.}$$

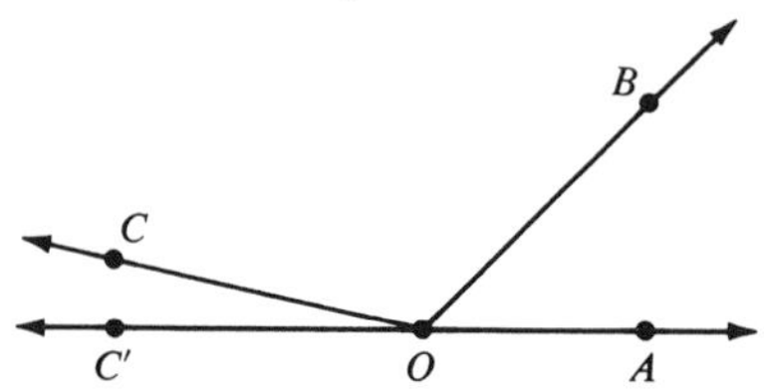

PROOF

Statements	Reasons
1. There exists a point C' such that $C'\text{-}O\text{-}A$ $(O \in \overset{\circ\!-\!\circ}{AC'})$.	1. Corollary to the definition of betweenness of points.
2. $\overleftrightarrow{C'A} = \overleftrightarrow{OA}$.	2. A line may be named by any two distinct points in it (Corollary 2 to Postulate 2.
3. $B \notin \overleftrightarrow{OA}$.	3. Hypothesis ($A\underline{O}B$ is an angle) and definition of an angle.
4. $B \notin \overleftrightarrow{C'A}$.	4. Statements 2, 3, and substitution property of equality.
5. $\angle AOB$ and $\angle BOC'$ are a linear pair.	5. Statements 1, 4, and definition of linear pair of angles.
6. $m\angle AOB + m\angle BOC' = 180$.	6. Statement 5 and Theorem 6-10.
7. $m\angle AOB + m\angle BOC = 180$.	7. Hypothesis.
8. $m\angle BOC = m\angle BOC'$.	8. Statements 6, 7, substitution property of equality, and addition property of equality.
9. $C \in \overleftrightarrow{OB}/\sim A$.	9. Hypothesis and definition of adjacent angles.
10. $\overleftrightarrow{C'A} \neq \overleftrightarrow{OB}$.	10. Statement 4 and ST_3.
11. $\overset{\circ\!-\!\circ}{C'A} \times \overleftrightarrow{OB}(O)$.	11. Statements 1, 10, and the corollary to Theorem 3-1.
12. $C' \in \overleftrightarrow{OB}/\sim A$.	12. Statement 11 and Theorem 5-7.
13. $\overrightarrow{OC} \subseteq \overleftrightarrow{OB}/\sim A$, $\overrightarrow{OC'} \subseteq \overleftrightarrow{OB}/\sim A$.	13. Statements 9, 12, and Corollary 2 to Theorem 5-6.
14. $\overrightarrow{OC} = \overrightarrow{OC'}$.	14. Statements 8, 13, and angle construction theorem.
15. $\overrightarrow{OC} \cup \overrightarrow{OB} = \overrightarrow{OC'} \cup \overrightarrow{OB}$.	15. Statement 14 and ST_1.
16. $\overrightarrow{OC} \cup \overrightarrow{OB} = \angle COB$, $\overrightarrow{OC'} \cup \overrightarrow{OB} = \angle C'OB$.	16. Hypothesis ($B\underline{O}C$ is an angle), statement 5 ($B\underline{O}C'$ is an angle), and the definition of angle.
17. $\angle COB = \angle C'OB$.	17. Statements 15, 16, and substitution property of equality.
18. $\angle AOB$ and $\angle COB$ are a linear pair.	18. Statements 5, 17, and substitution property of equality.

DEFINITIONS *An angle whose measure is* 90 *is called a* right angle. *An angle whose measure is less than* 90 *is called an* acute angle. *An angle whose measure is greater than* 90 *is called an* obtuse angle.

DEFINITION *If the sum of the measures of two angles is* 90, *then the two angles are* complementary *and each is the* complement *of the other.*

The proof of the following theorem is required in the exercises.

THEOREM 6-12 *If two angles of a linear pair have the same measure, then each is a right angle.*

DEFINITION *Two lines whose union contains a right angle are called* perpendicular lines.

We indicate that L_1 and L_2 are perpendicular by writing $L_1 \perp L_2$ or $L_2 \perp L_1$. We say that L_1 is perpendicular to L_2 or L_2 is perpendicular to L_1.

The definition of perpendicular lines has two corollaries whose proofs are considered in the exercises.

COROLLARY 1 TO THE DEFINITION OF PERPENDICULAR LINES *If two lines are perpendicular, then they intersect in a single point.*

Note: $L_1 \perp L_2(P) \longleftrightarrow (L_1 \times L_2(P) \wedge L_1 \perp L_2)$.

COROLLARY 2 TO THE DEFINITION OF PERPENDICULAR LINES *If two lines are perpendicular to the same line at distinct points, then these lines are distinct.*

DEFINITION *Two sets, each of which is a segment or a ray or a line, are perpendicular to each other if the line that contains one of the sets is perpendicular to the line that contains the other.*

Clearly, perpendicular sets can intersect in at most one point and need not have any point in common. The drawings below suggest two possibilities for two perpendicular segments. Observe that in the drawings we sometimes indicate that two sets are perpendicular by drawing a small square at the vertex of the right angle.

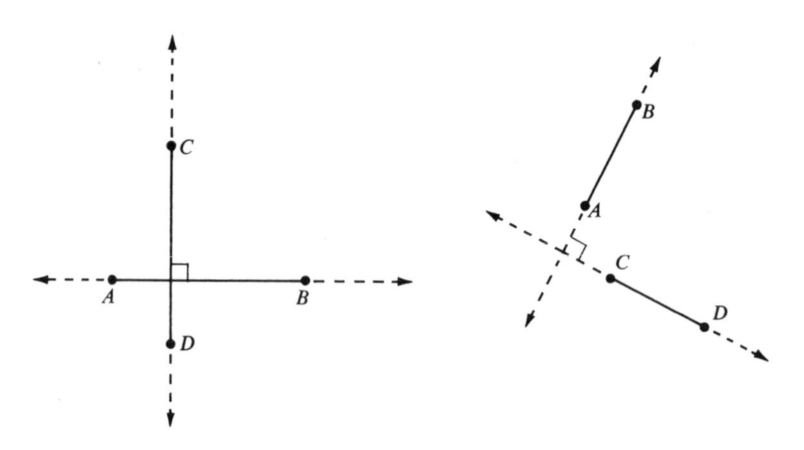

In Chapter 4 we agreed to call two segments congruent if they had the same length. We apply the same idea to angles.

DEFINITION *Two angles (whether distinct or not) that have the same measure are called* congruent angles *and each is said to be congruent to the other.*

We write $\angle RST \cong \angle ABC$ to express the fact that $\angle RST$ and $\angle ABC$ are congruent. Thus

$$\angle \alpha \cong \angle \beta \longleftrightarrow m\, \angle \alpha = m\, \angle \beta.$$

Clearly, $\angle\alpha = \angle\beta \longrightarrow m\angle\alpha = m\angle\beta$. However, the converse of this implication is not true. Why?

We complete this section by stating a number of theorems whose proofs are considered in the exercises.

THEOREM 6-13 *Any two right angles are congruent to each other.*

THEOREM 6-14 *If two angles are both congruent and supplementary, then each is a right angle.*

THEOREM 6-15 *The congruence relation for angles is an equivalence relation.*

THEOREM 6-16 *If two angles are congruent, then their supplements are congruent.*

THEOREM 6-17 *If two angles are congruent, then their complements are congruent.*

THEOREM 6-18 *If two angles are complementary, then each is an acute angle.*

Vertical Angles

Two lines that intersect in a single point form four angles as shown in the diagram. Observe that

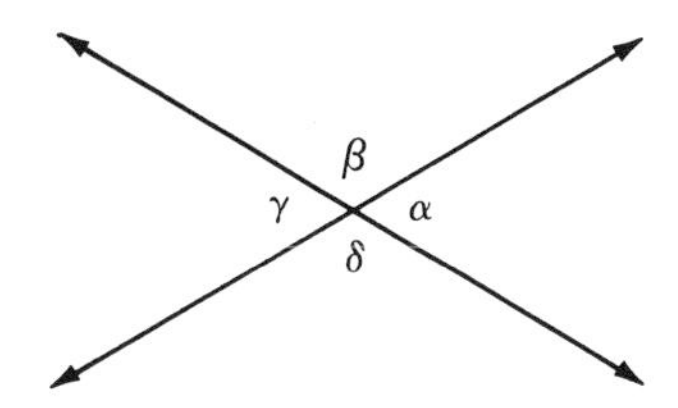

(1) $\angle\alpha$ and $\angle\gamma$ are nonadjacent angles.
(2) Each of these angles forms a linear pair with either $\angle\beta$ or $\angle\delta$.
(3) The sides of $\angle\alpha$ are opposite rays to the sides of $\angle\gamma$.

Similar statements could be made about $\angle\beta$ and $\angle\delta$. Angles that have the third property are called vertical angles.

DEFINITION *Two angles are vertical if the sides of one are opposite rays to the sides of the other.*

This definition can be stated more succinctly as follows:

$(\angle AOX$ and $\angle BOY$ are vertical angles$) \longleftrightarrow (A\text{-}O\text{-}B \wedge X\text{-}O\text{-}Y \wedge \overleftrightarrow{AB} \neq \overleftrightarrow{XY})$.

Since each of the vertical angles α and γ forms a linear pair with $\angle\beta$, each is supplementary to $\angle\beta$. Thus we have $m\angle\alpha + m\angle\beta = 180$ and $m\angle\gamma + m\angle\beta =$

180. It follows that $m\angle\alpha = m\angle\gamma$, and this in turn implies that $\angle\alpha \cong \angle\gamma$. Thus we have the following theorem.

THEOREM 6-19 *Vertical angles are congruent.*

Perpendiculars

The angles formed when two lines intersect in a single point are described in the following lemma.

LEMMA 2 *If $L_1 \times L_2(O)$, then the set of angles in $L_1 \cup L_2$ is the set of four angles each having vertex O, one side in L_1 and the other side in L_2.*

THEOREM 6-20 *If an angle is a right angle, then each of the other three angles contained in the union of the two lines that contain its sides is a right angle.*

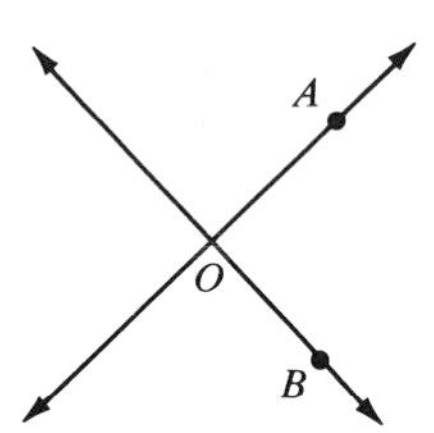

PROOF

If $\underline{AOB}$ is a right angle, then $\overleftrightarrow{AO} \perp \overleftrightarrow{OB}$ by our definition of perpendicular sets. If $\overleftrightarrow{AO} \perp \overleftrightarrow{OB}$, then we know that $\overleftrightarrow{AO} \times \overleftrightarrow{OB}(O)$, and our lemma tells us that there are four angles in $\overleftrightarrow{OA} \cup \overleftrightarrow{OB}$, each having vertex O with one side in $\overleftrightarrow{OB}$ and the other in $\overleftrightarrow{OA}$. $\angle AOB$ is one of these angles. Moreover, if $\overleftrightarrow{AO} \perp \overleftrightarrow{OB}$, we know from our definition that one of these four angles is a right angle. Therefore, by Theorem 6-10, all four are right angles.

COROLLARY TO THEOREM 6-20 *$\underline{AOB}$ is a right angle $\longleftrightarrow \overleftrightarrow{OA} \perp \overleftrightarrow{OB}$.*

THEOREM 6-21 (UNIQUE PERPENDICULAR THEOREM) *For each point in a line in a plane there is one and only one line in the given plane that contains the given point and is perpendicular to the given line.*

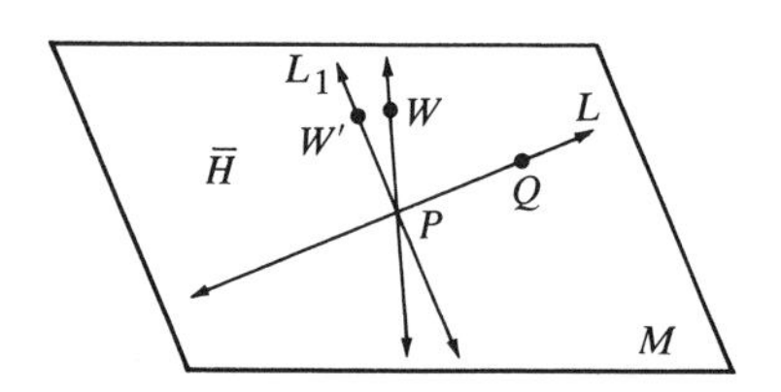

PROOF OF EXISTENCE

Since $L \subseteq M$, we know that M is the union of opposite closed half planes with common edge L. Let $\overline{H}$ be one of these. There is a point Q in L different from P. By the angle-construction theorem, there is a unique ray PW in H such that $\angle QPW$ is a right angle. Therefore, $\overleftrightarrow{PW}$ is perpendicular to L, contains P, and lies in M.

PROOF OF UNIQUENESS

To prove uniqueness we shall show that if L_1 is another line that contains P, lies in M, and is perpendicular to L, then $L_1 = \overleftrightarrow{PW}$. Since $L_1 \perp L$, we know that $L_1 \times L(P)$. From this and the fact that $L_1 \subseteq M$, we know that the intersection of L_1 with H is the interior of a ray PW'. Since $L_1 \perp L$, we know by the corollary to Theorem 6-20 that $W'PQ$ is a right angle. Therefore $m\angle QPW' = 90$. Thus $\angle QPW$ and $\angle QPW'$ are two angles with the same measure and a common side with the interiors of the other two sides in the same half plane. It follows from the angle-construction theorem that $\overrightarrow{PW} = \overrightarrow{PW'}$. Therefore $\overleftrightarrow{PW} = \overleftrightarrow{PW'}$. But, $\overleftrightarrow{PW'} = L_1$. Therefore, $\overleftrightarrow{PW} = L_1$, and our proof of uniqueness is complete.

DEFINITION *A line perpendicular to a segment at its midpoint is a* perpendicular bisector *of the segment.*

Observe that a segment has infinitely many perpendicular bisectors in space, but only one in a given plane that contains the segment.

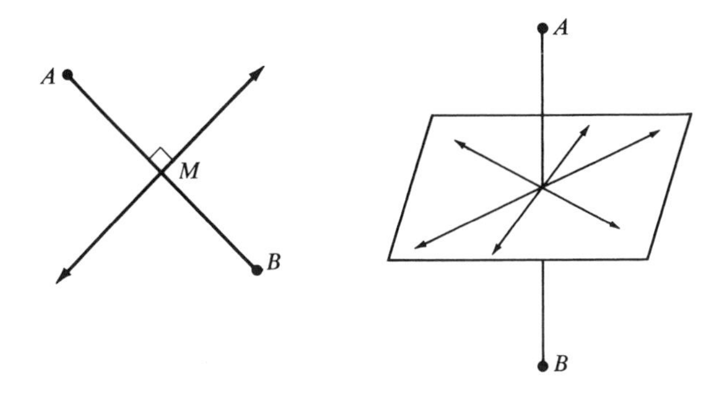

Exercises

1. Prove the corollary to Theorem 6-10.
2. Prove Theorem 6-12.
3. Prove Corollary 1 to the definition of perpendicular lines.
4. Prove Corollary 2 to the definition of perpendicular lines.
5. Prove Theorem 6-13.
6. Prove Theorem 6-14.
7. Prove Theorem 6-15.

8. Prove Theorem 6-16.
9. Prove Theorem 6-17.
10. Prove Theorem 6-18.
11. Construct a flow-diagram proof of the following statement: If $A\underline{B}C$ and $C\underline{B}D$ are a linear pair of angles, then they are adjacent.
12. Prove: If $A\underline{B}C$ and $C\underline{B}D$ are adjacent angles, then their interiors do not intersect.
13. Construct a flow-diagram proof of the following statement:

$$\left.\begin{array}{l} A\underline{O}B \text{ and } B\underline{O}C \text{ are adjacent angles} \\ B\underline{O}C \text{ and } C\underline{O}A \text{ are adjacent angles} \end{array}\right\} \longrightarrow m\angle AOB + m\angle BOC + m\angle COA = 360.$$

14. Prove Lemma 2.
15. Prove the corollary to Theorem 6-20.

Triangles and Quadrilaterals

Just as an angle can be partially described as the union of two rays, a triangle can be partially described as the union of three segments. Each of the drawings at the right is the union of $\overline{AB}$, $\overline{BC}$, and $\overline{AC}$. We accept the first three drawings as representing triangles, but reject the last two because $B \in \overleftrightarrow{AC}$.

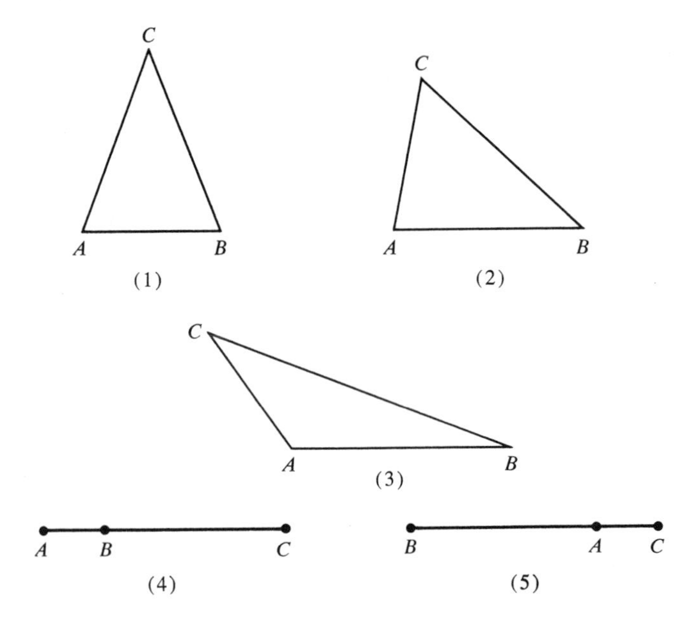

Notation: $ABC = \overline{AB} \cup \overline{BC} \cup \overline{CA}$.

DEFINITION *A triangle is the union of three segments joining three noncollinear points.*

The following symbolic statement of this definition is useful.

$$ABC \text{ is a triangle} \longleftrightarrow B \notin \overleftrightarrow{AC}.$$

Notation: If A, B, and C are three noncollinear points ($A \notin \overleftrightarrow{BC}$), the triangle they determine is denoted by the symbol $\triangle ABC$.

DEFINITIONS *Each of the points A, B, C, is a* vertex *of $\triangle ABC$; each of the segments AB, BC, CA is a* side *of $\triangle ABC$; each of the angles $\angle ABC$, $\angle BCA$, $\angle CAB$ is an angle of $\triangle ABC$. An angle of a triangle is said to be included between the two sides of the triangle that are contained in the angle. A side of a triangle is said to be included between the two angles of the triangle whose vertices are the endpoints of the side. A side of a triangle and the angle of the triangle whose vertex is not a point in that side are said to be opposite each other.*

Thus, in the diagram below, $\angle ABC$ is included between sides $\overline{AB}$ and $\overline{BC}$, side $\overline{AB}$ is included between $\angle BAC$ and $\angle ABC$, and $\angle BAC$ is opposite $\overline{BC}$ in $\triangle ABC$.

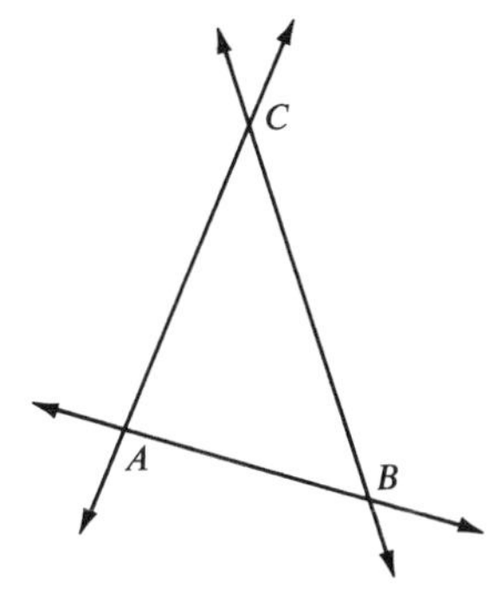

Observe that it is correct to say that the vertices of a triangle are in the triangle (thus $A \in \triangle ABC$) and the sides of a triangle are subsets of the triangle (thus $\overline{AB} \subseteq \triangle ABC$). However, it is not correct to say that the angles of a triangle are contained in the triangle. Thus the statement $\angle ABC \subseteq \triangle ABC$ is false. Why? Is it correct to say that a triangle is the intersection of its angles? Is $\triangle ABC = \triangle BCA$? Is $\triangle ABC = \triangle BAC$?

DEFINITION *The* interior of a triangle *is the intersection of half planes, each of which has a side of the triangle in its edge and contains the vertex of the angle opposite the side contained in the edge.*

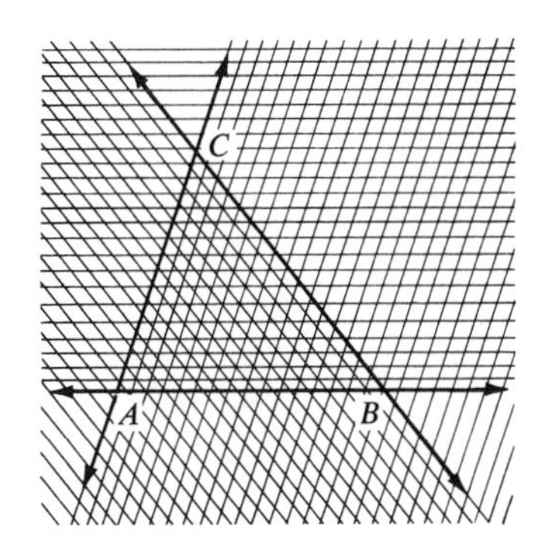

Thus the interior of $\triangle ABC = \overleftrightarrow{AB}/C \cap \overleftrightarrow{AC}/B \cap \overleftrightarrow{BC}/A$. Observe that the interior of a triangle could also be described as the intersection of the interior of the three angles of the triangle. Why?

DEFINITION *The set of points in the plane of a triangle that are not in the triangle nor in the interior of the triangle is called the* exterior of the triangle.

Since the interior of a triangle is, by definition, the intersection of convex sets, it is convex by Theorem 5-2. We have the following theorem.

THEOREM 6-22 *The interior of a triangle is a convex set.*

Having stated a definition of a triangle, we next define quadrilaterals. We accept the first three drawings below as representing quadrilaterals, but we reject the fourth because three of the points are collinear.

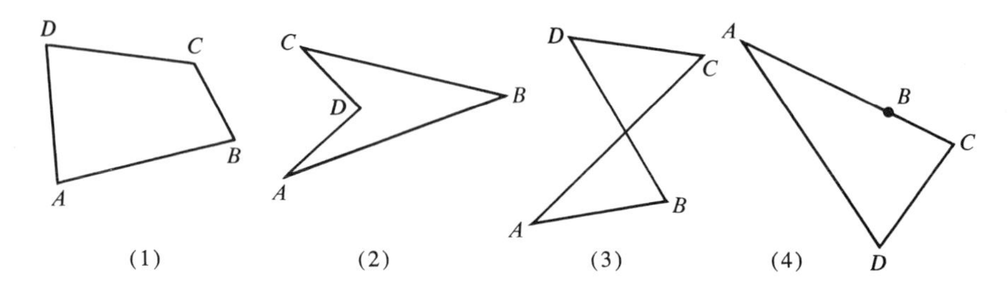

Notation: $ABCD = \overline{AB} \cup \overline{BC} \cup \overline{CD} \cup \overline{DA}$.

DEFINITION *ABCD is a* quadrilateral *if and only if points A, B, C, and D are coplanar and no three of them are collinear.*

Observe that the same four points can be vertices of different quadrilaterals as shown in the drawings below. Thus a quadrilateral is not determined by its

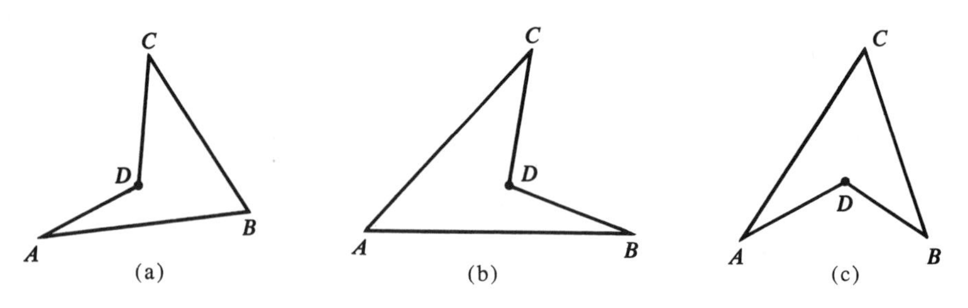

vertices; hence it cannot be described by a list of the vertices unless the list also indicates the sides of the quadrilateral. For this reason, when we list the vertices of a quadrilateral, we do it in such a way that (1) vertices whose names are adjacent in the list are endpoints of the same side, (2) the first and last vertices named in the list are endpoints of the same side.

A quadrilateral is determined by its sides. The quadrilateral determined by $\overline{AB}$, $\overline{BC}$, $\overline{CD}$, and $\overline{DA}$ is called quadrilateral $ABCD$. We sometimes use the symbol quad $ABCD$ to represent the phrase quadrilateral $ABCD$. Thus the quadrilaterals shown in the preceding drawings (a), (b), and (c) are respectively quad $ABCD$, quad $ABDC$, and quad $ADBC$. In general, is quad $RSTW =$ quad $STWR$? Is quad $RSTW =$ quad $TWRS$? Is quad $RSTW =$ quad $WTSR$? Name another quadrilateral that is equal to quad $RSTW$.

DEFINITIONS *The segments AB, BC, CD, and DA are called the* sides *of quadrilateral $ABCD$, and the points A, B, C, and D are called the* vertices *of quadrilateral $ABCD$. $|AB| + |BC| + |CD| + |DA|$ is called the* perimeter *of the quadrilateral $ABCD$. $\overline{AC}$ and $\overline{BD}$ are the* diagonals *of the quadrilateral $ABCD$.*

We close this section with an important theorem.

THEOREM 6-23 (CROSSBAR THEOREM) $P \in$ *interior* $\angle ABC \longrightarrow \overset{\circ\rightarrow}{BP} \succ\!\prec \overset{\circ\ \circ}{AC}$.

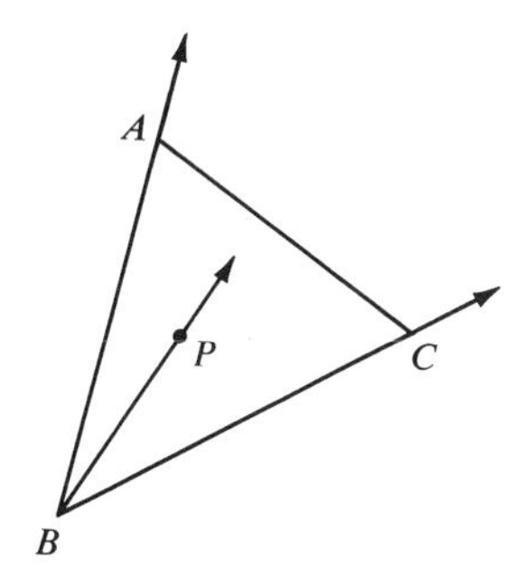

PLAN OF PROOF

We shall show that if $\overset{\circ\rightarrow}{BP}$ does not intersect $\overline{AC}$, then $\overleftrightarrow{BP}$ does not. This implies that A and C are in the same side of $\overleftrightarrow{BP}$, and, applying Theorem 6-4, we get a contradiction. Therefore, $\overset{\circ\rightarrow}{BP}$ intersects $\overline{AC}$, and since it does not contain either A or C, it must intersect $\overset{\circ\ \circ}{AC}$.

OUTLINE OF PROOF

There is a point Q such that Q-B-$P \longrightarrow \overset{\circ\rightarrow}{BQ}$ and $\overset{\circ\rightarrow}{BP}$ are opposite rays, $A \notin \overleftrightarrow{PQ}$ $\;\}\dashv$

$\nrightarrow$ $\overset{\circ\rightarrow}{BP}$ and $\overset{\circ\rightarrow}{BQ}$ lie in opposite half planes with edge $\overleftrightarrow{AB}$
$P \in$ interior $\angle ABC \longrightarrow P \in \overleftrightarrow{AB}/C \longrightarrow \overset{\circ\rightarrow}{BP} \subseteq \overleftrightarrow{AB}/C$ $\;\}\dashv$

$\nrightarrow$ $\overset{\circ\rightarrow}{BQ} \subseteq \overleftrightarrow{AB}/{\sim}C$
$\overset{\circ\ \circ}{AC} \subseteq \overleftrightarrow{AB}/C$
$\overleftrightarrow{AB}/C \cap \overleftrightarrow{AB}/{\sim}C = \emptyset$ $\;\}\dashv$

$$\left.\begin{array}{l}\nrightarrow \mathring{\overrightarrow{BQ}} \cap \overrightarrow{AC} = \emptyset \\ A \notin \mathring{\overrightarrow{BQ}}\end{array}\right\} \longrightarrow \left.\begin{array}{l}\mathring{\overrightarrow{BQ}} \cap \overrightarrow{AC} = \emptyset \\ \overline{AC} \subseteq \overrightarrow{AC}\end{array}\right\} \longrightarrow \left.\begin{array}{l}\mathring{\overrightarrow{BQ}} \cap \overline{AC} = \emptyset \\ \{B\} \cap \overline{AC} = \emptyset \\ \text{Suppose } \mathring{\overrightarrow{BP}} \cap \overline{AC} = \emptyset\end{array}\right\} \dashv$$

$$\nrightarrow (\mathring{\overrightarrow{BQ}} \cap \overline{AC}) \cup (\{B\} \cap \overline{AC}) \cup (\mathring{\overrightarrow{BP}} \cap \overline{AC}) = \emptyset \longrightarrow \overline{AC} \cap (\mathring{\overrightarrow{BQ}} \cup \{B\} \cup \mathring{\overrightarrow{BP}}) = \emptyset \dashv$$

$$\left.\begin{array}{l}\nrightarrow \overline{AC} \cap \overleftrightarrow{BP} = \emptyset \longrightarrow C \in \overleftrightarrow{BP}/A \\ \overrightarrow{BC} \neq \overrightarrow{BA}\end{array}\right\} \overset{*}{\longrightarrow} \begin{array}{l}\overrightarrow{BA} \text{ is between } \overrightarrow{BP} \text{ and } \overrightarrow{BC} \dashv \\ \overrightarrow{BC} \text{ is between } \overrightarrow{BP} \text{ and } \overrightarrow{BA} \dashv\!\!\!\dashv\end{array}$$

$$\left.\begin{array}{l}P \in \text{interior } \angle ABC \longrightarrow m\angle ABP + m\angle PBC = m\angle ABC \\ \nrightarrow m\angle ABP + m\angle ABC = m\angle PBC\end{array}\right\} \dashv$$

$$\left.\begin{array}{l}\nrightarrow\!\!\!\nrightarrow m\angle PBC + m\angle ABC = m\angle PBA \\ P \in \text{interior } \angle ABC \longrightarrow m\angle ABP + m\angle PBC = m\angle ABC\end{array}\right\} \dashv\!\!\!\dashv$$

$$\nrightarrow m\angle ABP = 0 \quad \text{(false)}$$

$$\nrightarrow\!\!\!\nrightarrow m\angle PBC = 0 \quad \text{(false).}$$

Therefore,

$$\left.\begin{array}{l}\mathring{\overrightarrow{BP}} \cap \overline{AC} \neq \emptyset \\ A \notin \overleftrightarrow{BP} \\ C \notin \overleftrightarrow{BP}\end{array}\right\} \longrightarrow \left.\begin{array}{l}\mathring{\overrightarrow{BP}} \cap \mathring{AC} \neq \emptyset \\ \overleftrightarrow{AC} \neq \overleftrightarrow{BP}\end{array}\right\} \longrightarrow \mathring{\overrightarrow{BP}} \times \mathring{AC}.$$

COROLLARY TO THEOREM 6-23 *If $P \in$ interior $\triangle ABC$, then $\mathring{\overrightarrow{BP}} \times \mathring{AC}(X)$ and B-P-X.*

Polygons

With the background provided by the last section, we are now ready to define a polygon of n sides.

Notation:

$P_1P_2P_3 \cdots P_{n-1}P_n = \overline{P_1P_2} \cup \overline{P_2P_3} \cup \cdots \cup \overline{P_{n-1}P_n} \cup \overline{P_nP_1}$ and $P = \{P_1, P_2, P_3, \ldots, P_{n-1}, P_n\}$, where $n \geqq 3$. Members of P are called *consecutive points* if they are endpoints of the same segment in $P_1P_2 \cdots P_n$. Two segments in $P_1P_2 \cdots P_n$ are called *consecutive segments* if they have a common endpoint. Three points in P are called *consecutive points* if they are endpoints of two consecutive segments.

DEFINITION *$P_1P_2P_3 \cdots P_n$ is a* polygon of n sides *(n-gon) if and only if $P_1, P_2, \ldots, P_n$ are distinct coplanar points and no three consecutive points in P are collinear.*

Notation: The polygon of n sides ($n \geqq 3$) determined by the segments $P_1P_2, P_2P_3, \ldots, P_{n-1}P_n, P_nP_1$ is denoted by the symbol polygon $P_1P_2 \cdots P_n$.

---

* See L-Exercise 4, page 143.

DEFINITIONS *The points* $P_1, P_2, \ldots, P_n$ *are the* vertices *of polygon* $P_1P_2 \cdots P_n$, *and the segments* $P_1P_2, P_2P_3, \ldots, P_nP_1$ *are its* sides. *Two vertices of polygon* $P_1P_2 \cdots P_n$ *that are endpoints of the same side are called* consecutive vertices. *Two vertices of a polygon that are not consecutive are called* non-consecutive vertices. *Two sides of polygon* $P_1P_2 \cdots P_n$ *that have a common endpoint are called* consecutive sides.

Side *AB* of polygon *ABCD* pictured below lies in the edge of a half plane containing the rest of the polygon. The same can be said for each of the other sides of the polygon. It may seem natural to say that *ABCD* is a convex polygon. We see, however, that this is not the case. If *ABCD* were a convex set of points, a segment joining any two of its points (*X* and *Y*, for example) would be contained in the polygon. Clearly, this is not true for $\overline{XY}$ in the drawing shown. Since we have no name for a polygon such as quad *ABCD*, we shall use the word *convax* to describe it. We adopt the following definition.

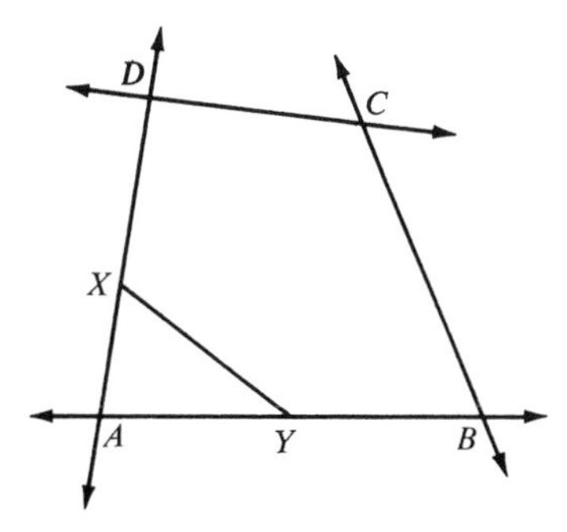

DEFINITION *A polygon is a* convax polygon *if and only if each of its sides lies in the edge of a half plane that contains the rest of the polygon.*

Using Theorems 5-3 and 5-6 we have the following.

COROLLARY TO THE DEFINITION OF CONVAX POLYGON *A polygon is convax if and only if for each pair of consecutive vertices a half plane whose edge is determined by these vertices contains all other vertices of the polygon.*

The figures below illustrate how each side of a convax quadrilateral lies in the edge of a half plane that contains all of the rest of the quadrilateral.

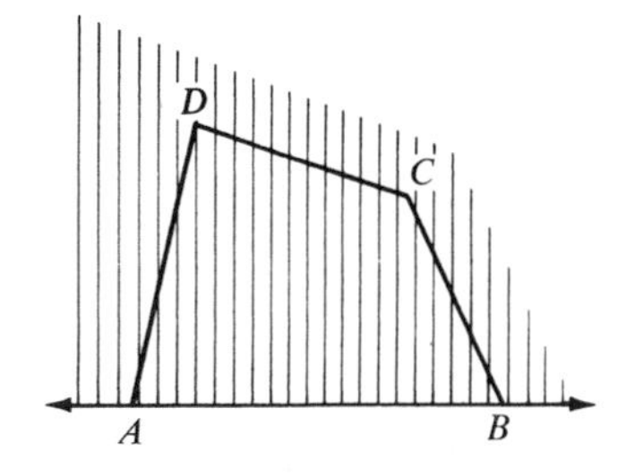

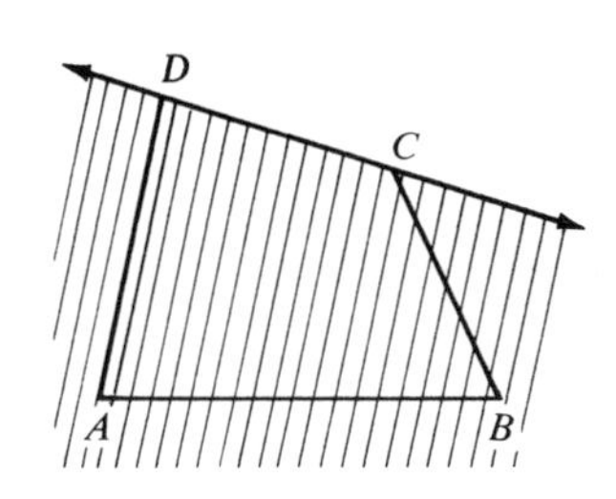

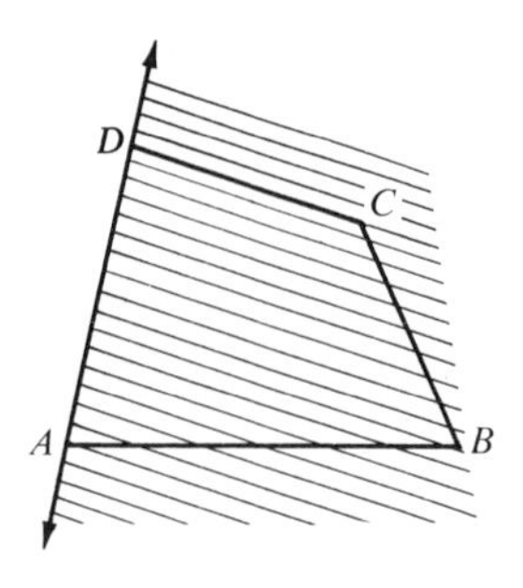

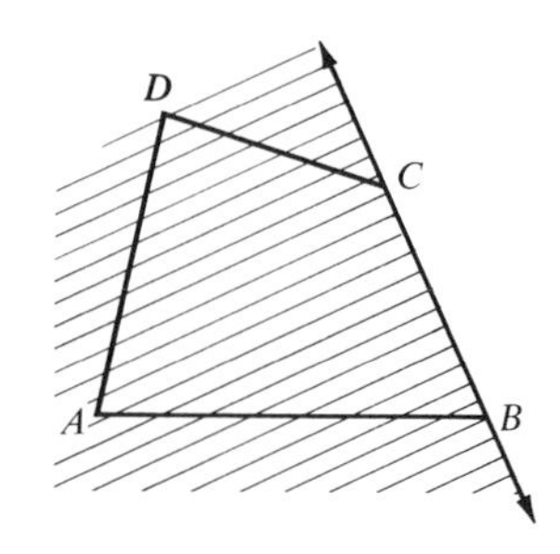

The next figures show a quadrilateral that is not convax.

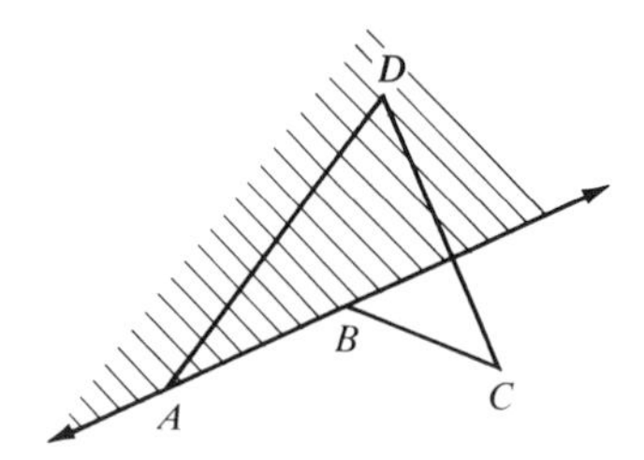

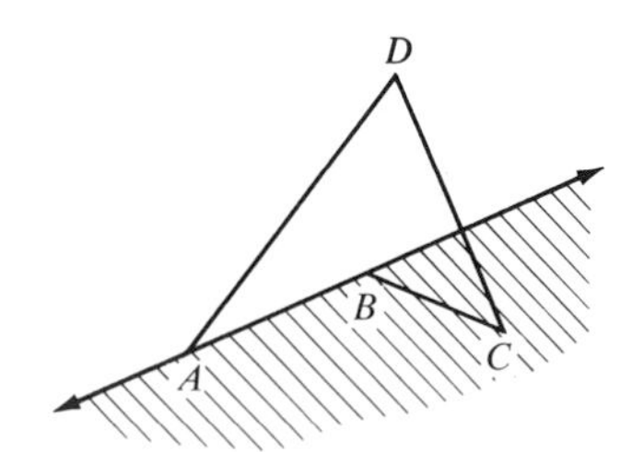

Is a triangle a convax polygon?

DEFINITION *The* interior *of a* convax polygon *is the intersection of all of the half planes each of which has a side of the polygon in its edge and contains the rest of the polygon.*

This definition enables us to prove the following theorem.

THEOREM 6-24 *The interior of a convax polygon is a convex set.*

DEFINITION *The* exterior *of a convax polygon is the set of points in the plane of the polygon that are not in the polygon and not in the interior of the polygon.*

DEFINITIONS *Any segment that joins two nonconsecutive vertices of a convax polygon is a* diagonal *of the convax polygon. Any angle that contains a pair of consecutive sides of a convax polygon is an* angle of the polygon. *Two angles of a convax polygon are* consecutive angles *of the polygon if and only if their intersection is a side of the polygon. If two angles of a convax polygon are not consecutive, they are* nonconsecutive angles.

Observe that we have defined the angles of a polygon only in the case where the polygon is convax. Accordingly, when we speak of the angles of a polygon, we are assuming the polygon is convax.

In the case of a quadrilateral, we can use the word *opposite* instead of the word *nonconsecutive,* as indicated by the following definition.

DEFINITIONS *Two vertices of a convax quadrilateral that are nonconsecutive are called* opposite vertices, *and each is said to be opposite the other. Two sides of a convax quadrilateral that are nonconsecutive are called* opposite sides, *and each is said to be opposite the other. Two angles of a convax quadrilateral that are nonconsecutive are called* opposite angles, *and each is said to be opposite the other.*

When we speak of opposite angles or opposite sides of a quadrilateral, we are assuming the quadrilateral is convax.

Exercises

1. (a) Can a point be in the exterior of a triangle and in the interior of one of the angles of the triangle?
   (b) Can a point be in the exterior of a triangle and not in the interior of any angle of the triangle?
   (c) Can a point be in an angle of the triangle and not be a point of the triangle?
   (d) Can a point be in two angles of a triangle and not be a point of the triangle?
   (e) Can a point be in the interior of two angles of a triangle and not be in the interior of the triangle?
2. Explain why polygon $P_1P_2P_3 \cdots P_n$ is equal to polygon $P_2P_3 \cdots P_nP_1$.
3. If $ABCDE$ is a convax polygon, explain why no three of the points A, B, C, D, E are collinear.

L 4. Prove:

$$\left.\begin{array}{l} B \in \overleftrightarrow{OA}/C \\ \overrightarrow{OB} \neq \overrightarrow{OC} \end{array}\right\} \longrightarrow \overrightarrow{OC} \text{ is between } \overrightarrow{OA} \text{ and } \overrightarrow{OB} \vee \overrightarrow{OB} \text{ is between } \overrightarrow{OC} \text{ and } \overrightarrow{OA}.$$

5. Prove: A triangle is not a convex set.

L 6. Prove: $\overrightarrow{BX}$ between $\overrightarrow{BA}$ and $\overrightarrow{BC} \longrightarrow C \in \overleftrightarrow{BX}/\sim A$.

7. Prove: A triangle is a convax polygon.

L 8. Prove: Every triangle separates its interior and its exterior.

L 9. Prove: If line L is in the plane of $\triangle ABC$ and contains a point of the interior of $\triangle ABC$, then L intersects $\triangle ABC$.

10. If line L intersects all of the sides of a triangle, then it contains at least one of the vertices of the triangle.
11. Use Postulate 13 (Pasch's postulate) to prove the crossbar theorem. (Suggestion: Using the drawing below, let T be a point such that T-B-C. Apply Pasch's postulate to noncollinear points T, A, and C.)

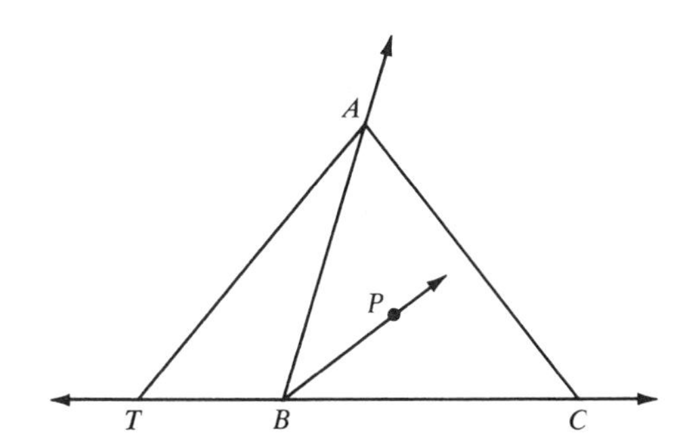

12. Prove: If the intersection of two distinct sides of a quadrilateral contains a point which is not a vertex, then the quadrilateral is not convex.

L 13. Prove: If $ABCD$ is a convex quadrilateral with the interior I, then
  (1) $\overrightarrow{AC}$ is between $\overrightarrow{AB}$ and $\overrightarrow{AD}$,
  (2) $\overset{\circ\!-\!\circ}{AC} \times \overset{\circ\!-\!\circ}{BD}(P) \wedge P \in I$.
  (3) B and D are in opposite sides of $\overleftrightarrow{AC}$,
  A and C are in opposite sides of $\overleftrightarrow{BD}$,
  (4) $\overleftrightarrow{AD} \times \overleftrightarrow{BC}(Q) \longrightarrow (B\text{-}C\text{-}Q \vee Q\text{-}B\text{-}C)$.

14. Prove:

$$\left.\begin{array}{l} ABCD \text{ is a convex quadrilateral} \\ X \in \overset{\circ\!-\!\circ}{DC} \\ Y \in \overset{\circ\!-\!\circ}{CB} \\ \overleftrightarrow{XY} \times \overleftrightarrow{AB}(T) \end{array}\right\} \longrightarrow (A\text{-}B\text{-}T \vee T\text{-}A\text{-}B).$$

15. Prove: The union of a convax quadrilateral and its interior is a convex set.

16. Prove: If $ABCDE$ is a convax polygon, then $ABCD$ is a convax polygon.

L 17. Prove: The union of a convax polygon and its interior is a convex set.

L 18. Prove: The interior of a convax polygon contains the interior of each of its diagonals and the interior of a convax polygon is not empty.

L 19. Prove: If $P_1P_2 \cdots P_n$ is a convax polygon ($n \geqq 4$), then the line containing a diagonal of the polygon intersects the polygon in exactly two points.

L 20. If $P_1P_2 \cdots P_n$ is a convax polygon ($n \geqq 4$) and R is any vertex other than P_1, P_2, P_3, then $\overrightarrow{P_2R}$ is between $\overrightarrow{P_2P_1}$ and $\overrightarrow{P_2P_3}$.

L 21. Prove: If $P_1P_2 \cdots P_n$ is a convax polygon ($n \geqq 4$), then $P_1P_3P_4 \cdots P_n$ is a convax polygon.

L 22. Prove: Every convax polygon separates its interior and its exterior.

References

Birkhoff and Beatley, *Basic Geometry*

Kay, *College Geometry*

Ringenberg, *College Geometry*

Chapter 7

Congruence and Inequalities

We know from school geometry that arguments based on congruence of triangles can be used to prove that certain pairs of "corresponding parts" are equal, or, more accurately, have equal measures. In this chapter we first refine our ideas about congruent triangles and then we capitalize on the fact that an *equality* derived from a congruence may help us to prove an *inequality*. Thus we can prove that $a > c$ if we know that $a > b$ and $b = c$. Indeed, this is precisely the proof pattern we use to prove the exterior-angle theorem, which is a very important geometric inequality to be derived later. As we proceed, we shall see that the idea of congruence is involved in many arguments designed to prove that two measures are not equal. Thus the study of congruence leads naturally to the study of geometric inequalities.

Congruent Triangles

In the physical world, our intuition tells us that two objects are congruent if they have the "same size and shape." In fact, our symbol for congruent ($\cong$) is derived from this idea, where "$=$" says "the same size" and "$\sim$" says "the same shape." If we want to determine whether two floor tiles are congruent, we can pick one of them up and place it on the other to see if they "fit" or "coincide." This method of testing is called "superposition." It is the method Euclid used in "proving" his SAS congruence theorem for triangles. This is unfortunate, because such arguments cannot serve as proofs in a formal development of geometry. Obviously, our triangles are not physical objects that can be moved about from place to place. The use of "proof by superposition" was one of several flaws in the otherwise magnificent logical structure that Euclid constructed. In our development we shall avoid the use of superposition by taking SAS as a postulate.

SAS: Given two triangles, not necessarily distinct, if two sides and the included angle of one are congruent respectively to two sides and the included angle of the other, then the triangles are congruent.

We have defined congruent segments as segments having the same length and congruent angles as angles having the same measure. We now define congruent triangles in terms of congruent segments and congruent angles. In order to do this, we need to consider the one-to-one correspondences that can be set up between the vertices of two triangles.

We shall use the symbol $\leftrightarrow$ for the phrase "corresponds to." Given two triangles, there are six different ways to establish a one-to-one correspondence between their vertices. For example, given triangles ABC and XYZ, we can pair their vertices as follows:

$A \leftrightarrow X$	$A \leftrightarrow X$	$A \leftrightarrow Y$	$A \leftrightarrow Z$	$A \leftrightarrow Y$	$A \leftrightarrow Z$
$B \leftrightarrow Y$	$B \leftrightarrow Z$	$B \leftrightarrow X$	$B \leftrightarrow X$	$B \leftrightarrow Z$	$B \leftrightarrow Y$
$C \leftrightarrow Z$	$C \leftrightarrow Y$	$C \leftrightarrow Z$	$C \leftrightarrow Y$	$C \leftrightarrow X$	$C \leftrightarrow X$

Since each angle has a unique vertex, any correspondence between the vertices of two triangles establishes a correspondence between the angles of the triangles. Since the sides of a triangle are determined by its vertices, a correspondence between the vertices of two triangles establishes a correspondence between their sides.

Notation: If ABC and XYZ are triangles, we shall use any of the symbols

$$A, B, C \leftrightarrow X, Y, Z; \quad A, C, B, \leftrightarrow X, Z, Y; \quad B, A, C, \leftrightarrow Y, X, Z;$$
$$B, C, A \leftrightarrow Y, Z, X; \quad C, A, B \leftrightarrow Z, X, Y; \quad C, B, A \leftrightarrow Z, Y, X$$

to denote the correspondence in which

$$\begin{array}{lll} A \leftrightarrow X, & \angle A \leftrightarrow \angle X, & \overline{AB} \leftrightarrow \overline{XY}, \\ B \leftrightarrow Y, & \angle B \leftrightarrow \angle Y, & \overline{BC} \leftrightarrow \overline{YZ}, \\ C \leftrightarrow Z, & \angle C \leftrightarrow \angle Z, & \overline{CA} \leftrightarrow \overline{ZX}. \end{array}$$

Observe that while $A, B, C \leftrightarrow X, Y, Z$ and $B, C, A \leftrightarrow Y, Z, X$ denote the same correspondence, $A, B, C \leftrightarrow X, Y, Z$ and $A, B, C \leftrightarrow Y, Z, X$ do not.

DEFINITIONS *Given a correspondence between two triangles, the angles that correspond are called* corresponding angles *and the sides that correspond are called* corresponding sides. *The sides and angles of the triangle are called the* parts *of the triangle.*

DEFINITION *If there exists a correspondence between the vertices of two triangles in which the corresponding angles are congruent and the corresponding sides are congruent, this correspondence is called a* congruent correspondence *or* congruence, *and the triangles are said to be* congruent.

Notation: If $A, B, C \leftrightarrow X, Y, Z$ is a congruent correspondence for triangles ABC and XYZ, we shall indicate this by writing $\triangle ABC \cong \triangle XYZ$, where the symbol $\cong$ is read "is congruent to." Clearly $\triangle ABC \cong \triangle XYZ \longleftrightarrow \triangle ACB \cong \triangle XZY \longleftrightarrow \triangle BAC \cong \triangle YXZ$, and so on, for any symbol where the letters X, Y, and Z are written in the same order as their corresponding letters A, B, and C, respectively.

From our definition we have $\triangle DEF \cong \triangle PQR \longleftrightarrow (\angle D \cong \angle P \wedge \angle E \cong \angle Q \wedge \angle F \cong \angle R \wedge \overline{DE} \cong \overline{PQ} \wedge \overline{EF} \cong \overline{QR} \wedge \overline{FD} \cong \overline{RP})$. Thus our definition of congruence for triangles asserts that the corresponding parts of congruent triangles are congruent.

The proof of the following theorem is required in the exercises.

THEOREM 7-1 *The congruence relation between triangles is an equivalence relation.*

We now generalize our definition of congruence so that it applies to convax polygons.

Notation: Given convax polygons $A_1A_2 \cdots A_n$ and $B_1B_2 \cdots B_n$ ($n \geqq 3$) the symbol $A_1, A_2, \ldots, A_n \leftrightarrow B_1, B_2, \ldots, B_n$ indicates a one-to-one correspondence between the vertices of these two polygons in which $\angle A_1 \leftrightarrow \angle B_1$, $\angle A_2 \leftrightarrow \angle B_2, \ldots,$ $\angle A_n \leftrightarrow \angle B_n$ and $\overline{A_1A_2} \leftrightarrow \overline{B_1B_2}$, $\overline{A_2A_3} \leftrightarrow \overline{B_2B_3}, \ldots, \overline{A_nA_1} \leftrightarrow \overline{B_nB_1}$.

DEFINITION *If there exists a correspondence between the vertices of two convax polygons in which the corresponding angles are congruent and the corresponding sides are congruent, this correspondence is called a* congruent correspondence *and the polygons are said to be* congruent.

Notation: If $A_1, A_2, \ldots, A_n \leftrightarrow B_1, B_2, \ldots, B_n$ is a congruent correspondence for polygons $A_1A_2 \cdots A_n$ and $B_1B_2 \cdots B_n$, we shall indicate this by writing poly $A_1A_2 \cdots A_n \cong$ poly $B_1B_2 \cdots B_n$.

DEFINITION *If each angle of a convax polygon is congruent to every other angle and each side is congruent to every other side, then the polygon is called a* regular polygon.

The Basic Congruence Postulate for Triangles

POSTULATE 16 (THE SAS POSTULATE) *Given two triangles, not necessarily distinct, if two sides and the included angle of one triangle are congruent respectively to two sides and the included angle of the other, then the triangles are congruent.*

DEFINITIONS *An* isosceles triangle *is a triangle having at least two congruent sides. Angles opposite congruent sides in an isosceles triangle are called* base angles; *a side common to two base angles is called a* base *and an angle opposite a base is called the* vertex angle corresponding to that base.

DEFINITION *An* equilateral triangle *is a triangle having three congruent sides.*

An important theorem about isosceles triangles is directly obtainable from Postulate 16. Consider a triangle ABC in which $\overline{AB} \cong \overline{BC}$. We have

$$\overline{AB} \cong \overline{BC} \xrightarrow{(1)} \left.\begin{matrix} \left\{\begin{matrix} \overline{AB} \cong \overline{CB} \\ \overline{BC} \cong \overline{BA} \end{matrix}\right. \\ (2)\ \angle B \cong \angle B \end{matrix}\right\} \xrightarrow{(3)} \triangle ABC \cong \triangle CBA \xrightarrow{(4)} \angle A \cong \angle C.$$

Observe that the reason for (3) is the SAS postulate and the reason for (4) is our definition of congruence for triangles. Since, according to our definition, $\triangle ABC$ is an isosceles triangle in which $\angle A$ and $\angle C$ are base angles, we have the following theorem.

THEOREM 7-2 (THE ISOSCELES-TRIANGLE THEOREM) *If a triangle is isosceles, then its base angles are congruent.*

From Theorem 4-9 we know that a segment has a unique midpoint. Theorem 6-21 tells us that, given a segment in plane M, there is in M a unique perpendicular to that segment at its midpoint. Thus every segment has a unique perpendicular bisector in each plane that contains it. The SAS postulate and a contrapositive of Theorem 7-2 enable us to prove the following theorem.

THEOREM 7-3 (THE PERPENDICULAR-BISECTOR THEOREM) *If a plane contains a segment, then the set of points that are in that plane and are equidistant from the endpoints of the segment is the perpendicular bisector of the segment in that plane.*

The proofs of Theorem 7-3 and its corollary are left as exercises.

COROLLARY TO THEOREM 7-3 *If two distinct points are each equidistant from the endpoints of a segment and are coplanar with it, they determine the perpendicular bisector of the segment in that plane.*

Exterior-Angle Theorem

We are now ready to define exterior angles for a triangle and to prove the exterior-angle theorem, which has many important consequences.

In the figure below, three lines intersect to form $\triangle ABC$. $\overleftrightarrow{AB}$ and $\overleftrightarrow{AC}$ form the linear pair of angles RAC and CAB. Also $\overleftrightarrow{AC}$ and $\overleftrightarrow{AB}$ form the linear pair of angles CAB and BAS. Similarly, each of the other angles of the triangle belong to two linear pairs.

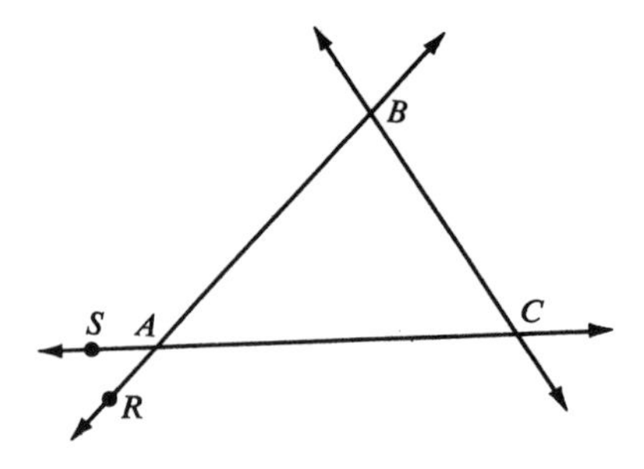

DEFINITION *An angle is an* exterior angle of a triangle *if and only if it forms a linear pair with an angle of the triangle.*

Thus, in the preceding figure, $\angle RAC$ and $\angle BAS$ are exterior angles of $\triangle ABC$. Sometimes, for emphasis, we speak of the angles of a triangle as *interior angles* of the triangle. Note that each exterior angle is adjacent to the interior angle with which it forms a linear pair. It is nonadjacent to the other interior angles of the triangle.

Recall that

$$(S\text{-}A\text{-}C \wedge B \notin \overleftrightarrow{AC}) \longleftrightarrow B\underline{A}S \text{ and } B\underline{A}C \text{ are a linear pair of angles.}$$

By our definition of exterior angle we have

$$B\underline{A}S \text{ is an exterior angle for } \triangle ABC \longleftrightarrow B\underline{A}S \text{ and } B\underline{A}C \text{ are a linear pair of angles.}$$

It follows that

$$B\underline{A}S \text{ is an exterior angle for } \triangle ABC \longleftrightarrow (S\text{-}A\text{-}C \wedge B \notin \overleftrightarrow{AC}).$$

THEOREM 7-4 (EXTERIOR-ANGLE THEOREM) *The measure of an exterior angle of a triangle is greater than the measure of either of the angles of the triangle that are nonadjacent to it.*

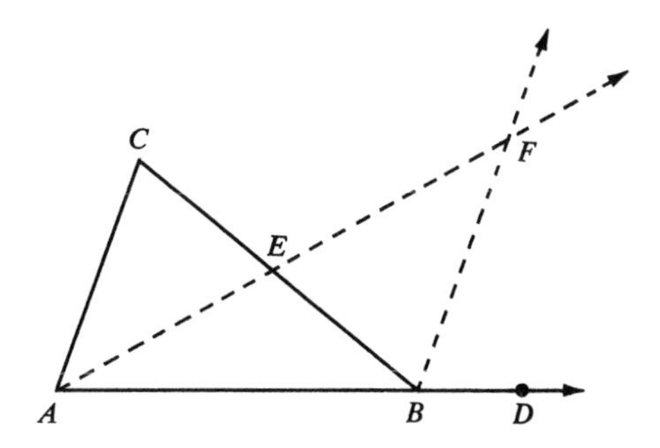

We must prove:

$$C\underline{B}D \text{ is an exterior angle of } \triangle ABC \longrightarrow \begin{cases} m\angle CBD > m\angle C \\ m\angle CBD > m\angle A. \end{cases}$$

PROOF

<table>
<tr><td>1. There is a point E such that E is the midpoint of $\overline{BC}$.</td><td>1. Theorem 4-9.</td></tr>
<tr><td>2. $|CE| = |BE|$ and $C\text{-}E\text{-}B$.</td><td>2. Statement 1 and definition of midpoint.</td></tr>
<tr><td>3. There is a point F in the interior of the ray opposite $\overrightarrow{EA}$ such that $|AE| = |EF|$.</td><td>3. Theorem 4-6 (point plotting theorem).</td></tr>
<tr><td>4. $A\text{-}E\text{-}F$.</td><td>4. Statement 3 and definition of opposite rays.</td></tr>
<tr><td>5. $A\text{-}B\text{-}D$ and $C \notin \overleftrightarrow{AB}$.</td><td>5. Hypothesis and definition of exterior angle of a triangle.</td></tr>
<tr><td>6. $B\underline{A}C$ is an angle.</td><td>6. Statement 5 and definition of angle.</td></tr>
<tr><td>7. $\overrightarrow{BF}$ is between $\overrightarrow{BC}$ and $\overrightarrow{BD}$.</td><td>7. Statements 2, 4, 5, 6, and Theorem 6-9.</td></tr>
</table>

Statements	Reasons
8. $m\angle CBD > m\angle CBF$.	8. Statement 7 and Corollary 1 to Theorem 6-4.
9. $\overline{CE} \cong \overline{BE}$ and $\overline{AE} \cong \overline{EF}$.	9. Statements 2, 3, and definition of congruent segments.
10. $A \notin \overleftrightarrow{BC}$.	10. Hypothesis (ABC is a triangle) and definition of triangle.
11. $\overleftrightarrow{AF} \neq \overleftrightarrow{BC}$.	11. Statement 10 and ST_3.
12. AEC and FEB are vertical angles.	12. Statements 2, 4, 11, and definition of vertical angles.
13. $\angle AEC \cong \angle FEB$.	13. Statement 12 and Theorem 6-19.
14. $C \notin \overleftrightarrow{AE}$ and $B \notin \overleftrightarrow{EF}$.	14. Statement 12 (AEC and FEB are angles) and definition of angle.
15. AEC and FEB are triangles.	15. Statement 14 and definition of triangle.
16. $\triangle AEC \cong \triangle FEB$.	16. Statements 9, 13, 15, and SAS postulate.
17. $\angle C \cong \angle FBC$.	17. Statement 16 and definition of congruent triangles.
18. $m\angle C = m\angle FBC$.	18. Statement 17 and definition of congruent angles.
19. $m\angle CBD > m\angle C$.	19. Statements 8, 18, and substitution property of equality.

To prove that $m\angle CBD > m\angle A$, we let G be the midpoint of $\overline{AB}$, let H be the point in $\overset{\circ}{\overrightarrow{CG}}$ such that $\overrightarrow{GC}$ and $\overrightarrow{GH}$ are opposite rays and $\overline{GC} \cong \overline{GH}$, and let $\overrightarrow{BK}$ be the ray opposite $\overrightarrow{BC}$. We then proceed as in the preceding proof. Finally, we show that since $m\angle ABK = m\angle CBD$, then $m\angle CBD > m\angle A$.

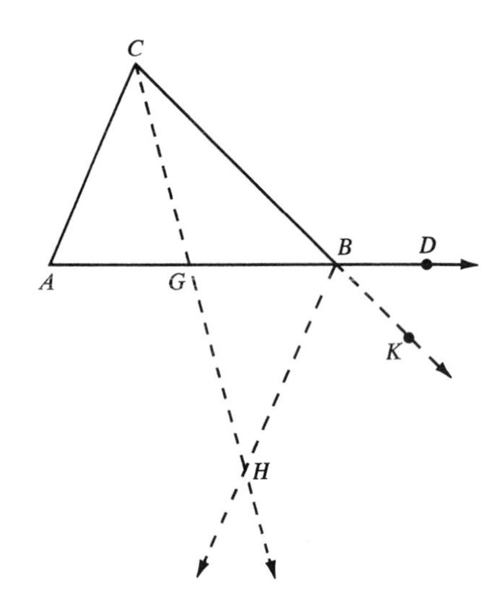

The proofs of the following corollaries to Theorem 7-4 are left as exercises.

COROLLARY 1 TO THEOREM 7-4 *The sum of the measures of any two angles of a triangle is less than* 180.

COROLLARY 2 TO THEOREM 7-4 *If one angle of a triangle is a right angle, then each of the other two angles of the triangle is acute.*

COROLLARY 3 TO THEOREM 7-4 *Each base angle of an isosceles triangle is acute.*

Exercises

1. Prove Theorem 7-1.
2. If A, B, and C are the vertices of triangle T_1 and X, Y, and Z are the vertices of triangle T_2, explain why the following implication is false:

$$T_1 \cong T_2 \longleftrightarrow (A, B, C \leftrightarrow X, Y, Z \text{ is a congruent correspondence}).$$

3. Prove Theorem 7-3.
4. Prove the corollary to Theorem 7-3.
5. Prove:
   (a) quad $ABCD \cong$ quad $BCDA \longrightarrow \overline{AC} \cong \overline{BD}$.
   (b) quad $ABCD \cong$ quad $CDAB \longrightarrow \angle DAC \cong \angle ACB$.
   (c) quad $ABCD \cong$ quad $ADCB \longrightarrow \overleftrightarrow{AC}$ is a perpendicular bisector of $\overline{BD}$.
   (d) quad $ABCD \cong$ quad $DCBA \longrightarrow \overline{AC} \cong \overline{BD}$.
   (e) quad $ABCD \cong$ quad $ADCB \longrightarrow \overrightarrow{AC}$ bisects $\angle DAB$.
   (f) poly $ABCDEF \cong$ poly $BCDEFA \longrightarrow DFB$ is an equilateral triangle.
6. Prove: Poly $A_1A_2 \cdots A_n \cong$ poly $A_2A_3 \cdots A_nA_1 \longrightarrow$ poly $A_1A_2 \cdots A_n$ is a regular polygon.
7. Prove: Poly $ABCDEF \cong$ poly $AFEDCB \longrightarrow \overleftrightarrow{AD}$ is a perpendicular bisector of $\overline{BF}$ and $\overline{EC}$.
8. Prove:
   (a) $(\triangle ABC \cong \triangle RST \wedge \overline{AB} \cong \overline{BC}) \longrightarrow \triangle ABC \cong \triangle TSR$.
   (b) $(\triangle ABC \cong \triangle RST \wedge \overline{AB} \neq \overline{BC}) \longrightarrow \triangle ABC \not\cong \triangle TSR$.
9. Write an indirect essay proof of Corollary 1 to Theorem 7-4.
10. Write a flow-diagram proof of Corollary 1 to Theorem 7-4.
11. Prove Corollaries 2 and 3 to Theorem 7-4.
12. Prove the isosceles-triangle theorem by proving

$$\left.\begin{array}{l} ABC \text{ is a triangle} \\ \overline{AB} \cong \overline{AC} \\ A\text{-}B\text{-}D \\ A\text{-}C\text{-}E \\ \overline{BD} \cong \overline{CE} \end{array}\right\} \longrightarrow \angle ABC \cong \angle ACB.$$

Triangle Inequalities

The exterior-angle theorem states a very important geometric inequality. We have already proved some other inequality statements. Some of these are listed here for reference.

(i) Corollary to Theorem 4-2:

$$A\text{-}B\text{-}C \longrightarrow (|AB| < |AC| \wedge |BC| < |AC|).$$

(ii) Corollary to Theorem 4-3:

$$(B \in \overleftrightarrow{AC} \wedge |AB| < |AC| \wedge |BC| < |AC|) \longrightarrow A\text{-}B\text{-}C.$$

(iii) Corollary 3 to Theorem 4-4:

$$(B \in \overrightarrow{AC} \wedge 0 < |AB| < |AC|) \longrightarrow B \in \overset{\circ\!-\!\circ}{AC}.$$

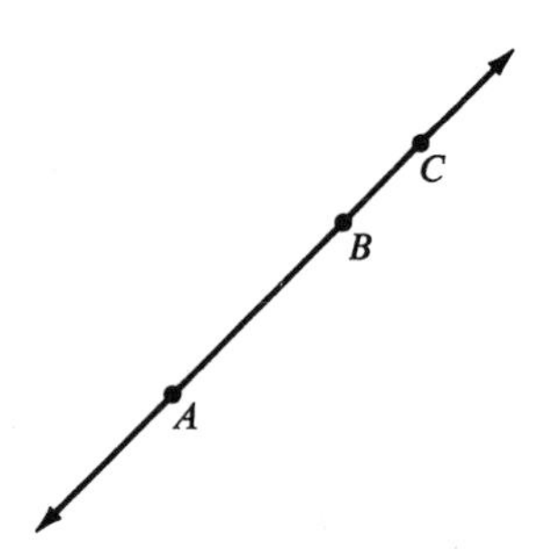

(iv) Corollary 1 to Theorem 6-4 and Theorem 6-6:

B is in the interior of $\angle AOC \dashv$
$$\leftrightarrow (m \angle AOB < m \angle AOC \wedge m \angle BOC < m \angle AOC).$$

(v) Corollary 2 to Theorem 6-4 and Theorem 6-6:

$$(B \in \overleftrightarrow{OA}/C \wedge m \angle AOB < m \angle AOC) \longrightarrow B \text{ is in the interior of } \angle AOC.$$

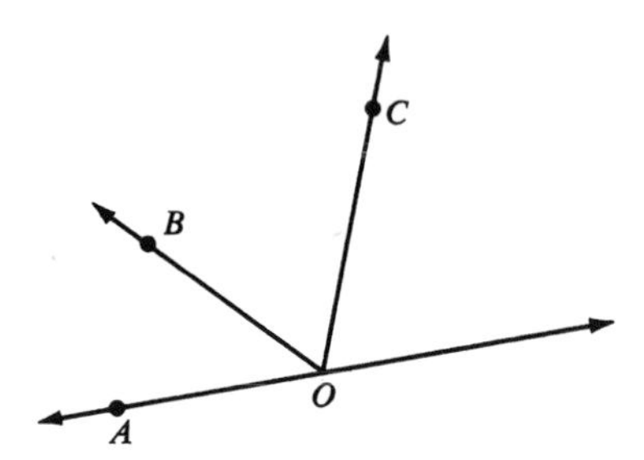

Other inequalities involving the parts of a triangle are suggested by modification of the hypotheses of previously proved theorems. For example, the isosceles-triangle theorem tells us that in $\triangle RST$

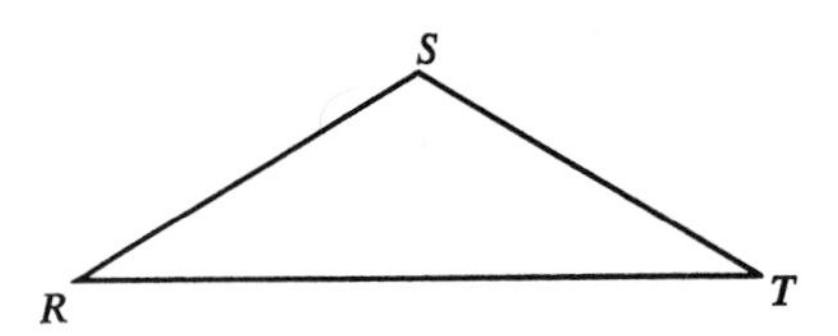

$$|ST| = |SR| \longrightarrow m \angle R = m \angle T.$$

If $|ST| > |SR|$, it seems reasonable to suppose that $m \angle R > m \angle T$. To prove this we must show

$$\left.\begin{array}{l} RST \text{ is a triangle} \\ |ST| > |SR| \end{array}\right\} \longrightarrow m \angle R > m \angle T.$$

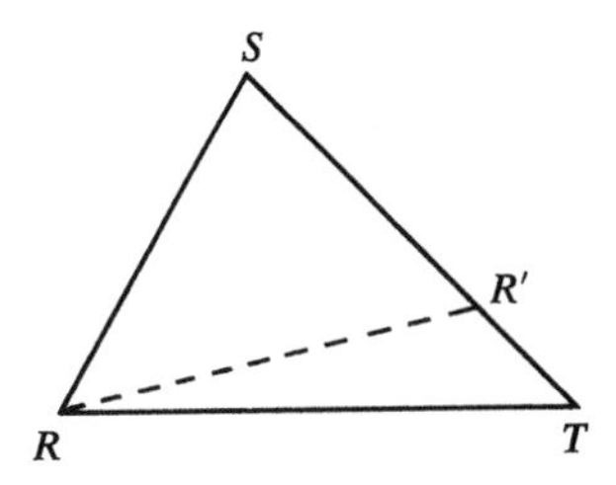

PROOF

Statement	Reason
1. $R \notin \overleftrightarrow{ST}$.	1. Hypothesis and definition of a triangle.
2. $R \neq S$.	2. Statement 1 and L-Exercise 6, page 68.
3. $\lvert RS \rvert > 0$.	3. Statement 2 and Postulate 10 (distance postulate).
4. There is a point $R' \in \overset{\circ\!\!\rightarrow}{ST}$ such that $\lvert R'S \rvert = \lvert RS \rvert$.	4. Theorem 4-6 (point-plotting theorem).
5. $\lvert R'S \rvert < \lvert ST \rvert$.	5. Hypothesis, statement 4, and substitution property of equality.
6. $R' \in \overset{\circ\!-\!\circ}{ST}$.	6. Statements 3, 4, 5, and (iii) above.
7. R' is in the interior of $\angle SRT$.	7. Statement 6 and Theorem 6-7.
8. $m\angle SRT > m\angle SRR'$.	8. Statement 7 and (iv) above.
9. $\overleftrightarrow{SR'} = \overleftrightarrow{R'T} = \overleftrightarrow{ST}$.	9. Statement 6 and Corollary 2 to Postulate 2.
10. $R \notin \overleftrightarrow{SR'} \wedge R \notin \overleftrightarrow{R'T}$.	10. Statements 1, 9, and substitution property of equality.
11. RSR' is an isosceles triangle.	11. Statements 4, 10 ($R \notin \overleftrightarrow{SR'}$), and definition of isosceles triangle.
12. $m\angle SRR' = m\angle SR'R$.	12. Statement 11, Theorem 7-2, and definition of congruent angles.
13. $m\angle SRT > m\angle SR'R$.	13. Statements 8, 12, and substitution property of equality.
14. $\angle SR'R$ is an exterior angle of $\triangle RR'T$.	14. Statements 6, 10 ($R \notin \overleftrightarrow{R'T}$), and definition of exterior angle of a triangle.
15. $m\angle SR'R > m\angle T$.	15. Statement 14 and Theorem 7-4.
16. $m\angle SRT > m\angle T$.	16. Statement 13, 15, and transitive property of order.

We have proved the following theorem.

THEOREM 7-5 *If two sides of a triangle are not congruent, then the angles opposite these sides are not congruent and the larger angle is opposite the longer side.*

At this point we apply a tautology called the "law of converses" to establish a converse of Theorem 7-5. We introduce this tautology by means of an example.

Suppose that (1) every person who enters a stadium passes through exactly one of the three gates A, B, C; (2) if a person enters by gate A, then his hand is stamped with an X; (3) if a person enters by gate B, then his hand is stamped with a Y; (4) if a person enters by gate C, then his hand is stamped with a Z.

At this point we might ask what we know about a person in the stadium whose hand is marked with a Z. We might be tempted to say that this person must have entered by gate C. However, we cannot be sure of this. For all we know, some of the people who entered gate A might have been stamped with a Z in addition to being stamped with an X. In order to be sure of an answer to our question, we need one more condition, namely (5) every person who enters the stadium is stamped with only one mark. Now we can be certain that, under the conditions stated, each of the converses of statements (2), (3), and (4) is true.

Since this example illustrates an important proof pattern known as the law of converses, we generalize it by employing symbols. If we let a represent the statement "a person enters the stadium by gate A," b represent the statement "a person enters the stadium by gate B," and so on, and x represent the statement "a person bears an X mark," and so on, our argument has this form:

$$\left.\begin{array}{l} (1)\ a \veebar b \veebar c \\ (2)\ a \longrightarrow x \\ (3)\ b \longrightarrow y \\ (4)\ c \longrightarrow z \\ (5)\ x \veebar y \veebar z \end{array}\right\} \longrightarrow \left\{\begin{array}{l} x \longrightarrow a \\ y \longrightarrow b \\ z \longrightarrow c. \end{array}\right.$$

Letting A represent the conjunction of premises (1), (2), (3), (4), and (5), we seek to prove among other things that $A \longrightarrow (x \longrightarrow a)$. We can do this by proving that $(x \wedge A) \longrightarrow a$. (See Exercise 7, page 34.)

$$\left.\begin{array}{l} \left.\begin{array}{l} b \longrightarrow y \\ \left.\begin{array}{l} x \\ x \veebar y \veebar z \end{array}\right\} \longrightarrow \left\{\begin{array}{l} \sim y \\ \sim z \end{array}\right. \\ c \longrightarrow z \end{array}\right\} \longrightarrow \begin{array}{l} \sim b \\ \sim c \end{array} \\ a \veebar b \veebar c \end{array}\right\} \longrightarrow a.$$

Observe that we did not use premise (2) in this proof. However, the presence of premise (2) does not affect the proof. The conclusions $y \longrightarrow b$ and $z \longrightarrow c$ can be established by similar proofs, each of which would appeal to premise (2).

A close examination of the law of converses will reveal that we can obtain another tautology from it by replacing (1) with $a \vee b \vee c$ and premise (5) with the statement "Not more than one of the statements x, y, and z is true."

Now we consider the following theorem, which is a converse of Theorem 7-5.

THEOREM 7-6 *If two angles of a triangle are not congruent, then the sides opposite these angles are not congruent and the longer side is opposite the larger angle.*

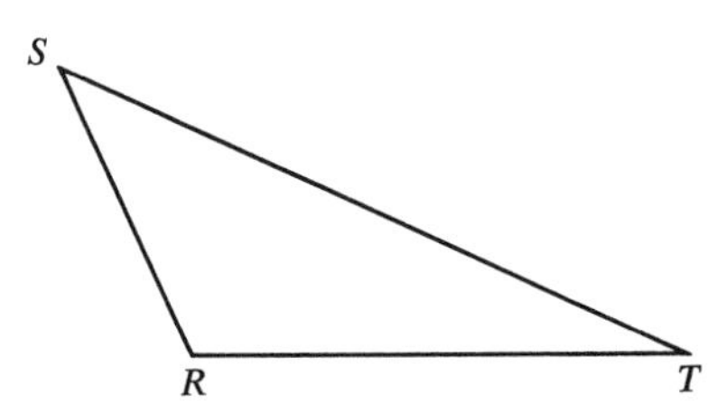

In terms of our drawing we must prove that in $\triangle RST, m\angle R > m\angle T \dashv$ $\nrightarrow |ST| > |RS|$. We proceed to apply the law of converses as follows: RST is a triangle $\dashv$

$$\left.\begin{array}{l} (1)\ |ST| > |RS| \veebar |ST| = |RS| \veebar |ST| < |RS| \\ \nrightarrow \left\{\begin{array}{l} (2)\ |ST| > |RS| \longrightarrow m\angle R > m\angle T \\ (3)\ |ST| = |RS| \longrightarrow m\angle R = m\angle T \\ (4)\ |ST| < |RS| \longrightarrow m\angle R < m\angle T \end{array}\right. \\ (5)\ m\angle R > m\angle T \veebar m\angle R = m\angle T \veebar m\angle R < m\angle T \end{array}\right\} \overset{(6)}{\dashv}$$

$$\nrightarrow \left\{\begin{array}{l} m\angle R > m\angle T \longrightarrow |ST| > |RS| \\ m\angle R = m\angle T \longrightarrow |ST| = |RS| \\ m\angle R < m\angle T \longrightarrow |ST| < |RS|. \end{array}\right.$$

Reasons: (1) Trichotomy, (2) Theorem 7-5, (3) Theorem 7-2, (4) Theorem 7-5, (5) trichotomy, and (6) law of converses.

We have proved:

$$RST \text{ is a triangle} \longrightarrow (m\angle R > m\angle T \longrightarrow |ST| > |RS|).$$

Since $a \longrightarrow (b \longrightarrow c)$ is equivalent to $(a \wedge b) \longrightarrow c$, we have proved that

$$(RST \text{ is a triangle} \wedge m\angle R > m\angle T) \longrightarrow |ST| > |RS|.$$

The verbal statement of this implication is equivalent to Theorem 7-6.

Theorem 7-6 has several important corollaries. Before we state these, we pause to observe that in the course of proving Theorem 7-6 by the law of converses, we also established the following implication:

$$(RST \text{ is a triangle} \wedge m\angle R = m\angle T) \longrightarrow |ST| = |RS|.$$

Thus we have a proof of the converse of the isosceles-triangle theorem as an unexpected bonus derived from our proof of Theorem 7-6.

THEOREM 7-7 *If two angles of a triangle are congruent, then the sides opposite these angles are congruent.*

DEFINITIONS *A* right triangle *is a triangle having one right angle. The side opposite the right angle in a right triangle is called the* hypotenuse, *and the other two sides are called the* legs.

We now state five corollaries to Theorem 7-6.

COROLLARY 1 TO THEOREM 7-6 *The hypotenuse of a right triangle is the longest side of the triangle.*

COROLLARY 2 TO THEOREM 7-6 *The shortest segment joining a point P to a line L not containing P is the segment PX, where $X \in L$ and $\overline{PX} \perp L$.*

COROLLARY 3 TO THEOREM 7-6 *The length of the line segment joining a point* *in the interior of one side of a triangle to the opposite vertex is less than the length of at least one of the other two sides.*

COROLLARY 4 TO THEOREM 7-6 *If ABC is an isosceles triangle with base $\overline{AB}$ and X is a point in $\overleftrightarrow{AB}$, then*

(1) $|CX| < |CB| \longleftrightarrow A\text{-}X\text{-}B$,
(2) $|CX| = |CB| \longleftrightarrow X \in \{A, B\}$,
(3) $|CX| > |CB| \longleftrightarrow A\text{-}B\text{-}X$ *or* $B\text{-}A\text{-}X$.

COROLLARY 5 TO THEOREM 7-6 *If the length of one side of a triangle is greater than or equal to the length of each of the other sides, then the line perpendicular to the line containing this side from the opposite vertex intersects this line in a point that is an interior point for the side.*

The following theorem can be proved by applying Theorem 6-6 and Theorem 7-2.

THEOREM 7-8 (THE TRIANGLE INEQUALITY) *The sum of the lengths of any two sides of a triangle is greater than the length of the third side.*

We must prove

$$RST \text{ is a triangle} \longrightarrow \begin{cases} |RT| < |RS| + |ST| \\ |RS| < |RT| + |TS| \\ |TS| < |RS| + |RT|. \end{cases}$$

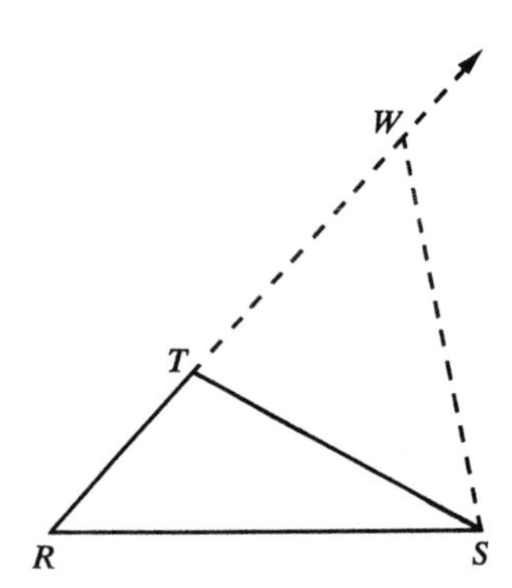

PROOF

1. In the interior of the ray opposite $\overrightarrow{TR}$ there is a point W such that $\lvert TW\rvert = \lvert TS\rvert$.	1. Theorem 4-6 (point-plotting theorem).
2. $R\text{-}T\text{-}W$ ($T \in \overset{\circ\!-\!\circ}{RW}$)	2. Statement 1 and definition of opposite rays.
3. $S \notin \overleftrightarrow{RT}$.	3. Hypothesis and definition of a triangle.
4. $\overleftrightarrow{RT} = \overleftrightarrow{RW}$, $\overleftrightarrow{RT} = \overleftrightarrow{TW}$.	4. Statement 2 and Corollary 2 to Postulate 2.

Statement	Reason
5. $S \notin \overleftrightarrow{RW}, S \notin \overleftrightarrow{TW}$.	5. Statements 3, 4, and substitution property of equality.
6. RSW is a triangle, TSW is a triangle.	6. Statement 5 and definition of a triangle.
7. T is in the interior of $\angle RSW$.	7. Statement 2 and Theorem 6-7.
8. $m\angle RSW > m\angle TSW$.	8. Statement 7 and (iv), page 152.
9. $m\angle W = m\angle TSW$.	9. Statements 1, 6, and Theorem 7-2 (isosceles-triangle theorem).
10. $\mathrm{m}\angle \mathrm{RSW} > \mathrm{m}\angle \mathrm{W}$.	10. Statements 8, 9, and substitution property of equality.
11. $\lvert RW\rvert > \lvert RS\rvert$.	11. Statements 6, 10, and Theorem 7-6.
12. $\lvert RW\rvert = \lvert RT\rvert + \lvert TW\rvert$.	12. Statement 2 and Theorem 4-2.
13. $\lvert RW\rvert = \lvert RT\rvert + \lvert TS\rvert$.	13. Statements 1, 12, and substitution property of equality.
14. $\lvert RT\rvert + \lvert TS\rvert > \lvert RS\rvert$.	14. Statements 11, 13, and substitution property of equality.

The other two inequalities in our conclusion can be proved in a similar manner.

Hinge Theorems—Inequalities Involving Two Triangles

We now apply the SAS postulate and the law of converses to prove the hinge theorems and the SSS theorem. For triangles ABC and $A'B'C'$ the SAS postulate implies

$$\left.\begin{array}{l}\overline{AB} \cong \overline{A'B'}\\ \overline{BC} \cong \overline{B'C'}\end{array}\right\} \longrightarrow (\angle B \cong \angle B' \longrightarrow \overline{AC} \cong \overline{A'C'}).$$

It seems reasonable to suppose that if $\angle B \not\cong \angle B'$, then $\overline{AC} \not\cong \overline{A'C'}$. In fact, we might even conjecture that

$$\left.\begin{array}{l}\overline{AB} \cong \overline{A'B'}\\ \overline{BC} \cong \overline{B'C'}\end{array}\right\} \longrightarrow (m\angle B > m\angle B' \longrightarrow \lvert AC\rvert > \lvert A'C'\rvert).$$

This conjecture is expressed in the following theorem.

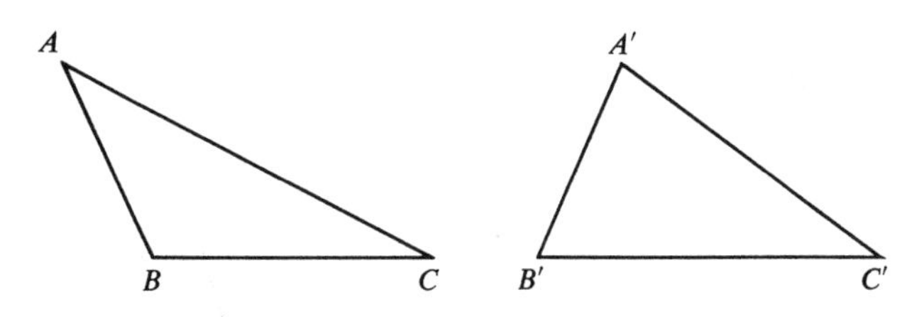

THEOREM 7-9 *If two sides of one triangle are congruent respectively to two sides of a second triangle and if the measure of the included angle of the first triangle is greater than the measure of the included angle of the second, then the third side of the first triangle is greater than the third side of the second triangle.*

We must prove:

$$\left.\begin{array}{l}(1)\ ABC \text{ is a triangle}\\ (2)\ A'B'C' \text{ is a triangle}\\ (3)\ \overline{AB} \cong \overline{A'B'}\\ (4)\ \overline{BC} \cong \overline{B'C'}\\ (5)\ m\angle ABC > m\angle A'B'C'\end{array}\right\} \longrightarrow |AC| > |A'C'|.$$

PROOF

By the angle-construction theorem there is a ray BW with $W \in \overleftrightarrow{BC}/A$ such that $m\angle CBW = m\angle A'B'C'$. Hence $m\angle CBW < m\angle CBA$. According to (v), page 152, we have W in the interior of $\angle CBA$. By the point-plotting theorem there is a point $D \in \overset{\circ}{\overrightarrow{BW}}$ such that $|BD| = |A'B'|$. Then by the SAS postulate $\triangle A'B'C' \cong \triangle DBC$. Either $D \in \overleftrightarrow{AC}$, as shown in figure (a), or $D \notin \overleftrightarrow{AC}$, as shown in figures (b) and (c).

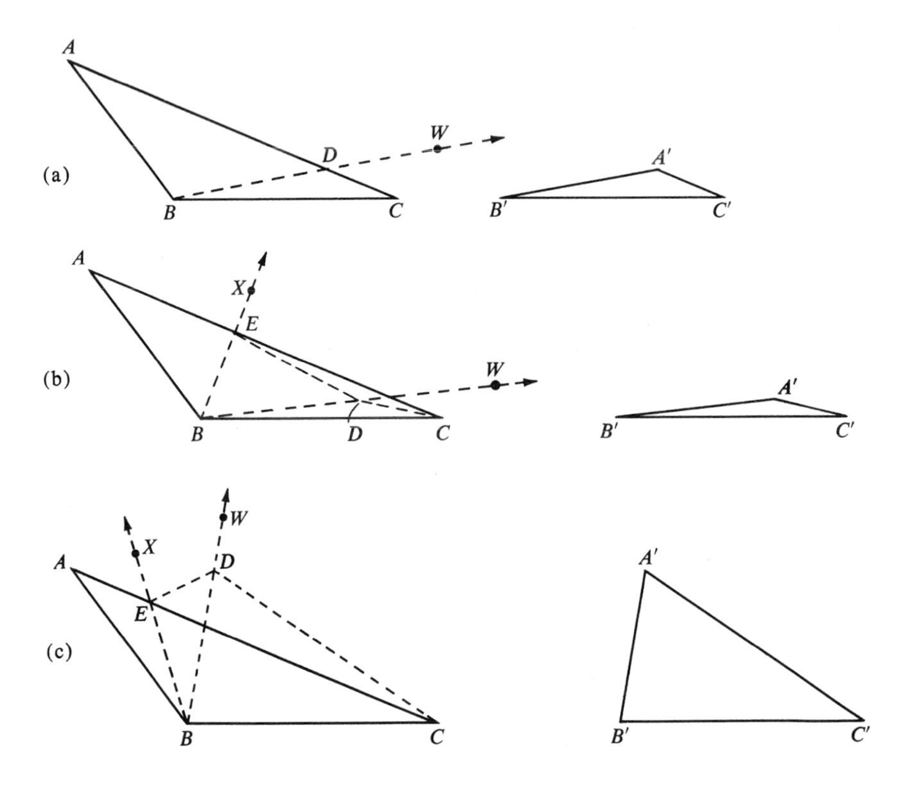

Case 1: $D \in \overleftrightarrow{AC}$. Since W is in the interior of $\angle CBA$, $\overset{\circ}{\overrightarrow{BW}} \times \overset{\circ\ \circ}{AC}$ by the crossbar theorem. Thus $D \in \overset{\circ\ \circ}{AC}$, and by (i), page 151, we have $|AC| > |DC|$. Since $|CD| = |A'C'|$, we have $|AC| > |A'C'|$, and our proof is complete.

Case 2: $D \notin \overleftrightarrow{AC}$. According to Theorem 6-5 there is a ray BX that bisects $\angle DBA$. Since $\overset{\circ}{\overrightarrow{BX}}$ lies in the interior of $\angle ABC$, we can again apply the crossbar theorem to prove that $\overset{\circ\!\!\longrightarrow}{BX}$ intersects $\overset{\circ\!\!-\!\!\circ}{AC}$ in a point we call E. $\triangle AEB \cong \triangle DEB$ by the SAS postulate. Hence $\overline{AE} \cong \overline{ED}$. Since $E \in \overset{\circ\!\!-\!\!\circ}{AC}$, we know that (a) $|AC| = |AE| + |EC|$. Since $D \notin \overleftrightarrow{EC}$, we know that EDC is a triangle. Therefore, we apply the triangle-inequality theorem (Theorem 7-8) to prove that (b) $|DE| + |EC| > |DC|$. Statements (a) and (b) imply that $|AC| > |DC|$. Since $|DC| = |A'C'|$, we have $|AC| > |A'C'|$.

The most useful of the five converses of Theorem 7-9 is the one obtained by exchanging the conclusion and premise (5).

$$\left.\begin{array}{l}(1)\ ABC \text{ is a triangle}\\(2)\ A'B'C' \text{ is a triangle}\\(3)\ \overline{AB} \cong \overline{A'B'}\\(4)\ \overline{BC} \cong \overline{B'C'}\\(5)\ |AC| > |A'C'|\end{array}\right\} \longrightarrow m\angle ABC > m\angle A'B'C'.$$

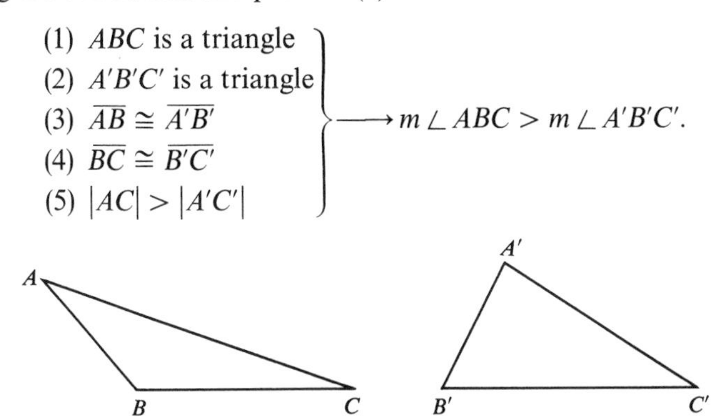

Using Theorems 7-7 and 7-9 and trichotomy, we have

$$\left.\begin{array}{l}(1)\\(2)\\(3)\\(4)\end{array}\right\} \longrightarrow \left.\begin{array}{l} m\angle B > m\angle B' \veebar m\angle B = m\angle B' \veebar m\angle B < m\angle B' \\ \left\{\begin{array}{l} m\angle B > m\angle B' \longrightarrow |AC| > |A'C'| \\ m\angle B = m\angle B' \longrightarrow |AC| = |A'C'| \\ m\angle B < m\angle B' \longrightarrow |AC| < |A'C'| \end{array}\right. \\ |AC| > |A'C'| \veebar |AC| = |A'C'| \veebar |AC| < |A'C'| \end{array}\right\} \dashv$$

$$\leftrightarrow \left\{\begin{array}{l} |AC| > |A'C'| \longrightarrow m\angle B > m\angle B' \\ |AC| = |A'C'| \longrightarrow m\angle B = m\angle B' \\ |AC| < |A'C'| \longrightarrow m\angle B < m\angle B'. \end{array}\right.$$

Thus we have proved the following implication: (ABC and $A'B'C'$ are triangles $\wedge\ \overline{AB} \cong \overline{A'B'} \wedge \overline{BC} \cong \overline{B'C'}$) $\longrightarrow$ ($|AC| > |A'C'| \longrightarrow m\angle B > m\angle B'$).

Since this implication is equivalent to (ABC and $A'B'C'$ are triangles $\wedge\ \overline{AB} \cong \overline{A'B'} \wedge \overline{BC} \cong \overline{B'C'} \wedge |AC| > |A'C'|$) $\longrightarrow m\angle B > m\angle B'$, we have proved the following theorem.

***THEOREM* 7-10** *If two sides of a triangle are congruent respectively to two sides of a second triangle and if the third side of the first triangle is longer than the third side of the second triangle, then the angles opposite these third sides are unequal, with the greater angle opposite the greater side.*

When we examine the proof of Theorem 7-10, we discover that the law of converses has again given us an important bonus. Our proof also establishes a second implication—namely, (ABC and $A'B'C'$ are triangles $\wedge\ \overline{AB} \cong \overline{A'B'} \wedge \overline{BC} \cong \overline{B'C'} \wedge \overline{AC} \cong \overline{A'C'}) \longrightarrow \angle B \cong \angle B'$. However, the conjunction of $\angle B \cong \angle B'$ with $\overline{AB} \cong \overline{A'B'}$ and $\overline{BC} \cong \overline{B'C'}$ proves that $\triangle ABC \cong \triangle A'B'C'$ by the SAS postulate. Thus, in the course of proving Theorem 7-10 by the law of converses, we have proved the SSS theorem, which can be verbalized as follows:

THEOREM 7-11 (SSS THEOREM) *If the three sides of one triangle are congruent respectively to the three sides of a second triangle, then the triangles are congruent.*

Exercises

1. Prove: If the midrays of angles R and S of triangle RST intersect at point O and $|ST| > |RT|$, then $|SO| > |RO|$.
2. Prove:

$$\left.\begin{array}{l} ABC \text{ is a triangle} \\ A\text{-}T\text{-}B \\ A\text{-}S\text{-}C \\ \overline{SC} \cong \overline{BT} \end{array}\right\} \longrightarrow (|AB| > |AC| \longleftrightarrow |SB| > |CT|).$$

3. Prove Corollary 1 to Theorem 7-6.
4. Prove Corollary 2 to Theorem 7-6.
5. Prove Corollary 3 to Theorem 7-6.
6. Use Corollary 3 to Theorem 7-6 and the law of converses to prove Corollary 4 to Theorem 7-6.
7. Prove Corollary 5 to Theorem 7-6.
8. Prove Theorem 7-8 by means of the crossbar theorem and the exterior-angle theorem.
9. Construct flow-diagram proofs for the following:

(a) $$\left.\begin{array}{l} BCD \text{ is a triangle} \\ |BD| > |DC| \\ A\text{-}B\text{-}C \end{array}\right\} \longrightarrow m\angle ABD > m\angle DBC.$$

(b) $$\left.\begin{array}{l} ABC \text{ is a triangle} \\ B\text{-}A\text{-}D \\ |AB| > |BC| \end{array}\right\} \longrightarrow m\angle DCB > m\angle BAC.$$

10. Prove:

(a) $$\left.\begin{array}{l} ACD \text{ is a triangle} \\ \overline{AB} \perp \overline{CD}(B) \\ |BC| < |BD| \end{array}\right\} \longrightarrow |AC| < |AD|.$$

(b) $$\left.\begin{array}{l} ACD \text{ is a triangle} \\ \overline{AB} \perp \overline{CD}(B) \\ |AC| < |AD| \end{array}\right\} \longrightarrow |BC| < |BD|.$$

11. Prove that the sum of the measures of the diagonals of a convex quadrilateral is greater than one-half the perimeter of the quadrilateral.

12. Prove: The perimeter of a quadrilateral is greater than the sum of the measures of its diagonals.
13. Prove: $O \in$ interior of triangle $RST \longrightarrow |RO| + |OS| < |RT| + |TS|$.
14. Prove: If O is in the interior of $\triangle ABC$, then $|AB| + |BC| + |CA| > |CO| + |BO| + |AO| > \frac{1}{2}(|AB| + |BC| + |CA|)$.
15. In triangles ABC and $A'B'C$, $\angle C$ and $\angle C'$ are right angles, $\overline{BC} \cong \overline{B'C'}$ and $|AB| > |A'B'|$. Prove: $m\angle A < m\angle A'$. Verbalize this statement.
16. Prove the SSS theorem by using SAS and the isosceles-triangle theorem.

Converses of the SAS Postulate

According to the SAS postulate

$$(1)\qquad \left.\begin{array}{l}\overline{AC} \cong \overline{A'C'}\\ \angle A \cong \angle A'\\ \overline{AB} \cong \overline{A'B'}\end{array}\right\} \longrightarrow \triangle ABC \cong \triangle A'B'C.$$

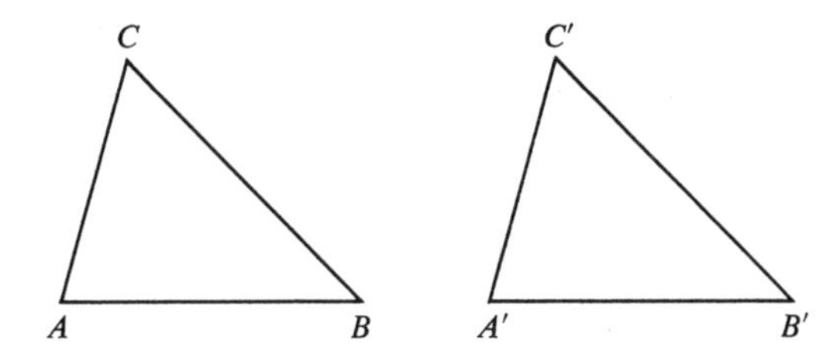

We convey this idea by saying that two sides and the included angle *determine* a triangle. It is natural to inquire what other sets of points determine a triangle. In view of our definition of congruent triangles, statement (1) is clearly equivalent to

$$(2)\qquad \left.\begin{array}{l}a\colon \overline{AC} \cong \overline{A'C'}\\ b\colon \angle A \cong \angle A'\\ c\colon \overline{AB} \cong \overline{A'B'}\end{array}\right\} \longrightarrow \left\{\begin{array}{l}x\colon \angle B \cong \angle B'\\ y\colon \overline{BC} \cong \overline{B'C'}\\ z\colon \angle C \cong \angle C'.\end{array}\right.$$

We observe that (2) is an implication in which both hypothesis and conclusion are conjunctive statements.

DEFINITION *If the hypothesis of a given implication is the conjunction of m statements and the conclusion is the conjunction of n statements, then any implication that can be formed by exchanging any number of statements in the conclusion with the same number of statements in the hypothesis is a* converse *of the given implication.*

Applying the definition, we see that (2) has nineteen converses. We can indicate each of these by merely listing the three statements that appear in its hypothesis, with the understanding that the remaining three statements appear in its conclusion. Thus (a, y, c) represents the implication

$$\left.\begin{matrix} a \\ y \\ c \end{matrix}\right\} \longrightarrow \left\{\begin{matrix} x \\ b \\ z. \end{matrix}\right.$$

We know that this converse is true because its verbal statement is equivalent to SSS (Theorem 7-11). Each of the converses (x, y, c) and (a, y, z) is true because the verbal statement of each is the SAS postulate. A careful study of our nineteen converses show that they can be classified under only six distinct verbal statements.

SSS: (a, y, c).
SAS: (x, y, c), (a, y, z).
ASA: (x, b, c), (a, b, z), (x, y, z).
SAA: (a, b, x), (z, b, c), (x, b, y), (x, z, c), (a, x, z), (y, b, z).
SSA: (a, x, c), (y, b, c), (a, z, c), (a, x, y), (y, z, c), (a, b, y).
AAA: (x, b, z).

We now proceed to investigate the converses of the SAS postulate. The process of investigating the converses of an implication involves all of the following operations:

1. Form all possible converses by applying the definition.
2. Verbalize each converse. Determine how many distinct verbal converses are available.
3. Make a decision about the truth of each of the distinct verbal converses. This decision must be proved. The falsity of a converse can be established by devising a counterexample.

Having established the first two converses of SAS, we proceed to investigate the remaining four.

THEOREM 7-12 (ASA THEOREM) *Given two triangles, not necessarily distinct, if two angles and the included side of one triangle are congruent respectively to two angles and the included side of the other, then the triangles are congruent.*

We must prove:

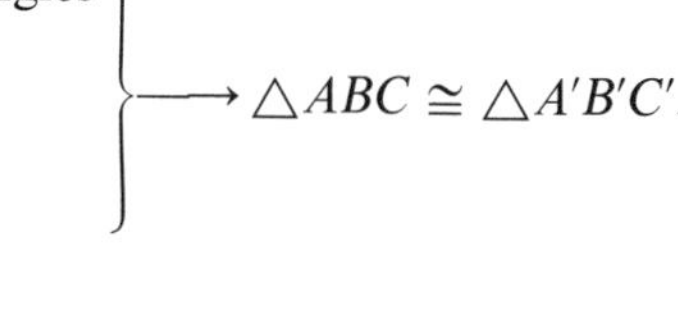

$$\left.\begin{matrix} (1)\ ABC \text{ and } A'B'C' \text{ are triangles} \\ (2)\ \overline{AB} \cong \overline{A'B'} \\ (3)\ \angle CAB \cong \angle C'A'B' \\ (4)\ \angle ABC \cong \angle A'B'C' \end{matrix}\right\} \longrightarrow \triangle ABC \cong \triangle A'B'C'.$$

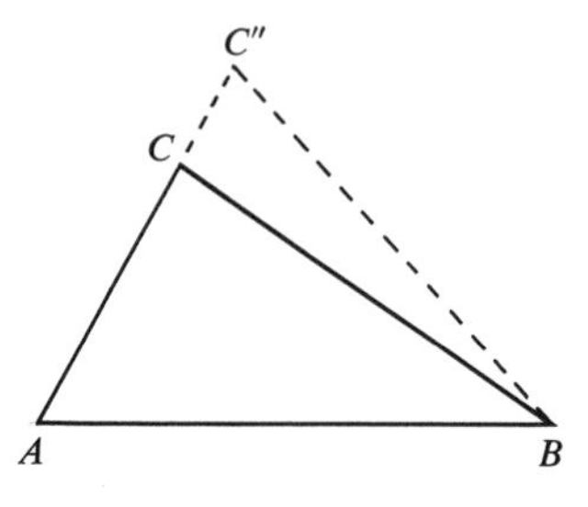

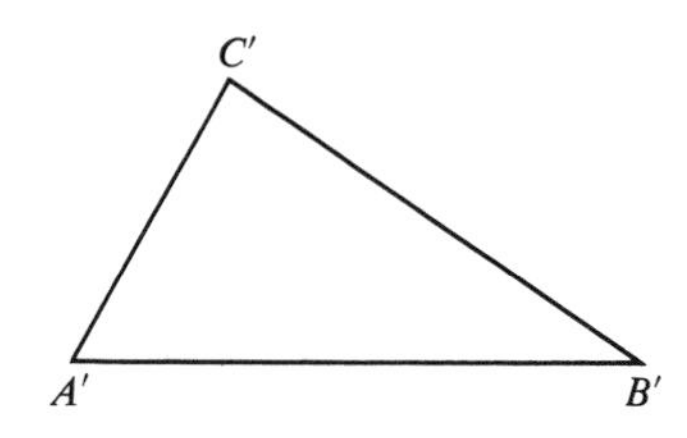

OUTLINE OF PROOF

By the point-plotting theorem there exists a point C'' in $\overset{\circ\!\!\rightarrow}{AC}$ such that $|AC''| = |A'C'|$. From this and premises (2) and (3), it follows that $\triangle ABC'' \cong \triangle A'B'C'$ by SAS. Hence, $\angle ABC'' \cong \angle A'B'C'$. This statement and premise (4) imply that $\underline{\angle ABC'' \cong \angle ABC}$. We can now prove indirectly that $|AC| = |AC''|$, for, if not, we must have either A-C-C'' or A-C''-C. Why? If A-C-C'', then C is in the interior of $\angle ABC''$, and it follows that $m\angle ABC'' > m\angle ABC$. If A-C''-C, then C'' is in the interior of $\angle ABC$, and we must have $m\angle ABC > m\angle ABC''$. In either case, $m\angle ABC'' \neq m\angle ABC$, which is equivalent to $\underline{\angle ABC'' \not\cong \angle ABC}$. The contradiction provided by the two underlined statements proves that $|AC| = |AC''|$ or, equivalently, that $\overline{AC} \cong \overline{AC''}$. Therefore, $\triangle ABC \cong \triangle ABC''$ by SAS, and, from the transitive property of congruence for triangles, we have $\triangle ABC \cong \triangle A'B'C'$.

THEOREM 7-13 (SAA THEOREM) *If a correspondence is established between the vertices of two triangles such that two angles and the side opposite one of them in one triangle are congruent to the corresponding parts of the other triangle, then the correspondence is a congruence.*

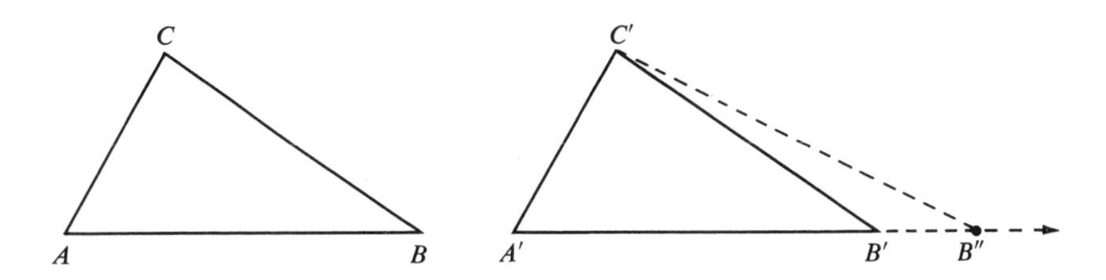

We must prove:

$$\left.\begin{array}{l} (1)\ ABC \text{ and } A'B'C' \text{ are triangles} \\ (2)\ \overline{AC} \cong \overline{A'C'} \\ (3)\ \angle CAB \cong \angle C'A'B' \\ (4)\ \angle ABC \cong \angle A'B'C' \end{array}\right\} \longrightarrow \triangle ABC \cong \triangle A'B'C'.$$

OUTLINE OF PROOF

There exists a point B'' in $\overset{\circ\!\!\rightarrow}{A'B'}$ such that $|AB| = |A'B''|$. Hence, $\triangle ABC \cong \triangle A'B''C'$ by SAS. It follows that $\angle ABC \cong \angle A'B''C'$. From this and premise (4), we have $\underline{\angle A'B'C' \cong \angle A'B''C'}$. We can prove indirectly that $|A'B'| = |A'B''|$ for, if not, we must have either A'-B'-B'' or A'-B''-B'. In either case, the exterior-angle theorem tells us that in $\angle A'B'C' \neq m\angle A'B''C'$ or, equivalently, that $\underline{\angle A'B'C' \not\cong \angle A'B''C'}$. The contradiction indicated by the underlined statements proves that $|A'B'| = |A'B''|$. From this statement and $|AB| = |A'B''|$ it follows that $\overline{AB} \cong \overline{A'B'}$. The conjunction of this statement with premises (2) and (3) implies that $\triangle ABC \cong \triangle A'B'C'$ by SAS.

COROLLARY TO THEOREM 7-13 (THE HL THEOREM) *If the hypotenuse and leg of one right triangle are congruent respectively to the hypotenuse and leg of another right triangle, then the two triangles are congruent.*

Proof of this corollary is requested in the exercises.

Although we might suspect that AAA does not represent a congruence theorem, the AAA statement is undecidable on the basis of our present postulate set. We shall investigate this matter in Chapter 8 after assuming the parallel postulate for Euclidean geometry (Postulate 17) and again in Chapter 10 after considering an alternate to the parallel postulate. We can at this time present a counterexample to show that SSA does not represent a congruence theorem.

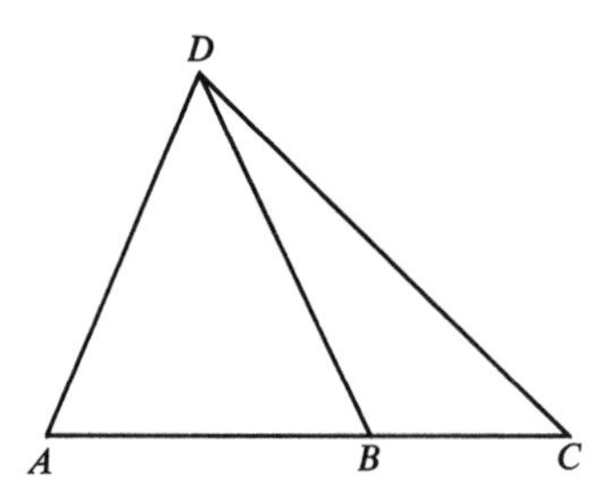

We know that $\overline{AB}$ has a perpendicular bisector in any plane that contains it. If D is any point in a perpendicular bisector of $\overline{AB}$ except the midpoint of $\overline{AB}$, then ADB is an isosceles triangle with $\overline{AD} \cong \overline{BD}$. Select a point C so that A-B-C. Then it is possible to prove each of the following statements: (1) $|DC| > |DB|$, (2) $|CA| > |CB|$, (3) $|DC| > |BC|$, and (4) $m\angle DAC < m\angle CBD$. (You are asked to supply the details in Exercise 3 below.) From these four statements it follows that *none* of the following correspondences is a congruent correspondence.

$C, D, A \leftrightarrow C, D, B$	$C, D, A \leftrightarrow D, C, B$
$C, D, A \leftrightarrow C, B, D$	$C, D, A \leftrightarrow B, C, D$
$C, D, A \leftrightarrow D, B, C$	$C, D, A \leftrightarrow B, D, C$

Therefore, $\triangle CDA \not\cong \triangle CDB$, even though two sides and a nonincluded angle of one are congruent to two sides and a nonincluded angle of the other.

Segments Related to a Triangle

We know that a segment has a unique midpoint and that two distinct points determine the segment joining them. Therefore we can define a median of a triangle as follows:

DEFINITION A median of a triangle *is a segment that joins a vertex of the triangle to the midpoint of the opposite side.*

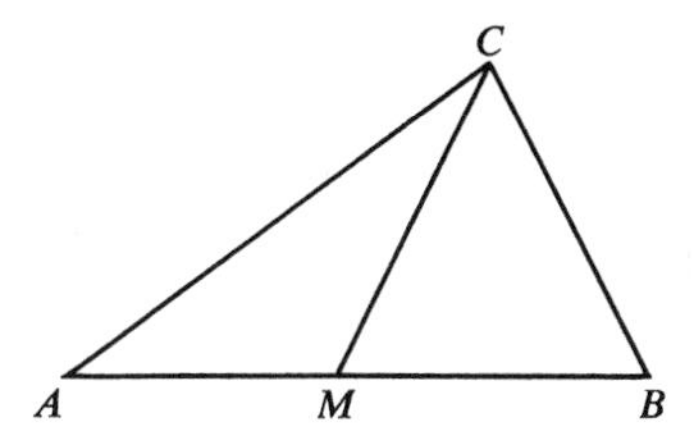

A symbolic form of this definition is

$\overline{CM}$ is a median of $\triangle ABC \longleftrightarrow ABC$ is a triangle and M is the midpoint of $\overline{AB}$.

We know that each angle of a triangle has a unique midray. According to the crossbar theorem, the midray of an angle of a triangle intersects the interior of the opposite side of the triangle. Therefore, there is a unique point in the interior of each side of a triangle that is the intersection of that side and the midray of the opposite angle.

DEFINITION *A* bisector of an angle of a triangle *is a segment that joins a vertex of the triangle to the point where the midray of this angle intersects the opposite side.*

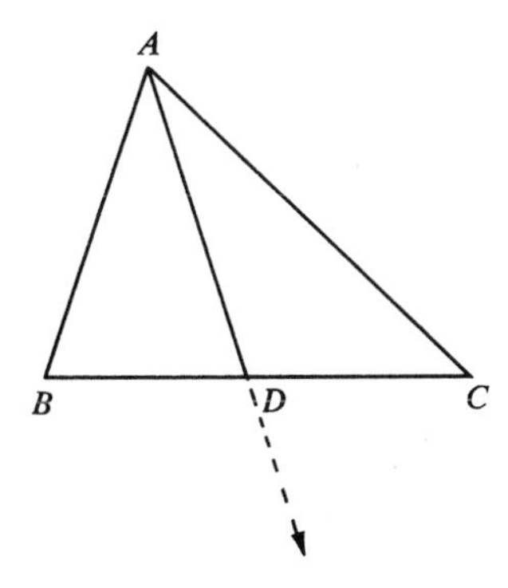

Thus we have:

$\overline{AD}$ is the bisector of $\angle BAC$ of $\triangle BAC \longleftrightarrow \overrightarrow{AD}$ is the midray of $\angle BAC$ and $D \in \overset{\circ\!-\!\circ}{BC}$.

The proof of the following theorem is considered in the exercises.

THEOREM 7-14 *The bisector of a vertex angle of an isosceles triangle is perpendicular to the base and bisects the base.*

Perpendicular to a Line from a Point

Before we define an altitude of a triangle, it is appropriate for us to prove

THEOREM 7-15 *Given a line and a point not in the line, there is exactly one line that contains the given point and is perpendicular to the given line.*

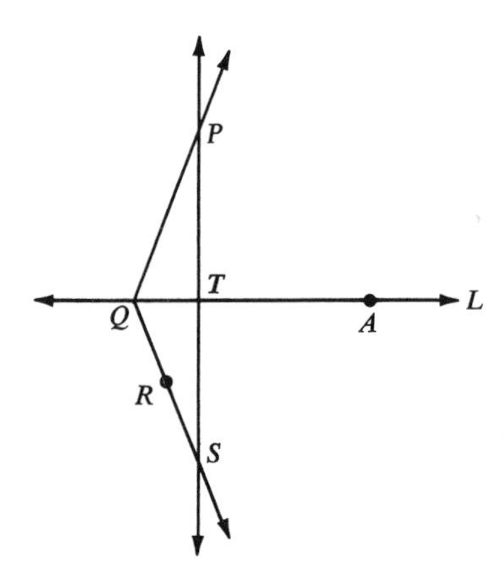

OUTLINE OF PROOF

We choose two distinct points Q and A in L. If $\overleftrightarrow{QP} \perp L$, then we have shown the existence of a line containing P perpendicular to L. If $\overleftrightarrow{QP}$ is not perpendicular to L, then $\angle AQP$ is either acute or obtuse. We shall assume that $\angle AQP$ is acute (otherwise we would consider $\angle A'QP$, where A' is a point such that A-Q-A'). By the angle-construction theorem there is a ray QR such that $\mathring{\overrightarrow{QR}} \subseteq L/{\sim}P$ and $m\angle AQR = m\angle AQP$. By the point-plotting theorem there is a point S in $\mathring{\overrightarrow{QR}}$ such that $|QS| = |QP|$. $S \in L/{\sim}P$ and, therefore, $\overset{\circ\,\circ}{SP} \times L(T)$. We can show that $S \notin \mathring{\overrightarrow{QP}}$, for if $S \in \mathring{\overrightarrow{QP}}$, we would have P-Q-S, and $P\underline{Q}A$ and $S\underline{Q}A$ would be a linear pair of angles that are congruent. As a result each would be a right angle. But $\angle AQP$ is acute. Since $S \notin \overleftrightarrow{QP}$ and $|QS| = |QP|$, we have an isosceles triangle SQP. $T \in \overset{\circ\,\circ}{PS}$ and, therefore, $\overrightarrow{QT}$ is between $\overrightarrow{QP}$ and $\overrightarrow{QS}$. Also, $m\angle AQR = m\angle AQP$. Thus $\overrightarrow{QT}$ bisects $\angle SQP$, and $\overline{QT}$ is the bisector of the vertex angle of isosceles $\triangle SQP$. By Theorem 7-14, $\overline{QT} \perp \overline{PS}$; hence, $\overleftrightarrow{QT} \perp \overleftrightarrow{PS}$. But $\overleftrightarrow{QT} = L$, and we have shown the existence of a line containing P and perpendicular to L.

To prove uniqueness, let L_1 be a line containing P such that $L_1 \perp L(X)$.

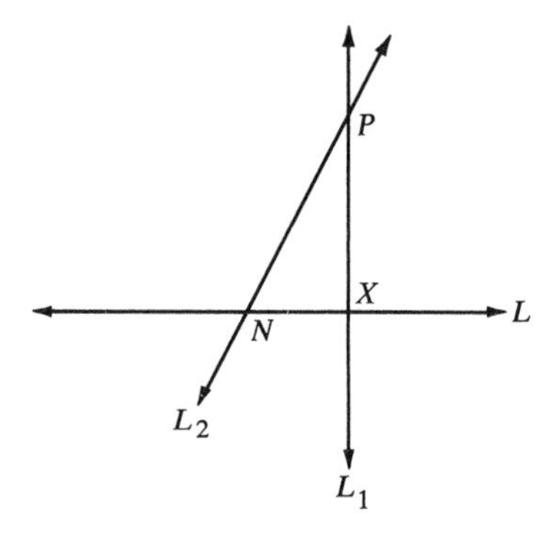

Let L_2 be any line distinct from L_1 and containing P. We must prove that L_2 is not perpendicular to L. Either L_2 intersects L or it does not. If $L_2 \cap L = \emptyset$, then L_2 is not perpendicular to L. If $L_2 \times L(N)$, then $N \neq X$ and PXT is a triangle with right angle at X. By Corollary 2 to Theorem 7-4 $\angle PNX$ is an acute angle; hence L_2 is not perpendicular to L.

DEFINITION *The* perpendicular from a point (not in a line) to the line *is a segment that joins the point to a point in the line and is perpendicular to the line.*

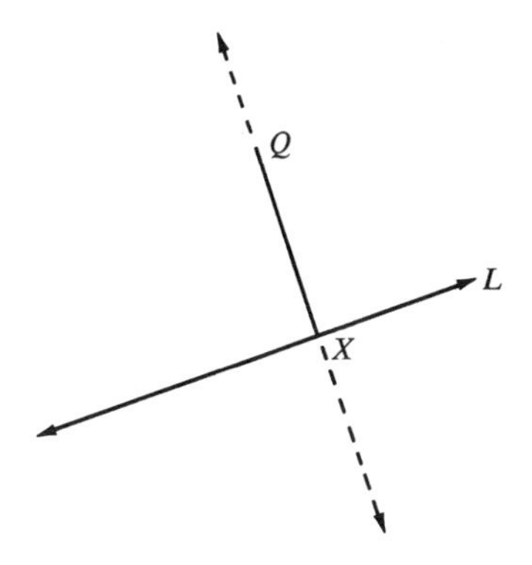

Stated symbolically, we have

$$(Q \notin L \wedge X \in L \wedge \overline{QX} \perp L) \longrightarrow (\overline{QX} \text{ is the perpendicular from } Q \text{ to } L).$$

DEFINITION *The point of intersection of a line with the perpendicular to the line from a point not in the line is called the* foot *of the perpendicular to the line from the point.*

DEFINITIONS *If a point is not in a line, its* distance from the line *is the length of the perpendicular from the point to the line. If a point is in a line, its distance from the line is zero.*

DEFINITION *The* altitude to a side of a triangle *is the perpendicular to the line containing that side from the opposite vertex. Such a segment is an* altitude of the triangle.

THEOREM 7-16 *The median to a base of an isosceles triangle is the altitude to that base and the bisector of the vertex angle corresponding to that base.*

A proof of Theorem 7-16 is requested in the exercises.

Fallacious Arguments Based on Diagrams

Let us try to prove the following statement: If two altitudes of a triangle are congruent, then the triangle is isosceles.

We must prove:

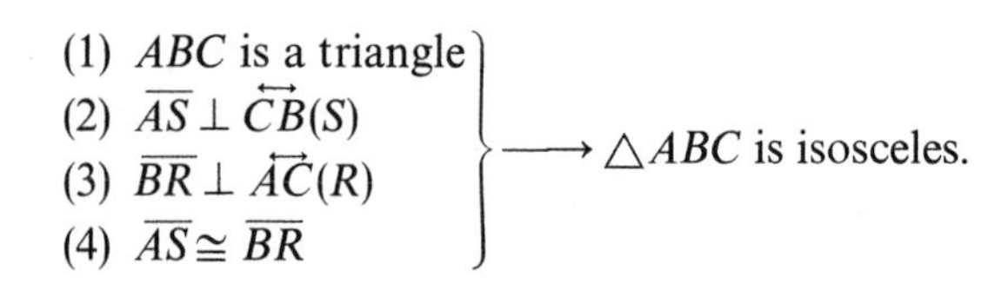

(1) ABC is a triangle
(2) $\overline{AS} \perp \overleftrightarrow{CB}(S)$
(3) $\overline{BR} \perp \overrightarrow{AC}(R)$
(4) $\overline{AS} \cong \overline{BR}$

$\longrightarrow \triangle ABC$ is isosceles.

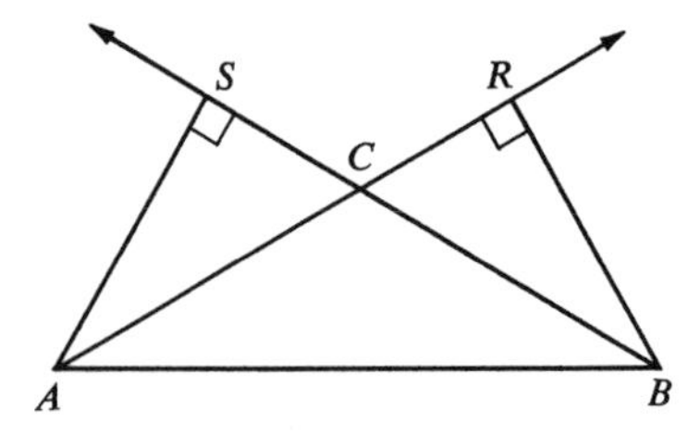

Discussion: In order to prove that $\triangle ABC$ is isosceles, we shall prove that $\overline{AC} \cong \overline{BC}$. Clearly, $\triangle ABS \cong \triangle BAR$ by the HL theorem. It follows that $\angle RAB \cong \angle SBA$. This gives us two congruent angles in $\triangle ABC$. Therefore, $\triangle ABC$ is isosceles by the converse of the isosceles-triangle theorem (Theorem 7-7).

Do you see the defect in this argument? Our reliance on the drawing has led us to make two unwarranted assumptions, namely that $\angle SBA = \angle CBA$ and that $\angle RAB = \angle CAB$. In an axiomatic development of geometry the question is not whether these statements are true but rather whether we can prove that they are implied by other statements that we have proved or explicitly assumed. We have failed to do this, because the truth of these statements is so strongly suggested by our drawing that we assumed them to be true without realizing that we were making unwarranted assumptions. Yet, if R were in the interior of the ray opposite $\overrightarrow{AC}$, we could not have $\angle RAB = \angle CAB$. Therefore, we must prove that R is in $\overset{\circ\rightarrow}{AC}$ and that S is in $\overset{\circ\rightarrow}{BC}$. The following lemma is useful in this situation.

LEMMA

$$\left.\begin{array}{l} m\angle BAC < 90 \\ \overline{BX} \perp \overrightarrow{AC}(X) \end{array}\right\} \longrightarrow X \in \overset{\circ\rightarrow}{AC}.$$

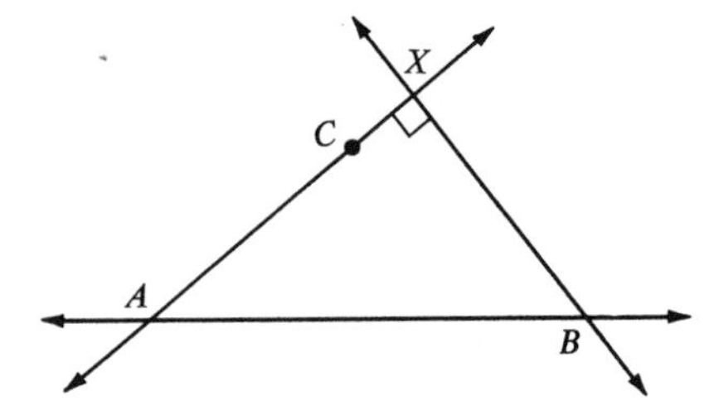

In the exercises you are asked to prove this lemma and to use it to complete the proof of the original statement.

One might argue that our reliance on the drawing in the previous example was not too serious, because, after all, our conclusion was correct. As indicated earlier, this has nothing to do with the validity of an argument in an axiomatic development of geometry. Nevertheless, it shall be noted that there are situations in which reliance on the drawing can lead to wrong or ridiculous conclusions. As an example, consider our "proof" of the following statement: All triangles are isosceles.

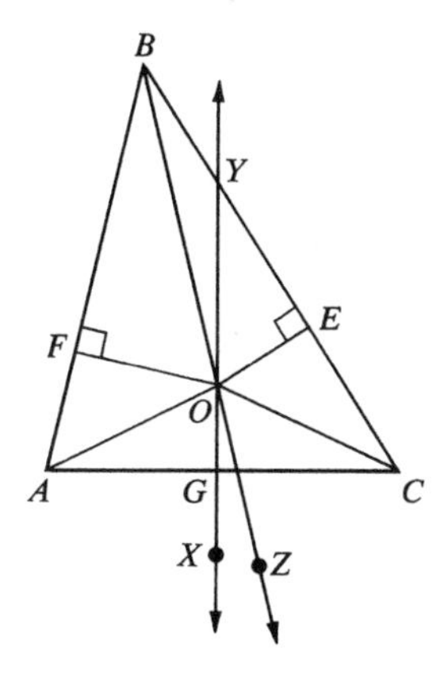

Given any triangle ABC, let $\overleftrightarrow{XY}$, the perpendicular bisector of $\overline{AC}$ in pl ABC, and $\overrightarrow{BZ}$, the midray of $\angle ABC$, intersect at O. Let E, F, and G be the feet of the perpendiculars from O to $\overline{BC}$, $\overline{AB}$, and $\overline{AC}$, respectively. Then $\triangle FOB \cong \triangle EOB$ by the SAA theorem and, therefore, $\underline{\overline{BF} \cong \overline{BE}}$ and $\overline{FO} \cong \overline{OE}$. Since O is in the perpendicular bisector of $\overline{AC}$, we know that $\overline{AO} \cong \overline{OC}$. It follows that right triangles AOF and COE are congruent by the HL theorem. This implies that $\underline{\overline{FA} \cong \overline{EC}}$. The conjunction of the underlined statements proves that $\overline{AB} \cong \overline{BC}$. Since we have proved that any triangle has two congruent sides, it follows that all triangles are isosceles.

The defect in this "proof' of an obviously false statement is not easy to explain. However, it is evident that we were led astray by the appearance of our diagram. Thus we were led to assume certain incidence relationships that were not true. Our proof depends on the tacit assumption that $E \in \overset{\circ\!-\!\circ}{BC}$ and $F \in \overset{\circ\!-\!\circ}{AB}$. It can be proved that this conjunctive statement is false.

Exercises

1. Prove the corollary to Theorem 7-13.
2. (a) Prove the following lemma:

$$\left.\begin{array}{l} m\angle BAC < 90 \\ \overline{BX} \perp \overleftrightarrow{AC}(X) \end{array}\right\} \longrightarrow X \in \overset{\circ\!\!\rightarrow}{AC}.$$

   (b) Use the lemma in part (a) to complete the proof of the statement: If two altitudes of a triangle are congruent, then the triangle is isosceles.

3. Complete the proof that SSA is false by establishing the four inequalities listed on page 164.
4. Prove Theorem 7-14.

In the next two exercises, investigate the converses of each of the implications obtained by separating Theorem 7-16 into two parts.

5. Given that ABC is a triangle, prove the implication

$$\left.\begin{array}{l}(1)\ \overline{AC} \cong \overline{BC} \\ (2)\ M \text{ is the midpoint of } \overline{AB}\end{array}\right\} \longrightarrow \overline{CM} \perp \overleftrightarrow{AB}(M).$$

and investigate the converses.
6. Given that ABC is a triangle, prove the implication

$$\left.\begin{array}{l}(1)\ \overline{AC} \cong \overline{BC} \\ (2)\ M \text{ is the midpoint of } \overline{AB}\end{array}\right\} \longrightarrow \overline{CM} \text{ is the bisector of } \angle ACB \text{ of } \triangle ABC$$

and investigate the converses.
7. If $A \neq B$ and $A \in L/B$, find the point X in L such that $|AX| + |XB|$ is as small as possible.
8. Prove:

$$\left.\begin{array}{l}ABC \text{ and } A'B'C' \text{ are triangles} \\ \overline{AB} \cong \overline{BC} \\ \overline{A'B'} \cong \overline{B'C'} \\ \angle B \cong \angle B' \\ |AB| > |A'B'|\end{array}\right\} \longrightarrow |AC| > |A'C'|.$$

9. Given a triangle ABC each of whose angles is acute, find points X, Y, and Z in $\overset{\circ\circ}{AB}$, $\overset{\circ\circ}{BC}$, and $\overset{\circ\circ}{CA}$, respectively, such that $|XY| + |YZ| + |ZX|$ is as small as possible.

In each of the following two exercises, the argument given to prove the implication is defective. Find the errors in the arguments. Show, in fact, that the implications are false by providing counterexamples.

10. $$\left.\begin{array}{l}ABC \text{ is a triangle} \\ |AX| = |BY| \\ |AC| < |BC|\end{array}\right\} \longrightarrow |AY| < |BX|.$$

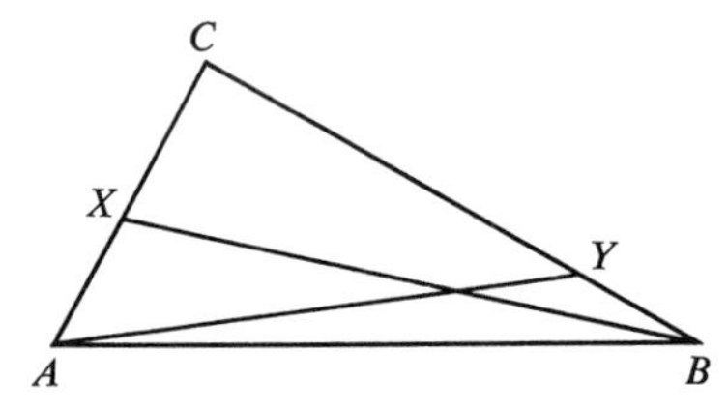

Argument:

$$\left.\begin{array}{l}\left.\begin{array}{l}ABC \text{ is a triangle} \\ |AC| < |BC|\end{array}\right\} \longrightarrow m\angle XAB > m\angle YBA \\ \qquad\qquad\qquad\qquad\quad \overline{AB} \cong \overline{BA} \\ \qquad\qquad\qquad\qquad\quad \overline{AX} \cong \overline{BY}\end{array}\right\} \longrightarrow |AY| < |BX|.$$

11. $$\left.\begin{array}{l}ABC \text{ and } A'B'C' \text{ are triangles} \\ |AB| = |A'B'| \\ m\angle B = m\angle B' \\ |BC| > |B'C'|\end{array}\right\} \longrightarrow |AC| > |A'C'|.$$

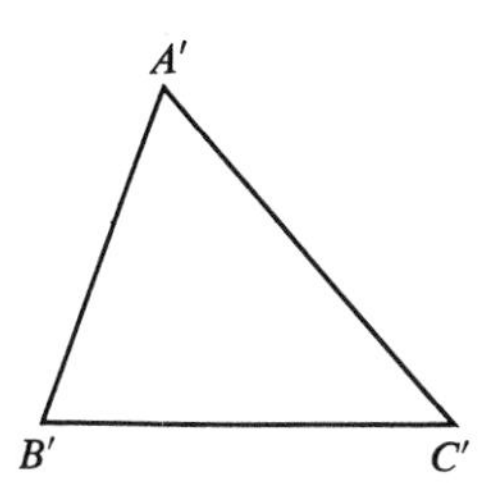

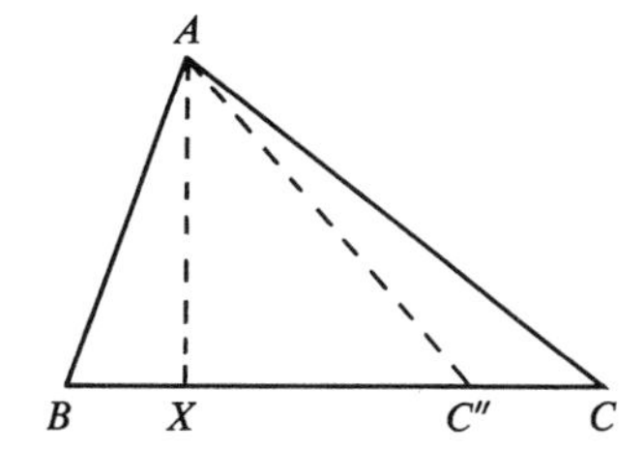

Argument: There is a line through A perpendicular to $\overleftrightarrow{BC}$ that intersects $\overleftrightarrow{BC}$ at X. There is a point C'' in $\mathring{\overrightarrow{BC}}$ such that $|BC''| = |B'C'|$. Since $|BC| > |B'C'|$, we have $|BC| > |BC''|$ with C'' in $\mathring{\overrightarrow{BC}}$. It follows that $C'' \in \overset{\circ\!-\!\circ}{BC}$, so that B-C''-C. $\angle AC''C$ is an exterior angle for $\triangle AXC''$. Therefore, $m\angle AC''C > 90$, and so it is the largest angle in $\triangle AC''C$. Thus $|AC| > |AC''|$. However, triangles ABC'' and $A'B'C'$ are congruent by SAS; hence $|AC''| = |A'C'|$. Therefore, $|AC| > |A'C'|$.

12. What is wrong with the following statement for the SAA theorem? Given two triangles, if two angles and a nonincluded side of one are congruent respectively to two angles and a nonincluded side of the other, then the triangles are congruent.

References

Alder, *Modern Geometry*

Courant and Robbins, *What is Mathematics*

Dubnov, *Mistakes in Geometric Proofs*

Fetisov, *Proof in Geometry*

Golos, *Foundations of Euclidean and Non-Euclidean Geometry*

Hemmerling, *Fundamentals of College Geometry*

Kay, *College Geometry*

Moise, *Elementary Geometry from an Advanced Standpoint*

Chapter 8

Parallelism

In this chapter we introduce the parallel postulate (Postulate 17) and consider some of its important consequences. We begin by stating some definitions.

Skew Lines and Parallel Lines

DEFINITION *Two lines are* skew lines *if and only if they are noncoplanar.*

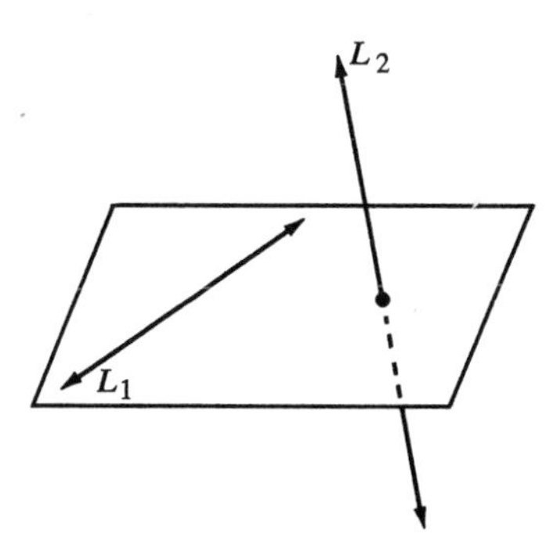

If L_1 and L_2 are two lines, we may state this definition as follows:

$$L_1 \text{ and } L_2 \text{ are skew lines} \longleftrightarrow L_1 \text{ and } L_2 \text{ are noncoplanar.}$$

DEFINITION *Two lines are* parallel *if and only if they are coplanar and do not intersect.*

We may use the symbol $L_1 \parallel L_2$ to represent the words "Line L_1 is parallel to L_2" and the symbol cop to represent the word "coplanar."

Thus the preceding definition may be stated as follows:

$$L_1 \parallel L_2 \longleftrightarrow (L_1 \text{ cop } L_2 \wedge L_1 \cap L_2 = \emptyset).$$

Using the symbol $\nparallel$ for "is not parallel to" and recalling that contradictions of equivalent statements are equivalent, we know that the following statement is true:

$$L_1 \nparallel L_2 \longleftrightarrow (L_1 \sim \text{cop } L_2 \vee L_1 \cap L_2 \neq \emptyset).$$

According to the definition of parallel lines, two parallel lines are contained in some plane. We now prove that there is only one plane that contains two parallel lines.

***THEOREM* 8-1** *Two parallel lines are contained in exactly one plane.*

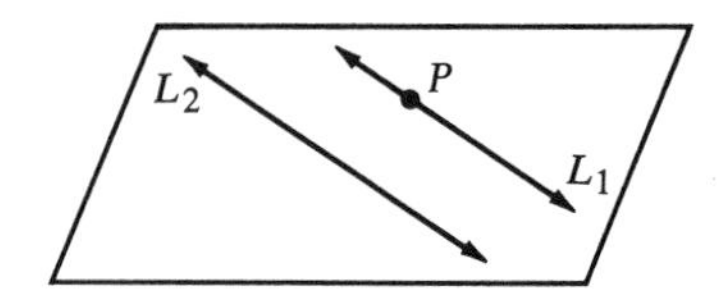

PROOF

If lines L_1 and L_2 are parallel lines, we know by definition that there is a plane that contains them. It remains for us to prove that there is a unique plane that contains $L_1 \cup L_2$. Let P be a point in L_1. Then $\underline{P \notin L_2}$, because two parallel lines have no common point. If each of the planes M_1 and M_2 contains $L_1 \cup L_2$, then each plane contains $P \cup L_2$; that is, $\underline{P \cup L_2 \subseteq M_1}$ and $\underline{P \cup L_2 \subseteq M_2}$. Since a line and a point not in it are contained in exactly one plane (Theorem 3-12), the underlined statements imply that $M_1 = M_2$, and our proof of uniqueness is complete.

Since two parallel lines lie in exactly one plane, we can say that two parallel lines determine a plane.

In our proof of Theorem 8-1, we observed that the plane determined by $L_1 \cup L_2$ (which, of course, contains L_1) is the same as the plane determined by $L_2 \cup P$. Therefore, the plane determined by L_2 and P contains L_1, and we have the following corollary to Theorem 8-1.

COROLLARY 1 TO *THEOREM* 8-1 *If two lines are parallel, then each lies in the plane determined by one of its points and the other line.*

The proof of Corollary 2 to Theorem 8-1 is required in the exercises.

COROLLARY 2 TO *THEOREM* 8-1 *Two lines are parallel if and only if one of the two lines lies in the half plane that contains one of its points and whose edge is the other line.* [$\overleftrightarrow{AB} \parallel \overleftrightarrow{CD} \longleftrightarrow \overleftrightarrow{AB} \subseteq \overleftrightarrow{CD}/A$.]

***THEOREM* 8-2** *If two lines are distinct, coplanar, and not parallel, then their intersection is a single point.*

DEFINITION *A* line is parallel to a plane *if it does not intersect the plane.*

THEOREM 8-3 *If two planes intersect in a line, then any line that is in one of the planes and is parallel to the other plane is parallel to the line of intersection.*

Proofs of Theorems 8-2 and 8-3 are requested in the exercises.

Transversals

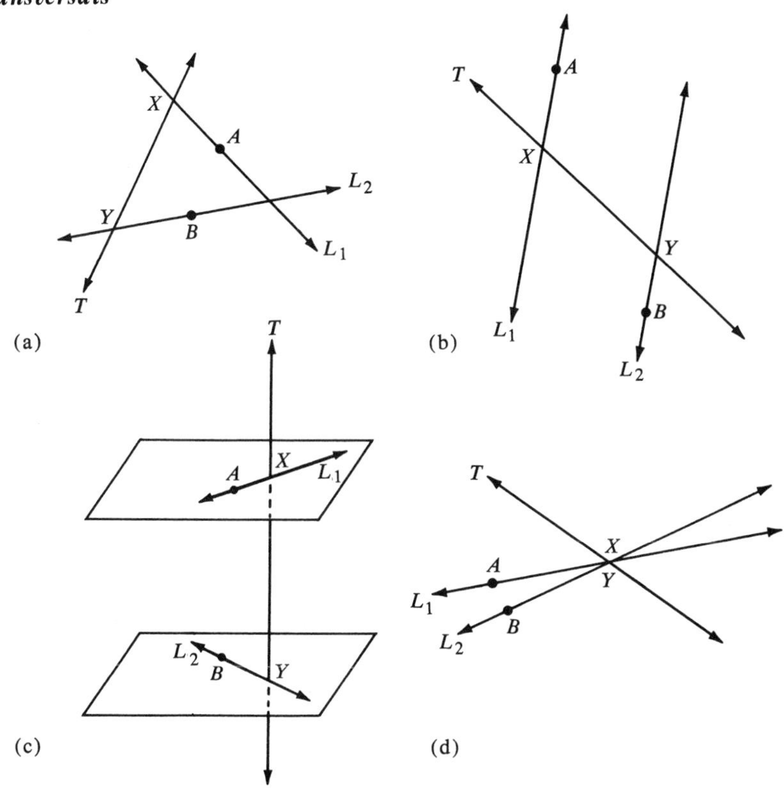

In figures (a) and (b) the line T is a transversal for the lines L_1 and L_2. However, we shall define the word "transversal" so that T is not a transversal for L_1 and L_2 in figures (c) and (d). We want to eliminate the possibility that L_1 and L_2 are noncoplanar [figure (c)] and the possibility that the point of intersection of T and L_1 is the same as the point of intersection of T and L_2 [figure (d)]. These considerations suggest the following definition of a transversal for two lines.

DEFINITION $\overleftrightarrow{XY}$ *is a transversal for* $\overleftrightarrow{XA}$ *and* $\overleftrightarrow{YB}$ *if and only if each of the following statements is true:*

(1) $\overleftrightarrow{XY} \cap \overleftrightarrow{XA} = \{X\}$,
(2) $\overleftrightarrow{XY} \cap \overleftrightarrow{YB} = \{Y\}$,
(3) $X \neq Y$,
(4) $\overleftrightarrow{XA}$ cop $\overleftrightarrow{YB}$.

It follows that L is a transversal for L_1 and L_2 if and only if L intersects $L_1 \cup L_2$ in distinct points and L_1 and L_2 are coplanar.

COROLLARY TO THE DEFINITION OF TRANSVERSAL

(1) $A \in \overleftrightarrow{XY}/\sim B \longrightarrow \overleftrightarrow{XY}$ *is a transversal for* $\overleftrightarrow{XA}$ *and* $\overleftrightarrow{YB}$, *and*
$A \in \overleftrightarrow{XY}/B \longrightarrow \overleftrightarrow{XY}$ *is a transversal for* $\overleftrightarrow{XA}$ *and* $\overleftrightarrow{YB}$.
(2) *If* $\overleftrightarrow{XY}$ *is a transversal for* $\overleftrightarrow{XA}$ *and* $\overleftrightarrow{YB}$, *then* $\overleftrightarrow{XY}$, $\overleftrightarrow{XA}$, *and* $\overleftrightarrow{YB}$ *are coplanar.*
(3) *If* $\overleftrightarrow{XY}$ *is a transversal for* $\overleftrightarrow{XA}$ *and* $\overleftrightarrow{YB}$, *then* $\overleftrightarrow{XY} \neq \overleftrightarrow{XA} \wedge \overleftrightarrow{XY} \neq \overleftrightarrow{BY} \wedge \overleftrightarrow{BY} \neq \overleftrightarrow{AX}$, *and no point common to any two of the lines* $\overleftrightarrow{XY}$, $\overleftrightarrow{XA}$, *and* $\overleftrightarrow{YB}$ *is in the other line.*

A transversal for two given lines forms eight angles with the two lines. Certain pairs of these angles are important in our study, and we provide names to describe them.

DEFINITION *If points A and B are in opposite half planes with common edge $\overleftrightarrow{XY}$, then $A\underline{X}Y$ and $B\underline{Y}X$ are* alternate interior *angles.*

Thus $A \in \overleftrightarrow{XY}/\sim B \longleftrightarrow A\underline{X}Y$ and $B\underline{Y}X$ are alternate interior angles.

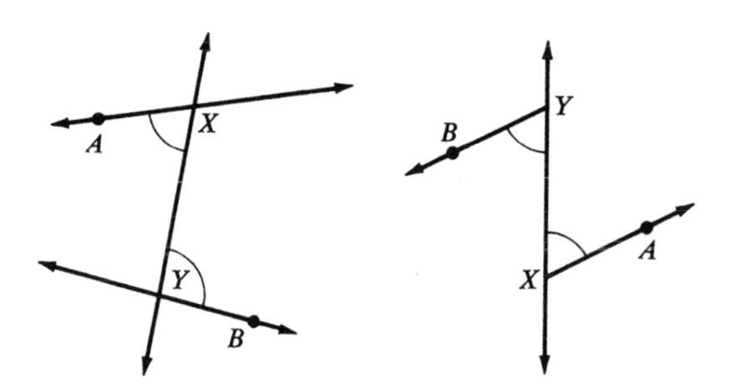

DEFINITION *If points A and B are in the same half plane with common edge $\overleftrightarrow{XY}$, then $A\underline{X}Y$ and $B\underline{Y}X$ are* consecutive interior angles.

Thus $A \in \overleftrightarrow{XY}/B \longleftrightarrow A\underline{X}Y$ and $B\underline{Y}X$ are consecutive interior angles.

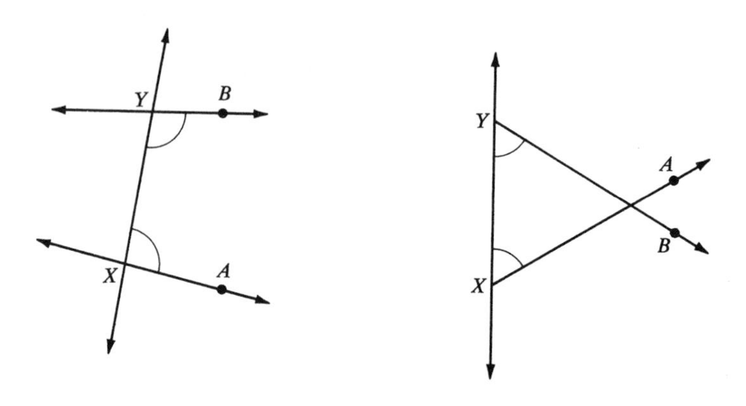

DEFINITION *If points A and B are in the same half plane with edge* $\overleftrightarrow{XY}$ *and X-Y-C or Y-X-C, then* $A\underline{X}C$ *and* $B\underline{Y}C$ *are* corresponding angles.

$[A \in \overleftrightarrow{XY}/B \wedge (X\text{-}Y\text{-}C \vee Y\text{-}X\text{-}C)] \longleftrightarrow A\underline{X}C$ and $B\underline{Y}C$ are corresponding angles.

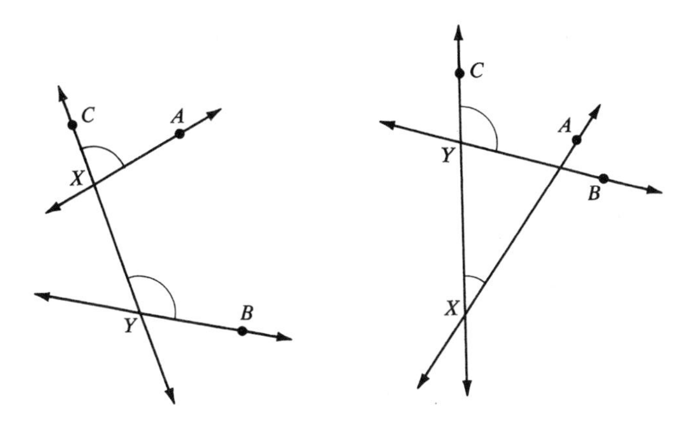

Exercises

1. If $D \notin$ pl ABC, then name all pairs of skew lines determined by A, B, C, and D.

L 2. Prove: If lines L_1 and L_2 are skew, then they do not intersect.

3. Prove: If $\overleftrightarrow{AB}$ is skew to $\overleftrightarrow{CD}$, then $\overleftrightarrow{AC}$ is skew to $\overleftrightarrow{BD}$.

4. (a) Is "parallel" an equivalence relation for lines?
   (b) Is "skew" an equivalence relation for lines?

5. Prove Corollary 2 to Theorem 8-1.

L 6. Prove Theorem 8-2:

$$\left.\begin{array}{l} L_1 \neq L_2 \\ L_1 \text{ cop } L_2 \\ L_1 \nparallel L_2 \end{array}\right\} \longrightarrow L_1 \times L_2.$$

Verbalize each of its contrapositives and investigate its converses.

7. Prove Theorem 8-3.

8. Prove the corollary to the definition of transversal.

L 9. Prove:

$$\left.\begin{array}{l} A\underline{X}Y \text{ and } B\underline{Y}X \text{ are alternate} \\ \text{interior angles.} \\ A\text{-}X\text{-}A'. \\ B\text{-}Y\text{-}B'. \end{array}\right\} \longrightarrow \left\{\begin{array}{l} A'\underline{X}Y \text{ and } B'\underline{Y}X \text{ are alternate} \\ \text{interior angles.} \\ A'\underline{X}Y \text{ and } B\underline{Y}X \text{ are consecutive} \\ \text{interior angles.} \end{array}\right.$$

(b) Let the statment $\angle AXY \cong \angle BYX$ be annexed to the hypothesis of the implication in part (a). Prove: $\angle A'XY \cong \angle B'YX$ and $\angle A'XY$ and $\angle BYX$ are supplementary.

10. Prove:

$$\left.\begin{array}{l} A\underline{X}B \text{ and } A\underline{Y}C \text{ are corresponding angles} \\ A\underline{X}B \text{ and } A\underline{Z}D \text{ are corresponding angles} \\ Y \neq Z \end{array}\right\} \longrightarrow \left\{\begin{array}{l} A\underline{Y}C \text{ and } A\underline{Z}D \text{ are} \\ \text{corresponding angles.} \end{array}\right.$$

L 11. Prove:

$$\left.\begin{array}{l} A\underline{X}Y \text{ and } X\underline{Y}B \text{ are alternate} \\ \text{interior angles.} \\ A\underline{X}Y \text{ and } X\underline{W}C \text{ are alternate} \\ \text{interior angles.} \\ X\text{-}Y\text{-}W. \end{array}\right\} \longrightarrow \left\{\begin{array}{l} X\underline{Y}B \text{ and } X\underline{W}C \text{ are} \\ \text{corresponding angles.} \end{array}\right.$$

Proving Lines Parallel

THEOREM 8-4 (ALTERNATE-INTERIOR-ANGLES THEOREM) *If a pair of alternate interior angles are congruent when two lines are cut by a transversal, then the lines are parallel.*

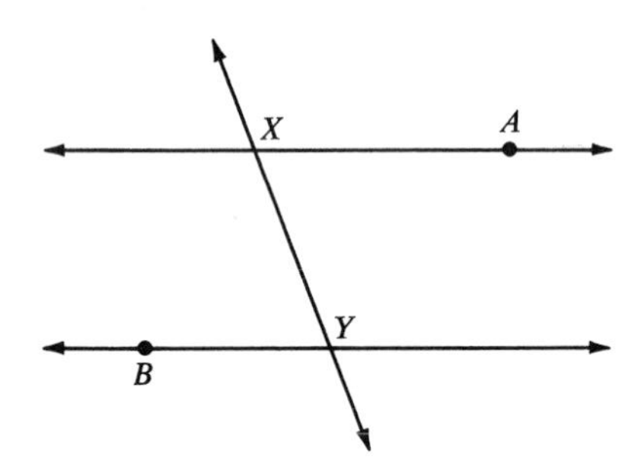

We must prove:

$$\left.\begin{array}{l} (1)\ A\underline{X}Y \text{ and } B\underline{Y}X \text{ are alternate interior angles} \\ (2)\ \angle AXY \cong \angle BYX \end{array}\right\} \longrightarrow \overleftrightarrow{XA} \parallel \overleftrightarrow{YB}.$$

We shall prove this theorem by proving the second contrapositive:

$$\left.\begin{array}{l} (1)\ A\underline{X}Y \text{ and } B\underline{Y}X \text{ are alternate interior angles} \\ (3)\ \overleftrightarrow{XA} \nparallel \overleftrightarrow{YB} \end{array}\right\} \longrightarrow \angle AXY \ncong \angle BYX.$$

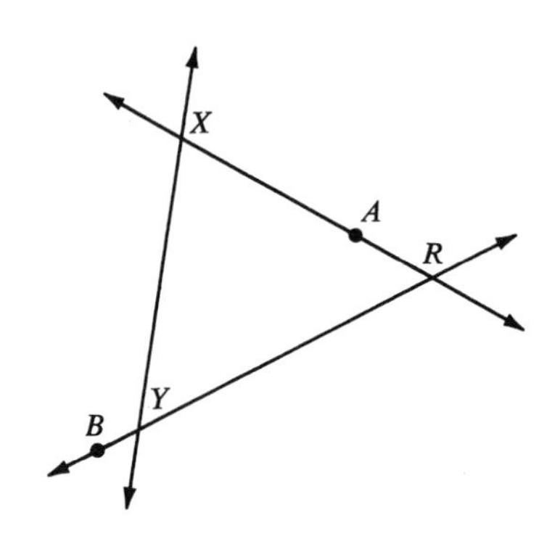

PROOF

Since AXY and BYX are alternate interior angles, we know $A \in \overleftrightarrow{XY}/{\sim}B$ and $\overleftrightarrow{XY}$ is a transversal for $\overleftrightarrow{XA}$ and $\overleftrightarrow{YB}$. By the corollary to the definition of transversal, we know that the three lines are coplanar and distinct. The non-parallel lines $\overleftrightarrow{XA}$ and $\overleftrightarrow{YB}$ must intersect in a point R, and $R \notin \overleftrightarrow{XY}$. Therefore, $R \in \overleftrightarrow{XY}/A$ or $R \in \overleftrightarrow{XY}/B$. Suppose $R \in \overleftrightarrow{XY}/A$; then $R \in \overleftrightarrow{XY}/{\sim}B$ and we have R-Y-B. Since $X \notin \overleftrightarrow{YR}$, we know that $\angle BYX$ is an exterior angle for $\triangle XYR$. Therefore, $m\angle BYX > m\angle RXY$. It follows by the definition of congruent angles that $\angle BYX \not\cong \angle RXY$. But $\angle RXY = \angle AXY$, so $\angle BYX \not\cong \angle AXY$, and our proof is complete for the case where $R \in \overleftrightarrow{XY}/A$. Similar reasoning applies when $R \in \overleftrightarrow{XY}/B$. Therefore, the proof of the theorem is complete.

The proofs of the following corollaries are left as exercises.

COROLLARY 1 TO THEOREM 8-4 (CORRESPONDING-ANGLES COROLLARY) *If a pair of corresponding angles are congruent when two lines are cut by a transversal, then the lines are parallel.*

COROLLARY 2 TO THEOREM 8-4 (CONSECUTIVE-INTERIOR-ANGLES COROLLARY) *If a pair of consecutive interior angles are supplementary when two lines are cut by a transversal, then the lines are parallel.*

COROLLARY 3 TO THEOREM 8-4 *If two given distinct lines are perpendicular to the same line and all three lines are coplanar, then the two given lines are parallel.*

$$\left.\begin{array}{l} L_1 \perp L \\ L_2 \perp L \\ L_1 \neq L_2 \\ L_1, L_2, \textit{and } L \textit{ are coplanar} \end{array}\right\} \longrightarrow L_1 \parallel L_2.$$

We are now in a position to prove an important existence theorem.

THEOREM 8-5 *If L is a line and point $P \notin L$, then there is at least one line that contains P and is parallel to L.*

The proof of Theorem 8-5 is requested in the exercises.

The Parallel Postulate

If point P is in the line L, we know there is no line that contains P and is parallel to L. Theorem 8-5 states that if the point P is not in L, there is at least one line that contains P and is parallel to L. It seems obvious that there is only one such line. As we observed earlier, this statement is undecidable on the basis of the postulates we have stated thus far. We accept the following postulate, which characterizes Euclidean geometry.

POSTULATE 17 (PARALLEL POSTULATE) *For any point in space there is at most one line that contains the point and is parallel to a given line.*

Combining Postulate 17 and Theorem 8-5, we have the following important theorem.

THEOREM 8-6 *Through a given point not in a given line there is one and only one line parallel to the given line.*

Let us analyze Postulate 17. If each of the lines L_1 and L_2 contains the point P and each is parallel to line L, we must conclude that L_1 and L_2 are the same line. Thus we have

$$\left.\begin{array}{l}(1)\ P\in L_1\\ (2)\ P\in L_2\\ (3)\ L_1\parallel L\\ (4)\ L_2\parallel L\end{array}\right\}\longrightarrow L_1=L_2.$$

The fourth contrapositive is noteworthy.

$$C_4:\ \left.\begin{array}{l}(1)\ P\in L_1\\ (2)\ P\in L_2\\ (3)\ L_1\parallel L\\ (4)\ L_1\neq L_2\end{array}\right\}\longrightarrow L_2\nparallel L.$$

If two distinct lines have a common point and one of them is parallel to a given line, then the other is not parallel to that line.

In this form (C_4) Postulate 17 is a principal component in the proof of the following theorem.

THEOREM 8-7 *If a line intersects one of two parallel lines in a single point, then it intersects the other in a single point, provided all three lines are coplanar.*

We must prove the following implication:

$$\left.\begin{array}{l}L_1\times L_2(P)\\ L_1\parallel L_3\\ L_1, L_2, L_3 \text{ are cop}\end{array}\right\}\longrightarrow L_2\times L_3.$$

PROOF

$$\underline{L_1, L_2, L_3 \text{ are cop}}\longrightarrow L_2 \text{ cop } L_3$$

$$\underline{L_1\times L_2(P)}\longrightarrow\left\{\begin{array}{l}L_1\neq L_2\\ P\in L_2\\ P\in L_1\end{array}\right.$$

$$\left.\begin{array}{l}L_1\neq L_2\\ P\in L_2\\ P\in L_1\\ \underline{L_1\parallel L_3}\end{array}\right\}\longrightarrow L_2\nparallel L_3$$

$$\left.\begin{array}{l}P\in L_1\\ L_1\parallel L_3\end{array}\right\}\longrightarrow P\notin L_3;\qquad \left.\begin{array}{l}P\in L_2\\ P\notin L_3\end{array}\right\}\longrightarrow L_2\neq L_3$$

$$\left.\begin{array}{l}L_2 \text{ cop } L_3\\ L_2\nparallel L_3\\ L_2\neq L_3\end{array}\right\}\longrightarrow L_2\times L_3.$$

The reader may supply the reasons.

Now we use the parallel postulate to prove an important theorem, a converse of Theorem 8-4.

THEOREM 8-8 *If two lines are parallel and are cut by a transversal, then the alternate interior angles formed are congruent.*

We shall establish this theorem by proving a contrapositive: If a pair of alternate interior angles are not congruent when two lines are cut by a transversal, then the lines are not parallel.

We must prove:

$$\left.\begin{array}{l}(1)\ \underline{TQO} \text{ and } \underline{QOS} \text{ are alternate interior angles} \\ (2)\ \angle TQO \not\cong \angle QOS\end{array}\right\} \longrightarrow \overleftrightarrow{TQ} \nparallel \overleftrightarrow{OS}.$$

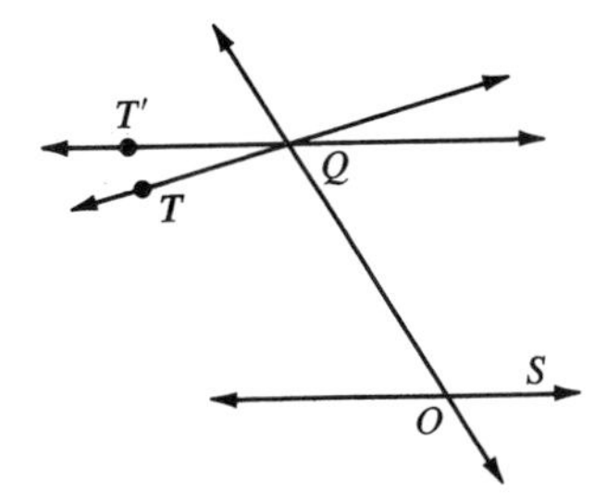

PROOF

Statement	Reason
1. $T \in \overleftrightarrow{OQ}/\sim S$.	1. Hypothesis and definition of alternate interior angles.
2. There is a point T' such that (a) $T' \in \overleftrightarrow{OQ}/\sim S$, (b) $m\angle T'QO = m\angle QOS$.	2. Angle-construction theorem.
3. $T' \in \overleftrightarrow{OQ}/T$.	3. Statements 1, 2(a), and plane separation lemma.
4. $\angle T'QO \cong \angle QOS$.	4. Statement 2(b) and definition of congruent angles.
5. $\underline{T'QO}$ and $\underline{QOS}$ are alternate interior angles.	5. Statement 2(a) and definition of alternate interior angles.
6. $\overleftrightarrow{T'Q} \parallel \overleftrightarrow{OS}$.	6. Statements 4, 5, and Theorem 8-4.
7. $\angle T'QO \not\cong \angle TQO$.	7. Hypothesis, Statement 4, and Theorem 6-15.
8. $\angle T'QO \neq \angle TQO$.	8. Statement 7 and "If two angles are equal, then they are congruent."
9. $\overrightarrow{QT'} \neq \overrightarrow{QT}$.	9. Statement 8 and (ii), page 117.
10. $\underline{TQT'}$ is an angle.	10. Statements 3, 9, and Corollary 2 to the definition of angle.
11. $\overleftrightarrow{QT} \neq \overleftrightarrow{QT'}$.	11. Statement 10 and definition of angle.
12. $\overleftrightarrow{QT} \nparallel \overleftrightarrow{OS}$.	12. Statements 6, 11, and parallel postulate.

The proofs of the following corollaries to Theorem 8-8 and the proof of Theorem 8-9 are left as exercises.

COROLLARY 1 TO THEOREM 8-8 *If two parallel lines are cut by a transversal, then each pair of corresponding angles are congruent.*

COROLLARY 2 TO THEOREM 8-8 *If two parallel lines are cut by a transversal, then each pair of consecutive interior angles are supplementary.*

COROLLARY 3 TO THEOREM 8-8 *If a transversal for two parallel lines is perpendicular to one of the two parallel lines, then it is perpendicular to the other.*

COROLLARY 4 TO THEOREM 8-8 *Two lines which are respectively perpendicular to two distinct intersecting lines intersect in a single point, provided all four lines are coplanar.*

THEOREM 8-9 *If two distinct lines are parallel to the same line, then they are parallel to each other, provided all three lines are coplanar.*

Exercises

1. Prove Corollary 1 to Theorem 8-4.
2. Prove Corollary 2 to Theorem 8-4.
3. Prove Corollary 3 to Theorem 8-4.
4. Write the other three contrapositives of Postulate 17.
5. Prove:

$$\left.\begin{array}{l} B\text{-}C\text{-}D \\ F \in \overrightarrow{BC}/E \\ \angle EBD \cong \angle FCD \\ \overrightarrow{BG} \text{ bisects } \angle EBD \\ \overrightarrow{CH} \text{ bisects } \angle FCD \end{array}\right\} \longrightarrow \overleftrightarrow{BG} \parallel \overleftrightarrow{CH}.$$

6. Prove:

$$\left.\begin{array}{l} ABCD \text{ is a convax quadrilateral} \\ \overline{AB} \cong \overline{DC} \\ \overline{AD} \cong \overline{BC} \end{array}\right\} \longrightarrow \left\{\begin{array}{l} \overleftrightarrow{DC} \parallel \overleftrightarrow{AB} \\ \overleftrightarrow{AD} \parallel \overleftrightarrow{BC}. \end{array}\right.$$

7. Prove: If the diagonals of a quadrilateral bisect each other, then the lines which contain the opposite sides are parallel.
8. Prove Theorem 8-5.
9. Prove:

$$\left.\begin{array}{r} \underline{XOC} \text{ and } \underline{YOC} \text{ are congruent adjacent obtuse angles} \\ \overline{OX} \cong \overline{OY} \\ \underline{BOC} \text{ and } \underline{AOC} \text{ are congruent adjacent acute angles} \\ \overline{OB} \cong \overline{OA} \\ A \in \overrightarrow{OC}/X \end{array}\right\} \longrightarrow \overleftrightarrow{BA} \parallel \overleftrightarrow{XY}.$$

10. Prove:

$$\left.\begin{array}{l} \overleftrightarrow{AB} \parallel \overleftrightarrow{CD} \\ A \in \overleftrightarrow{BD}/{\sim}C \\ \overline{AB} \cong \overline{CD} \end{array}\right\} \longrightarrow \overleftrightarrow{AD} \parallel \overleftrightarrow{BC}.$$

11. Prove Corollary 1 to Theorem 8-8.

12. Prove Corollary 2 to Theorem 8-8.
13. Prove Corollary 3 to Theorem 8-8.
14. Prove Corollary 4 to Theorem 8-8.
15. Write an indirect essay proof of Theorem 8-9.
16. Prove:

$$\left.\begin{array}{l}\overleftrightarrow{AB} \parallel \overleftrightarrow{CD} \\ \angle BAD \cong \angle DCB \\ C \in \overleftrightarrow{AD}/B\end{array}\right\} \longrightarrow \overleftrightarrow{AD} \parallel \overleftrightarrow{BC}.$$

Parallel Segments and Rays

We have stated that two coplanar lines are parallel if they do not intersect. However, there are coplanar segments that do not intersect that we do not regard as parallel segments. We think of parallel segments as subsets of parallel lines, as indicated by the following definitions.

DEFINITIONS *Two segments that are subsets of parallel lines are* parallel segments. *A* line is parallel to a segment *if it is parallel to the line that contains the segment.*

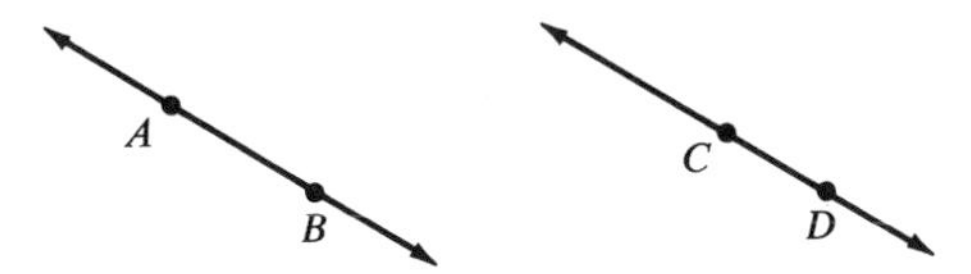

Thus $\overline{AB} \parallel \overline{CD} \longleftrightarrow \overleftrightarrow{AB} \parallel \overleftrightarrow{CD}$ and $\overleftrightarrow{AB} \parallel \overline{CD} \longleftrightarrow \overleftrightarrow{AB} \parallel \overleftrightarrow{CD}$.
The definition of parallel rays is somewhat more involved.

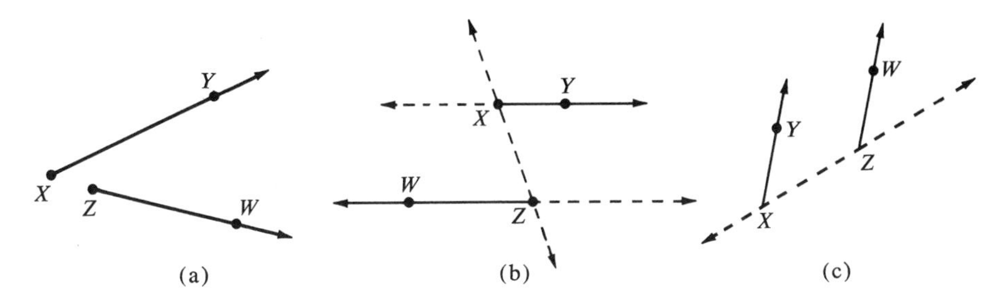

In figure (a) we have coplanar rays that do not intersect, but we reject the idea that these rays are parallel. In figure (b) rays XY and ZW are subsets of parallel lines XY and ZW, but they are not regarded as parallel because they seem to extend in opposite directions. We consider that two rays are parallel if and only if they are subsets of parallel lines and extend in the same direction, as shown in figure (c). We adopt the following definitions.

DEFINITION $\overrightarrow{XY} \parallel \overrightarrow{ZW} \longleftrightarrow (\overleftrightarrow{XY} \parallel \overleftrightarrow{ZW} \wedge Y \in \overleftrightarrow{XZ}/W)$.

DEFINITION $\overrightarrow{XY}$ *is antiparallel* $\overrightarrow{ZW} \longleftrightarrow (\overleftrightarrow{XY} \parallel \overleftrightarrow{ZW} \wedge Y \in \overleftrightarrow{XZ}/\sim W)$.

Notation: We use the symbol $\downharpoonleft\upharpoonright$ for "antiparallel." Thus

$$\overrightarrow{RS} \downharpoonleft\upharpoonright \overrightarrow{EF} \longleftrightarrow (\overleftrightarrow{RS} \parallel \overleftrightarrow{EF} \wedge S \in \overleftrightarrow{RE}/\sim F).$$

THEOREM 8-10 *Through a given point outside the line that contains a given ray there is exactly one ray* $\begin{Bmatrix}\text{parallel} \\ \text{antiparallel}\end{Bmatrix}$ *to the given ray and having the given point as endpoint.*

Thus, if $A \notin \overleftrightarrow{BC}$, then there is a unique ray AP that is $\begin{Bmatrix}\text{parallel} \\ \text{antiparallel}\end{Bmatrix}$ to $\overrightarrow{BC}$.

THEOREM 8-11 $\overrightarrow{AB} \parallel \overrightarrow{CD} \longleftrightarrow \overrightarrow{AB} \downharpoonleft\upharpoonright \overrightarrow{DC}$.

Proofs for Theorems 8-10 and 8-11 are considered in the exercises.

THEOREM 8-12 *If the sum of the measures of two consecutive interior angles is less than 180, then the half lines that are the interiors of the sides of these angles that do not lie in the transversal, intersect in a point.*

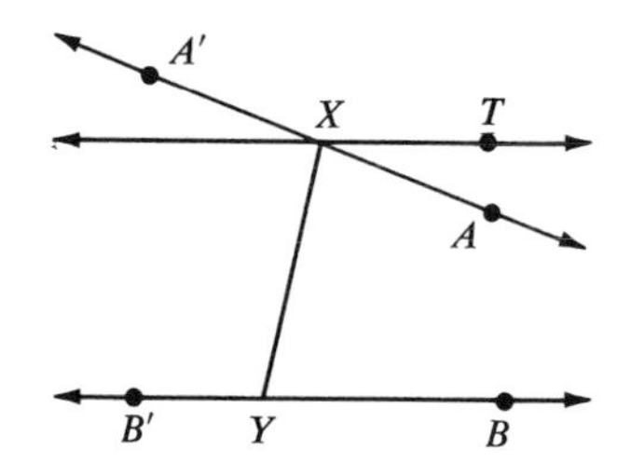

We must prove:

$$\left.\begin{array}{l}(1)\ B\underline{Y}X \text{ and } Y\underline{X}A \text{ are consecutive interior angles} \\ (2)\ m\angle BYX + m\angle YXA < 180\end{array}\right\} \longrightarrow \overset{\circ\rightarrow}{XA} \times \overset{\circ\rightarrow}{YB}.$$

OUTLINE OF PROOF

Since $B\underline{Y}X$ and $Y\underline{X}A$ are consecutive interior angles that are not supplementary, we know by the contrapositive of Corollary 2 to Theorem 8-8 that $\overleftrightarrow{XA} \times \overleftrightarrow{BY}(P)$. Since $P \in \overleftrightarrow{XA}$, it must be in $\overset{\circ\rightarrow}{XA}$ or $\overset{\circ\rightarrow}{XA'}$, where $\overrightarrow{XA'}$ is the ray opposite $\overrightarrow{XA}$. Similarly, P must be in $\overset{\circ\rightarrow}{YB}$ or $\overset{\circ\rightarrow}{YB'}$, where $\overrightarrow{YB'}$ is the ray opposite $\overrightarrow{YB}$. (We note that $P \neq X$ and $P \neq Y$.) We shall show that $P \in \overset{\circ\rightarrow}{XA}$ by showing that it cannot be in $\overset{\circ\rightarrow}{XA'}$, and we shall show that $P \in \overset{\circ\rightarrow}{YB}$ because it cannot be in $\overset{\circ\rightarrow}{YB'}$.

Clearly $\overrightarrow{XA} \nparallel \overrightarrow{YB}$. Therefore, by Theorem 8-10, there is a ray XT that is parallel to $\overrightarrow{YB}$. It follows that $m\angle YXT + m\angle XYB = 180$. From this statement and premise (2) we have $\underline{m\angle YXA < m\angle YXT}$. Moreover, $\underline{A \in \overleftrightarrow{XY}/T}$, because $A \in \overleftrightarrow{XY}/B$ by our definition of consecutive interior angles (premise 1) and $\overleftrightarrow{XY}/T = \overleftrightarrow{XY}/B$ by our definition of parallel rays ($\overrightarrow{XT} \parallel \overrightarrow{YB}$). The underlined statements imply that A is in the interior of $\angle YXT$. Thus $A \in XT/Y$; that is, $\overleftrightarrow{XT}/A = \overleftrightarrow{XT}/Y$. Since we have A'-X-A and $T \notin \overleftrightarrow{AA'}$, we know that A' is in the opposite side of $\overleftrightarrow{XT}$ from A. Thus $\overset{\circ}{\overrightarrow{XA'}} \subseteq \overleftrightarrow{XT}/\sim Y$. We also know by Corollary 2 to Theorem 8-1 that $\overleftrightarrow{YB} \subseteq \overleftrightarrow{XT}/Y$. Therefore, $\overset{\circ}{\overrightarrow{XA'}}$ cannot intersect $\overleftrightarrow{YB}$, because $\overset{\circ}{\overrightarrow{XA'}}$ and $\overleftrightarrow{YB}$ lie in opposite half planes with common edge $\overleftrightarrow{XT}$. Hence our point of intersection P is in $\overset{\circ}{\overrightarrow{XA}}$. It follows that $P \in \overleftrightarrow{XY}/B$. Therefore, $P \notin \overset{\circ}{\overrightarrow{YB'}}$, because $\overset{\circ}{\overrightarrow{YB'}} \subseteq \overleftrightarrow{XY}/\sim B$. Hence $P \in \overset{\circ}{\overrightarrow{YB}}$. Since P is common to $\overset{\circ}{\overrightarrow{XA}}$ and $\overset{\circ}{\overrightarrow{YB}}$, and $\overleftrightarrow{XA} \asymp \overleftrightarrow{BY}$, we have $\overset{\circ}{\overrightarrow{XA}} \asymp \overset{\circ}{\overrightarrow{YB}}(P)$.

Recall from Chapter 6 our proving (Theorem 6-7) that $S \subseteq I$, where I represents the interior of a given angle and S represents the set of interior points of segments having one endpoint in the interior of one side of the angle and the other endpoint in the interior of the other side. Using the parallel postulate, we now prove that $I \subseteq S$.

We must prove:

$$\left.\begin{array}{l} COD \text{ is an angle} \\ P \text{ is in the interior of } \angle COD \end{array}\right\} \longrightarrow \left\{\begin{array}{l} \text{There is a segment } AB \text{ such that} \\ A\text{-}P\text{-}B \wedge A \in \overset{\circ}{\overrightarrow{OC}} \wedge B \in \overset{\circ}{\overrightarrow{OD}}. \end{array}\right.$$

PLAN OF PROOF

There exists a line that contains P and is parallel to $\overleftrightarrow{OC}$. This line intersects $\overset{\circ}{\overrightarrow{OD}}$ in a single point T. We select a point B such that O-T-B and use Theorem 8-12 to prove that $\overset{\circ}{\overrightarrow{BP}} \asymp \overset{\circ}{\overrightarrow{OC}}$.

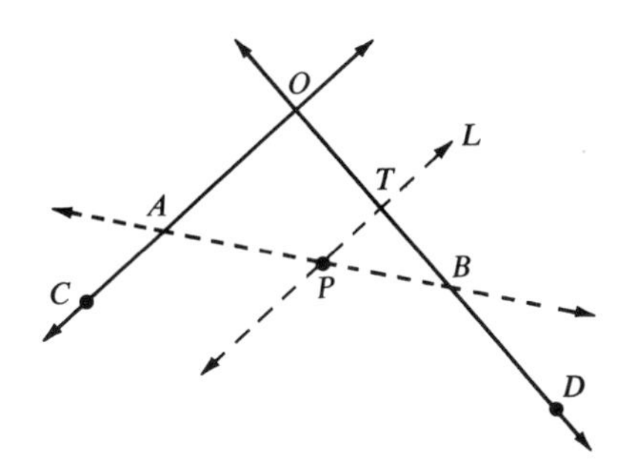

PROOF

In the plane of $\angle COD$ there exists a line L that contains P and is parallel to $\overleftrightarrow{OC}$. Since COD is an angle, we know that $\overleftrightarrow{OC}$ intersects $\overleftrightarrow{OD}$ in a single point O. According to Theorem 8-7, $\overleftrightarrow{OD}$ intersects L in a single point, which we shall call point T $[\overleftrightarrow{OD} \asymp L(T)]$. Thus $\underline{T \in \overleftrightarrow{OD}}$. Since P is in the interior of $\angle COD$, we know that $P \in \overleftrightarrow{OC}/D$ and $P \in \overleftrightarrow{OD}/C$. Therefore, $\overleftrightarrow{OC}/P = \overleftrightarrow{OC}/D$. By Corollary 1 to Theorem 8-1 we know that $L \subseteq \overleftrightarrow{OC}/P$. Hence $L \subseteq \overleftrightarrow{OC}/D$. Since $T \in L$, we have $\underline{T \in \overleftrightarrow{OC}/D}$. Using Corollary 3 to Theorem 5-8, we see that the underlined statements imply that $T \in \overset{\circ}{\overrightarrow{OD}}$. Therefore, $\overrightarrow{OT} = \overrightarrow{OD}$ and $\overleftrightarrow{OT} = \overleftrightarrow{OD}$. Thus

$$\left.\begin{array}{r} P \in \overleftrightarrow{OD}/C \longrightarrow P \in \overleftrightarrow{OT}/C \longrightarrow C\underline{O}T \text{ and } O\underline{T}P \text{ are consecutive interior angles} \\ \overleftrightarrow{OC} \parallel \overleftrightarrow{TP} \end{array}\right\} \dashv$$

$$\leftrightarrow \underline{m\angle COT + m\angle OTP = 180}.$$

Now there is a point B such that O-T-B, which implies that $\overleftrightarrow{OB} = \overleftrightarrow{OT}$ and $B \in \overline{OT}$. Since $\overrightarrow{OT} = \overrightarrow{OD}$, we have $B \in \overrightarrow{OD}$, and since $B \neq O$, we have $B \in \overset{\circ}{\overrightarrow{OD}}$. Moreover, $P \in \overleftrightarrow{OT}/C$ and $\overleftrightarrow{OB} = \overleftrightarrow{OT}$ implies $P \in \overleftrightarrow{OB}/C$ and $P \notin \overleftrightarrow{OT}$. Therefore,

$$\left.\begin{array}{l} P \notin \overleftrightarrow{OB} \\ O\text{-}T\text{-}B \end{array}\right\} \longrightarrow O\underline{T}P \text{ is an exterior angle for } \triangle PTB \dashv$$

$$\leftrightarrow m\angle TBP < m\angle OTP \dashv$$

$$\leftrightarrow \underline{m\angle COT + m\angle TBP < m\angle COT + m\angle OTP}.$$

The last two underlined statements imply that $\underline{m\angle COT + m\angle TBP < 180}$. Since $P \in \overleftrightarrow{OB}/C$, we note that $\underline{\underline{\angle COT \text{ and } \angle TBP \text{ are consecutive interior angles}}}$. According to Theorem 8-12, the statements with the double underlines imply that $\overset{\circ}{\overrightarrow{OC}} \times \overset{\circ}{\overrightarrow{BP}}(A)$ and hence $A \in \overset{\circ}{\overrightarrow{OC}}$ and $\overleftrightarrow{OA} = \overleftrightarrow{OC}$. $\overrightarrow{OD} \cap \overleftrightarrow{OC} = \{O\}$, and therefore $\overset{\circ}{\overrightarrow{OD}} \subseteq \overleftrightarrow{OC}/D$. However, $B \in \overset{\circ}{\overrightarrow{OD}}$; hence $B \in \overleftrightarrow{OC}/D$ or, equivalently, $\overleftrightarrow{OC}/B = \overleftrightarrow{OC}/D$. Thus,

$$P \in \overleftrightarrow{OC}/D \longrightarrow P \in \overleftrightarrow{OC}/B \longrightarrow \underline{P \in \overleftrightarrow{OA}/B}.$$

Since $A \in \overset{\circ}{\overrightarrow{BP}}$, we have $\overrightarrow{BA} = \overrightarrow{BP}$, and accordingly $P \in \overrightarrow{BA}$. But $B \neq P$, and therefore $\underline{P \in \overset{\circ}{\overrightarrow{BA}}}$. From the last two underlined statements, we have $P \in \overleftrightarrow{OA}/B \cap \overset{\circ}{\overrightarrow{BA}}$. By L-Exercise 9, page 112, $\overleftrightarrow{OA}/B \cap \overset{\circ}{\overrightarrow{BA}} = \overset{\circ}{\overline{BA}}$. Hence $P \in \overset{\circ}{\overline{BA}}$.

We have proved that any point in the interior of an angle is an interior point of some segment having one endpoint in the interior of one side of the angle and the other endpoint in the interior of the other side. Thus we have proved the following theorem.

THEOREM 8-13 *If I represents the interior of a given angle and S represents the set of interior points of segments having one endpoint in the interior of one side of the angle and the other endpoint in the interior of the other side, then $I \subseteq S$.*

In the exercises below, you are asked to prove that $I \subseteq S \longrightarrow$ parallel postulate. From this statement and Theorem 8-13 we see that, for the axiomatic system presented in Chapters 1 through 7, the parallel postulate and the statement $I \subseteq S$ are equivalent.

Exercises

L 1. Prove: $\overleftrightarrow{AB} \parallel \overleftrightarrow{CD} \longrightarrow B \notin \overleftrightarrow{AC}$.

2. Prove:

$$\left.\begin{array}{l} m\angle TQA > m\angle BOQ \\ B \in \overleftrightarrow{\mathring{O}\mathring{Q}}/A \\ O\text{-}Q\text{-}T \end{array}\right\} \longrightarrow \overrightarrow{\mathring{Q}A} \times \overrightarrow{\mathring{O}B}.$$

3. Prove:

$$\left.\begin{array}{l} \overrightarrow{\mathring{O}Y} \times \overrightarrow{\mathring{O}X} \\ X\text{-}Q\text{-}X' \\ Y\text{-}O\text{-}Y' \\ O\text{-}Q\text{-}W \end{array}\right\} \longrightarrow \left\{\begin{array}{l} m\angle Y'OQ > m\angle WQX' \\ \overrightarrow{\mathring{O}Y'} \cap \overrightarrow{\mathring{Q}X'} = \emptyset. \end{array}\right.$$

4. Prove:

$$\left.\begin{array}{l} S\text{-}O\text{-}Q\text{-}T \\ A \in \overleftrightarrow{ST}/B \\ m\angle TQA + m\angle BOS > 180 \end{array}\right\} \longrightarrow \overrightarrow{\mathring{Q}A} \times \overrightarrow{\mathring{O}B}.$$

5. Prove Theorem 8-10:
   (a) $A \notin \overleftrightarrow{BC} \longrightarrow$ There is a unique ray AP that is parallel to $\overrightarrow{BC}$.
   (b) $A \notin \overleftrightarrow{BC} \longrightarrow$ There is a unique ray AP that is antiparallel to $\overrightarrow{BC}$.
6. Prove Theorem 8-11.

L 7. Prove:

$$\left.\begin{array}{l} \overrightarrow{AB} \parallel \overrightarrow{CD} \\ X \in \mathring{C}\mathring{D} \end{array}\right\} \longrightarrow \left\{\begin{array}{l} \overrightarrow{AB} \parallel \overrightarrow{XD} \\ \overrightarrow{AB} \downarrow\!\uparrow \overrightarrow{XC}. \end{array}\right.$$

8. Prove:

$$\left.\begin{array}{l} \overrightarrow{AB} \parallel \overrightarrow{XY} \\ \overrightarrow{CD} \parallel \overrightarrow{XY} \\ X \in \overleftrightarrow{AC} \\ A \neq C \end{array}\right\} \longrightarrow \overrightarrow{AB} \parallel \overrightarrow{CD}.$$

L 9. Prove: $\overrightarrow{AB} \parallel \overrightarrow{CD} \longrightarrow \overrightarrow{BA} \parallel \overrightarrow{DC}$.

L 10. Prove: $\overrightarrow{AB} \parallel \overrightarrow{CD} \longrightarrow ABDC$ is a convax quadrilateral.

11. Prove:
   (a) If two rays are in distinct lines and are parallel to the same ray, then they are parallel to each other, provided that all three rays are coplanar.
   (b) If two rays are in distinct lines and are antiparallel to the same ray, then they are parallel to each other, provided that all three rays are coplanar.
12. Prove:

$$\left.\begin{array}{l} \overrightarrow{AB} \parallel \overrightarrow{CD} \\ O \in \text{interior } \angle BAC \\ O \in \text{interior } \angle ACD \end{array}\right\} \longrightarrow \overrightarrow{\mathring{A}O} \times \overrightarrow{\mathring{C}D}(T) \text{ such that } A\text{-}O\text{-}T.$$

13. Points X and Y are in I, the interior of $\triangle ABC$. Prove that $\overrightarrow{\mathring{B}X} \times \overrightarrow{\mathring{C}Y}(O)$ such that $O \in I$.

L 14. Let L be a line and let P be a point not in L such that there are two distinct lines L_1 and L_2 through P each of which is parallel to L. Prove that L is wholly contained in the interior of an angle that is a subset of $L_1 \cup L_2$.

L 15. Let I be the interior of a given angle and let S be the set of interior points of segments having one endpoint in the interior of one side of the angle and the other endpoint in the interior of the other side of the angle. Prove that $I \subseteq S$ implies the parallel postulate.

Sum of the Measures of the Angles of a Triangle

We are now in a position to prove an important theorem that characterizes Euclidean geometry.

THEOREM 8-14 *The sum of the measures of the angles of a triangle is* 180.

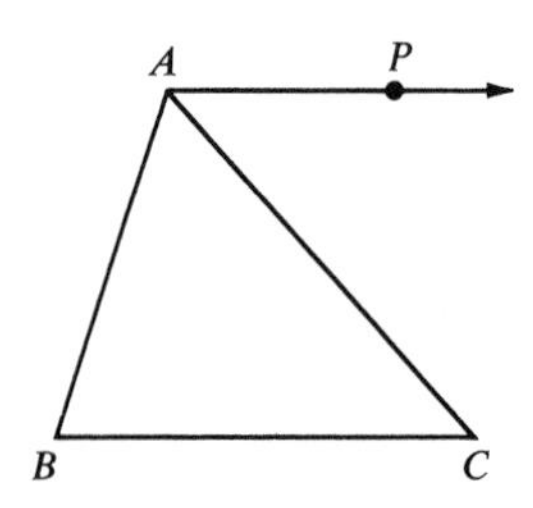

In terms of our drawing we must prove

$$ABC \text{ is a triangle} \longrightarrow m\angle BCA + m\angle CAB + m\angle ABC = 180.$$

PROOF

Statement	Reason
1. $A \notin \overleftrightarrow{CB}$.	1. Hypothesis and definition of triangle.
2. There is a ray AP such that $\overrightarrow{AP} \downarrow\uparrow \overrightarrow{CB}$.	2. Statement 1 and Theorem 8-10.
3. $\overleftrightarrow{AP} \parallel \overleftrightarrow{BC}$ and $P \in \overleftrightarrow{AC}/\sim B$.	3. Statement 2 and definition of antiparallel rays.
4. $P\underline{A}C$ and $A\underline{C}B$ are alternate interior angles.	4. Statement 3 and definition of alternate interior angles.
5. $m\angle PAC = m\angle ACB$.	5. Statements 3, 4, and Theorem 8-8.
6. $\overrightarrow{AP} \parallel \overrightarrow{BC}$.	6. Statement 2 and Theorem 8-11.
7. $P \in \overleftrightarrow{AB}/C$.	7. Statement 6 and definition of parallel rays.
8. $P\underline{A}B$ and $A\underline{B}C$ are consecutive interior angles.	8. Statement 7 and definition of consecutive interior angles.
9. $m\angle PAB + m\angle ABC = 180$.	9. Statements 3, 8, and Corollary 2 to Theorem 8-8.
10. $C \in \overleftrightarrow{AB}/P$.	10. Statement 7 and plane separation lemma.
11. $C \in \overleftrightarrow{AP}/B$.	11. Statement 3 and Corollary 2 to Theorem 8-1.
12. $C \in \overleftrightarrow{AP}/B \cap \overleftrightarrow{AB}/P$.	12. Statements 10, 11, and definition of intersection of sets.
13. $C \in$ interior $\angle PAB$.	13. Statement 12 and definition of interior of an angle.
14. $\overrightarrow{AC}$ is between $\overrightarrow{AP}$ and $\overrightarrow{AB}$.	14. Statement 13 and Theorem 6-6.
15. $m\angle PAC + m\angle CAB = m\angle PAB$.	15. Statement 14 and Theorem 6-4.
16. $m\angle ACB + m\angle CAB + m\angle ABC = 180$.	16. Statements 5, 9, 15, and substitution property of equality.

COROLLARY 1 TO THEOREM 8-14 *The measure of an exterior angle of a triangle is equal to the sum of the measures of its two nonadjacent interior angles.*

COROLLARY 2 TO THEOREM 8-14 *If two triangles have two pairs of corresponding angles congruent, then the third pair of angles are congruent.*

COROLLARY 3 TO THEOREM 8-14 *The measure of each angle of an equilateral triangle is* 60.

COROLLARY 4 TO THEOREM 8-14 *The sum of the measures of the angles of a convax quadrilateral is* 360.

The Sum of the Measures of the Angles of a Convax Polygon

We have noted (Theorem 8-14) that the sum of the measures of the angles of a triangle is 180 and the sum of the measures of the angles of a quadrilateral is 360 (Corollary 4 to Theorem 8-14). Thus if S_n represents the sum of the measures of the angles of a convax polygon having n sides, we see that $S_3 = 1 \cdot 180$ and $S_4 = 2 \cdot 180$. These facts suggest that $S_n = (n - 2)180$ for $n \geqq 3$. The proof involves mathematical induction.*

Having verified our formula for $n = 3$ and $n = 4$, we must now prove that

$$S_n = (n - 2)180 \longrightarrow S_{n\;1} = [(n + 1) - 2]\,180 = (n - 1)180.$$

Let our convax polygon having $n + 1$ sides be denoted by poly $P_1P_2P_3 \cdots P_{n+1}$. Then

$$S_{n+1} = m\angle P_{n+1}P_1P_2 + m\angle P_1P_2P_3 + m\angle P_2P_3P_4 + \cdots + m\angle P_nP_{n+1}P_1.$$

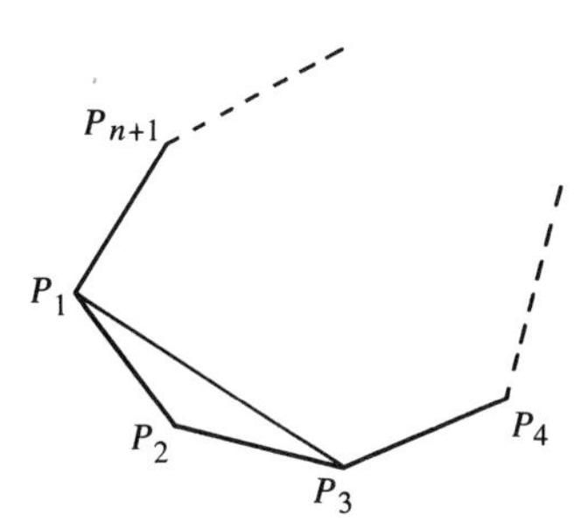

Since our polygon is convax, $\overrightarrow{P_1P_3}$ is between $\overrightarrow{P_1P_{n+1}}$ and $\overrightarrow{P_1P_2}$, and $\overrightarrow{P_3P_1}$ is between $\overrightarrow{P_3P_2}$ and $\overrightarrow{P_3P_4}$ (L-Exercise 20, page 144). Thus

$$m\angle P_{n+1}P_1P_2 = m\angle P_{n+1}P_1P_3 + m\angle P_3P_1P_2$$

and

$$m\angle P_2P_3P_4 = m\angle P_2P_3P_1 + m\angle P_1P_3P_4.$$

Substituting, we have

$$S_{n+1} = (m\angle P_{n+1}P_1P_3 + m\angle P_3P_1P_2) + m\angle P_1P_2P_3 + (m\angle P_2P_3P_1 + m\angle P_1P_3P_4) + \cdots + m\angle P_nP_{n+1}P_1.$$

* For a discussion of mathematical induction, see Appendix B.

Rearranging terms, we have

$$S_{n+1} = (m\angle P_3P_1P_2 + m\angle P_1P_2P_3 + m\angle P_2P_3P_1) + (m\angle P_{n+1}P_1P_3 + m\angle P_1P_3P_4 + \cdots + m\angle P_nP_{n+1}P_1)$$

$=$ sum of the measures of the angles of $\triangle P_1P_2P_3$ + sum of the measures of the angles of poly $P_1P_3P_4 \cdots P_{n+1}$.

By L-Exercise 21, page 144, poly $P_1P_3P_4 \cdots P_{n+1}$ is a convax polygon. Clearly, it has n sides. Therefore, the sum of the measures of the angles of poly $P_1P_3P_4 \cdots P_{n+1}$ is denoted by S_n. Therefore,

$$S_{n+1} = S_n + 180.$$

$$S_n = (n-2)180 \longrightarrow S_n + 180 = (n-2)180 + 180 \longrightarrow S_{n+1} = (n-1)180.$$

We have proved the following theorem.

THEOREM 8-15 *The sum of the measures of the angles of a convax polygon having n sides* ($n \geqq 3$) *is* $(n-2)180$.

COROLLARY 1 TO THEOREM 8-15 *The measure of each angle of a regular polygon having n sides is* $(n-2)180/n$.

DEFINITION *Let V be a vertex of a convax polygon. Either angle that forms a linear pair with the angle of the polygon whose vertex is V is called an exterior angle of the polygon.*

COROLLARY 2 TO THEOREM 8-15 *For any convax polygon, the sum of the measures of the exterior angles, one at each vertex, is* 360.

Exercises

1. Prove Corollaries 1, 2, and 3 to Theorem 8-14.
2. Prove Corollary 4 to Theorem 8-14.
3. Prove:
$$\left.\begin{array}{l} \overrightarrow{BA} \parallel \overrightarrow{CD} \\ \overrightarrow{CO} \text{ bisects } \angle DCB \\ \overrightarrow{BO} \text{ bisects } \angle ABC \end{array}\right\} \longrightarrow \overleftrightarrow{CO} \perp \overleftrightarrow{BO}.$$
4. Prove Corollary 1 to Theorem 8-15.
5. Prove Corollary 2 to Theorem 8-15.
6. Prove:
$$\left.\begin{array}{l} \overline{BA} \perp \overline{AC} \\ \overleftrightarrow{AD} \perp \overleftrightarrow{BC}(D) \end{array}\right\} \longrightarrow \left\{\begin{array}{l} \angle ABC \cong \angle DAC \\ \angle BCA \cong \angle BAD. \end{array}\right.$$
7. Prove:
$$\left.\begin{array}{l} |AC| = |CD| = |CB| \\ A\text{-}C\text{-}D \\ B \notin \overleftrightarrow{AD} \end{array}\right\} \longrightarrow m\angle ABD = 90.$$

8. Prove:

$$\left.\begin{array}{l} L_1 \perp L_2 \\ L_3 \perp L_1 \\ L_4 \perp L_2 \\ L_1, L_2, L_3, \text{ and } L_4 \text{ are coplanar} \\ L_3 \neq L_2 \text{ and } L_4 \neq L_1 \end{array}\right\} \longrightarrow L_3 \perp L_4.$$

9. Prove:

$$\left.\begin{array}{l} \overrightarrow{AB} \parallel \overrightarrow{CD} \\ O \in \overleftrightarrow{AC}/B \\ O \in \overleftrightarrow{AB}/C \\ O \in \overleftrightarrow{DC}/A \end{array}\right\} \longrightarrow m\angle AOC = m\angle BAO + m\angle DCO.$$

10. Poly $ABCDEFGH$ is a regular polygon; prove $\overline{DA} \parallel \overline{BC}$ and $\overline{FC} \parallel \overline{HA}$.

Quadrilaterals

The idea of parallelism plays a role in the definitions of certain convax quadrilaterals. We state the following definitions for future reference.

DEFINITIONS *A* trapezoid *is a convax quadrilateral in which exactly one pair of opposite sides are parallel. The two parallel sides are called* bases *of the trapezoid, and the line segment that joins the midpoints of the nonparallel sides is called the* median *of the trapezoid.*

An isosceles trapezoid *is a trapezoid in which the nonparallel sides are congruent.*

A parallelogram *is a convax quadrilateral in which both pairs of opposite sides are parallel. Either pair of parallel sides are referred to as* bases *of the parallelogram.*

An altitude *of a trapezoid or parallelogram corresponding to a given base is a line segment that joins a point in the line that contains a base to a point in the line that contains the opposite base and is perpendicular to both lines.*

A rectangle *is a parallelogram that has a right angle.*

A rhombus *is a parallelogram that has two consecutive sides congruent.*

A square *is a parallelogram that has a right angle and two consecutive sides congruent.*

Exercises

L 1. Prove: If a quadrilateral is a parallelogram, then
   (a) Each pair of opposite sides are congruent.
   (b) Each pair of opposite angles are congruent.
   (c) Its diagonals bisect each other.

L 2. Prove:
   (a) $\left.\begin{array}{l} \overrightarrow{AD} \parallel \overrightarrow{BC} \\ \overline{AD} \cong \overline{BC} \end{array}\right\} \longrightarrow ABCD$ is a parallelogram.
   (b) A parallelogram exists.

L 3. Prove: A convex quadrilateral is a parallelogram if
(a) each pair of opposite sides are congruent.
(b) each pair of opposite angles are congruent.
(c) its diagonals bisect each other.

L 4. Prove:
(a) If $ABCD$ is a parallelogram, then $\overrightarrow{AB} \parallel \overrightarrow{DC}$ and $\overrightarrow{AD} \parallel \overrightarrow{BC}$.
(b) If each pair of nonconsecutive sides of a quadrilateral are parallel, then the quadrilateral is a parallelogram.
(c) $\left.\begin{array}{l}\overline{AB} \parallel \overline{DC} \\ \overline{AD} \parallel \overline{BC}\end{array}\right\} \longrightarrow ABCD$ is a parallelogram.

L 5. Prove:
(a) $(\overrightarrow{AD} \parallel \overrightarrow{BC} \wedge \overline{AD} \not\cong BC) \longrightarrow ABCD$ is a trapezoid.
(b) A trapezoid exists.

L 6. If L_1 and L_2 are parallel lines and X and Y are distinct points in L_1, then the distance from X and Y to L_2 are equal.

L 7. Prove:

$$\left.\begin{array}{l} ABC \text{ is a triangle} \\ X \in \overset{\circ\!-\!\circ}{AC} \end{array}\right\} \longrightarrow \left\{\begin{array}{l} \text{There exists a ray } XT \parallel \overrightarrow{AB} \\ \overset{\circ\!\to}{XT} \times \overset{\circ\!-\!\circ}{BC}. \end{array}\right.$$

8. Prove:

$$\left.\begin{array}{l} ABC \text{ is a triangle} \\ M \text{ is the midpoint of } \overline{AB} \\ \overleftrightarrow{MX} \parallel \overleftrightarrow{AC} \end{array}\right\} \longrightarrow \left\{\begin{array}{l} \overleftrightarrow{MX} \times \overset{\circ\!-\!\circ}{BC}(T) \text{ such that} \\ \overline{BT} \cong \overline{TC}. \end{array}\right.$$

L 9. Prove:

$$\left.\begin{array}{l} ABC \text{ is a triangle} \\ M \in \overset{\circ\!-\!\circ}{AC} \\ N \in \overset{\circ\!-\!\circ}{BC} \\ M\text{-}N\text{-}T \end{array}\right\} \longrightarrow \left\{\begin{array}{l} \overleftrightarrow{MN} \neq \overleftrightarrow{CB} \\ \overleftrightarrow{MN} \neq \overleftrightarrow{CA} \\ C \notin \overleftrightarrow{MN} \\ T \in \overleftrightarrow{CB}/\sim M \\ C\underline{N}M \text{ and } B\underline{N}T \text{ are vertical angles.} \end{array}\right.$$

10. Prove:

$$\left.\begin{array}{l} ABC \text{ is a triangle} \\ M \text{ is the midpoint of } \overline{AC} \\ N \text{ is the midpoint of } \overline{BC} \end{array}\right\} \longrightarrow \left\{\begin{array}{l} \overline{MN} \parallel \overline{AB} \\ |MN| = \tfrac{1}{2}|AB|. \end{array}\right.$$

L 11. (a) Prove:

$$\left.\begin{array}{l} ABC \text{ is an equilateral triangle.} \\ M, N, P \text{ are the midpoints of} \\ \overline{AC}, \overline{BC}, \text{ and } \overline{AB}, \text{ respectively.} \end{array}\right\} \longrightarrow \left\{\begin{array}{l} MNP \text{ is an equilateral} \\ \text{triangle.} \end{array}\right.$$

(b) Use part (a) to prove that the statement "Two triangles are congruent provided that the three angles of one triangle are congruent respectively to the three angles of the other triangle" is false. (Observe that this provides a counterexample for the statement symbolized as AAA in Chapter 7.)

12. Prove:

$$\left.\begin{array}{l} ABCD \text{ is a trapezoid with bases } \overline{AB} \text{ and } \overline{DC} \\ |DC| < |AB| \\ \overrightarrow{CX} \parallel \overrightarrow{DA} \end{array}\right\} \longrightarrow \overset{\circ\!\to}{CX} \times \overset{\circ\!-\!\circ}{AB}.$$

13. Prove: $ABCD$ is an isosceles trapezoid with bases $\overline{AB}$ and $\overline{DC} \longrightarrow \angle DAB \cong \angle CBA$.

14. Prove:

$$\left.\begin{array}{l}\overline{AB} \text{ and } \overline{DC} \text{ are the bases of} \\ \text{trapezoid } ABCD. \\ N \in \overset{\circ\!-\!\circ}{CB}.\end{array}\right\} \longrightarrow \overrightarrow{DN} \times \overleftrightarrow{AB}(T) \text{ such that } A\text{-}B\text{-}T$$

15. Prove: The median of a trapezoid is parallel to the bases and its length is one-half the sum of the lengths of the bases.
16. Prove: If $ABCD$ is a convax quadrilateral and M, N, O, and P are the midpoints of $\overline{AB}$, $\overline{BC}$, $\overline{CD}$, and $\overline{DA}$, respectively, then $MNOP$ is a parallelogram.

L 17. Prove: If two coplanar angles have their sides respectively parallel, then the angles are congruent.

References

Golos, *Foundations of Euclidean and Non-Euclidean Geometry*
Hemmerling, *Fundamentals of College Geometry*
Moise, *Elementary Geometry from an Advanced Standpoint*
Prenowitz and Jordan, *Basic Concepts of Geometry*
Ringenberg, *College Geometry*

Chapter 9
Lines and Planes in Space

In this chapter we continue our study of incidence relations involving points, lines, and planes in space.

Lines Perpendicular to Planes

Before we define the phrase "a line and a plane are perpendicular to each other," we review some facts about intersecting lines and planes.

Postulate 8 tells us that if two distinct planes M_1 and M_2 intersect, their intersection is a line L. We use the symbol $M_1 \cap M_2(L)$ for the statement "Planes M_1 and M_2 intersect in a line that we shall call line L." If the intersection of two planes is a line, then the planes are distinct intersecting planes. Why? Thus for planes M_1 and M_2 we have

(i) $$(M_1 \cap M_2 \neq \emptyset \wedge M_1 \neq M_2) \longleftrightarrow M_1 \cap M_2(L).$$

Theorem 3-9 tells us that if a line L intersects a plane M not containing it, the intersection is a single point. We can call this point P. If we use the symbol $L \times M(P)$ for the statement "Line L and plane M intersect in a single point that we shall call P," Theorem 3-9 can be expressed as follows:

$$\left.\begin{matrix} L \cap M \neq \emptyset \\ L \nsubseteq M \end{matrix}\right\} \longrightarrow L \times M(P).$$

On the other hand, if the intersection of a line L and a plane M is a single point, we know that their intersection is not empty and that the line does not lie in the plane. Why? Thus for line L and plane M we have

(ii) $$(L \cap M \neq \emptyset \wedge L \nsubseteq M) \longleftrightarrow L \times M.$$

DEFINITIONS A line and a plane are perpendicular to each other *if and only if they intersect in a single point and any line that contains this point and lies in the plane is perpendicular to the given line.* A segment or a ray is perpendicular to a plane *if and only if the line that contains it is perpendicular to that plane.*

THEOREM 9-1 *If a line is perpendicular to each of two distinct intersecting lines at their point of intersection, then the line is perpendicular to the plane determined by them.*

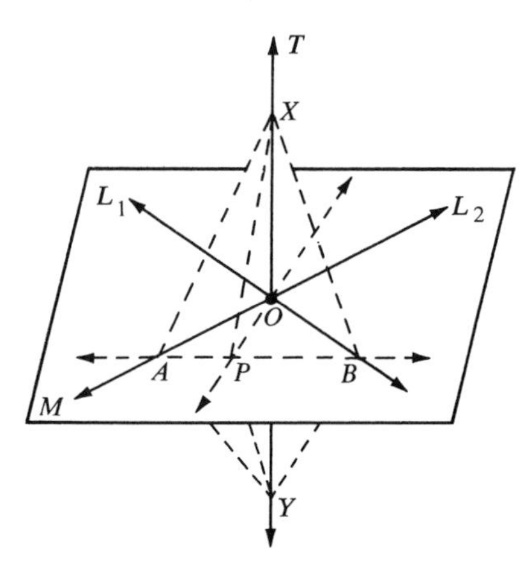

$$\left.\begin{array}{l} L_1 \times L_2(O) \\ \textit{Line } T \perp L_1(O) \\ \textit{Line } T \perp L_2(O) \\ L_1 \textit{ and } L_2 \textit{ determines plane } M \end{array}\right\} \longrightarrow T \perp M(O).$$

In order to prove this important theorem, we must apply the definition of a line perpendicular to a plane. If $\overleftrightarrow{OP}$ is any line in M, we must prove that $T \perp \overleftrightarrow{OP}$.

OUTLINE OF PROOF

If $\overleftrightarrow{OP} = L_1$ or $\overleftrightarrow{OP} = L_2$, this is true by the hypothesis. If $\overleftrightarrow{OP} \neq L_1$ and $\overleftrightarrow{OP} \neq L_2$, then P is in the interior of one of the four angles formed by L_1 and L_2. Theorem 8-13 tells us that there is a line L that contains P and intersects L_1 and L_2 in points A and B, respectively, so that A-P-B. Also AOB is a triangle. Select points X and Y in T so that X-O-Y and $|XO| = |YO|$. If we can prove that $|XP| = |YP|$, then it follows that $\overleftrightarrow{PO} \perp T$. Why?

To prove that $|XP| = |YP|$ we first observe that $\overleftrightarrow{AO}$ is a perpendicular bisector of $\overline{XY}$ and $\overleftrightarrow{BO}$ is a perpendicular bisector of $\overline{XY}$. Moreover, neither of the points X or Y is in plane M, because, if so, we would have two lines (L_1 and L_2) each perpendicular to the same line (T) at the same point O with all three lines in the same plane. Therefore, $X \notin \overleftrightarrow{AB}$, $Y \notin \overleftrightarrow{AB}$, $X \notin \overleftrightarrow{AP}$, and $Y \notin \overleftrightarrow{AP}$. It follows that ABX, ABY, APX, and APY are triangles. The rest of the argument can be flow-diagramed as follows.

$$\left.\begin{array}{r} \overleftrightarrow{OA} \text{ is a perpendicular bisector of } \overline{XY} \longrightarrow \overline{AX} \cong \overline{AY} \\ \overleftrightarrow{OB} \text{ is a perpendicular bisector of } \overline{XY} \longrightarrow \overline{BX} \cong \overline{BY} \\ \overline{AB} \cong \overline{AB} \end{array}\right\} \longrightarrow \triangle ABX \cong \triangle ABY \dashv$$

$$\left.\begin{array}{l}\leftrightarrow \angle XAP \cong \angle YAP \\ \overline{AP} \cong \overline{AP} \\ \overline{AX} \cong \overline{AY}\end{array}\right\} \longrightarrow \triangle XAP \cong \triangle YAP \longrightarrow \overline{XP} \cong \overline{YP} \longrightarrow \left.\begin{array}{l}|XP| = |YP| \\ |XO| = |YO|\end{array}\right\} \dashv$$

$$\leftrightarrow \overleftrightarrow{PO} \text{ is a perpendicular bisector of } \overline{XY} \longrightarrow \overleftrightarrow{PO} \perp T.$$

Thus we have proved that if $\overleftrightarrow{OP} \neq L_1$ and $\overleftrightarrow{OP} \neq L_2$ and $\overleftrightarrow{OP}$ is in M, then $\overleftrightarrow{OP} \perp T$. Recall that if $\overleftrightarrow{OP} = L_1$ or $\overleftrightarrow{OP} = L_2$, then $\overleftrightarrow{OP} \perp T$. The conjunction of these two underlined statements tell us that if $\overleftrightarrow{OP}$ is any line in M through O, then $\overleftrightarrow{OP} \perp T$. Therefore, $T \perp M$ according to the definition above, and our proof is complete.

The proofs of the following theorems are considered in the exercises.

THEOREM 9-2 *Through a given point there is a unique plane perpendicular to a given line.*

THEOREM 9-3 *All the perpendiculars to a given line at a given point in the line lie in the plane that is perpendicular to the line at that point.*

THEOREM 9-4 *Through a given point there is a unique line perpendicular to a given plane.*

Parallel Lines and Planes

Theorem 3-10 tells us that if a plane intersects two parallel planes, then the lines of intersection are coplanar and nonintersecting. Since lines that are coplanar and nonintersecting are parallel lines, we have the following theorem.

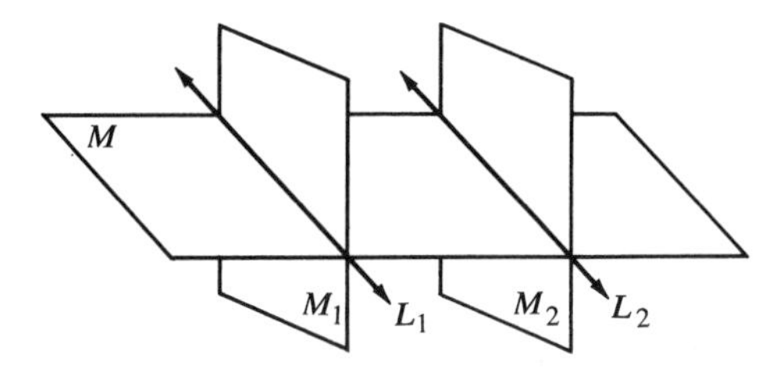

THEOREM 9-5 *If a plane intersects each of two parallel planes in a line, then these lines of intersection are parallel.*

Suppose that line L_1 intersects plane M in a single point O and that line L_2 is parallel to L_1. Since $L_1 \times M(O)$ we know that $L_1 \nsubseteq M$. L_1 and L_2 are parallel and by Theorem 8-1, L_1 and L_2 are contained in exactly one plane, which we call Z. We know that $Z \neq M$, since $L_1 \subseteq Z$ and $L_1 \nsubseteq M$. However, each of the planes contains point O and thus $Z \cap M \neq \emptyset$. Therefore, M and Z intersect in a line L. Because $L \subseteq M$ and $L_1 \nsubseteq M$, we have $L_1 \neq L$. O is contained in L_1 and L; therefore, $L_1 \times L(O)$. But $L_1 \parallel L_2$ and L_1, L, and L_2 are contained in plane Z.

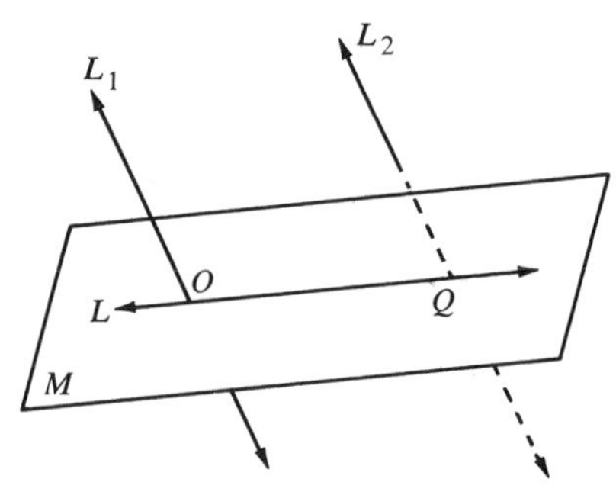

From the three underlined statements we have $L_2 \times L(Q)$ by Theorem 8-7. $Q \in L$ implies $Q \in M$ and therefore $L_2 \cap M \neq \emptyset$. We also know that $L_2 \neq L$ and $L = Z \cap M$; thus $L_2 \neq Z \cap M$. Since $L_2 \subseteq Z$, we must have $L_2 \nsubseteq M$. The double-underlined statements imply $L_2 \times M$ from (ii), page 195. We have proved the following theorem.

THEOREM 9-6 *If a plane intersects one of two parallel lines in a single point, then it intersects the other line in a single point.*

COROLLARY TO THEOREM 9-6 *If a plane contains exactly one of two parallel lines, then it is parallel to the other.*

By interchanging the words "plane" and "line" in Theorem 9-6, we have the following theorem, whose proof is considered in the exercises.

THEOREM 9-7 *If a line intersects one of two parallel planes in a single point, then it intersects the other plane in a single point.*

Recall that in Chapter 8 we stated that a line is parallel to a plane if and only if they do not intersect. As a result of our definition, we proved (Theorem 8-3) the following theorem, which we now restate as Theorem 9-8.

THEOREM 9-8 *If two planes intersect in a line, then any line that is in one of the planes and is parallel to the other plane is parallel to the line of intersection.*

The following corollaries to Theorem 9-8 are considered in the exercises.

COROLLARY 1 TO THEOREM 9-8 *If two lines intersect in a single point and each is parallel to a given plane, then the plane determined by these lines is parallel to the given plane.*

COROLLARY 2 TO THEOREM 9-8 *If a line is parallel to a plane, then any line that contains a point in the plane and is parallel to the given line lies in the plane.*

COROLLARY 3 TO THEOREM 9-8 *If two lines that intersect in a single point and lie in plane M_1 are parallel respectively to two lines that intersect in a single point and lie in a different plane M_2, then $M_1 \parallel M_2$.*

Exercises

1. Q, A, and B are points in plane M. P is a point not in M.
   (a) If $\overline{PQ} \perp \overline{QA}$ and $\overline{PQ} \perp \overline{QB}$, must $\overline{PQ} \perp M$? Why?
   (b) If Q, A, and B are noncollinear points and $\overline{PQ} \perp \overline{QA}$ and $\overline{PQ} \perp \overline{QB}$, must $\overline{PQ} \perp M$? Why?
2. (a) Suppose point $P \in \text{pl}\, M$ and $\overleftrightarrow{PR} \perp M$ and $\overleftrightarrow{PQ} \perp M$. What conclusion can you draw about $\overleftrightarrow{PR}$ and $\overleftrightarrow{PQ}$? Must $Q = R$?
   (b) Suppose A, B, and C are points such that $A \notin M$ and $B, C \in M$. If $\overleftrightarrow{AB} \perp M$ and $\overleftrightarrow{AC} \perp M$, must $\overleftrightarrow{AB} = \overleftrightarrow{AC}$? Must $B = C$?
3. P is a point and M is a plane such that $P \notin M$.
   (a) There is a line $\overleftrightarrow{PQ} \perp M$. Why? Let $\overleftrightarrow{PR}$ be any line perpendicular to $\overleftrightarrow{PQ}$ at P. Use an indirect argument to explain why $\overleftrightarrow{PR}$ is parallel to M.
   (b) How many lines are perpendicular to $\overleftrightarrow{PQ}$ at P? Why are they coplanar?
4. M_1 and M_2 are parallel planes, A, $B \in M_1$; C, $D \in M_2$; and $\overline{AD} \parallel \overline{BC}$. Prove $\overline{AD} \cong \overline{BC}$.
5. There are four parts to the proof of Theorem 9-2. We must prove both existence and uniqueness for each of the two cases:

   Case 1: The given point P is in the line L.
   Case 2: The given point P is not in line L.

   (a) Prove existence for Case 1. (b) Prove uniqueness for Case 1.
   (c) Prove existence for Case 2. (d) Prove uniqueness for Case 2.
6. Prove Theorem 9-3.
7. In proving Theorem 9-4 we must prove both existence and uniqueness when the given point P is in the given plane M (Case 1) and when the point P is not in the given plane (Case 2).
   (a) Prove existence for Case 1. (b) Prove uniqueness for Case 1.
   (c) Prove existence for Case 2. (d) Prove uniqueness for Case 2.
8. Prove: If two angles are not coplanar and the sides of one are respectively parallel to the sides of the other, then the planes determined by these angles are parallel.
9. Write an essay proof of the corollary to Theorem 9-6.
10. Prove Theorem 9-7.
11. Write an indirect essay proof of Corollary 1 to Theorem 9-8.
12. Write an indirect essay proof of Corollary 2 to Theorem 9-8.
13. Prove Corollary 3 to Theorem 9-8.

Parallel Lines and Planes in Space

If a line is perpendicular to one of two parallel planes, we can prove that it is perpendicular to the other. Thus for line L and planes M_1 and M_2 we can prove

$$\left.\begin{array}{l} L \perp M_1 \\ M_1 \parallel M_2 \end{array}\right\} \longrightarrow L \perp M_2.$$

OUTLINE OF PROOF

Let line L be perpendicular to plane M_1 at O. Then L intersects M_1 in a single point. According to Theorem 9-7, L must also intersect M_2 in a single point, which we shall call Q. Let P be any point not in L. Then P and L determine

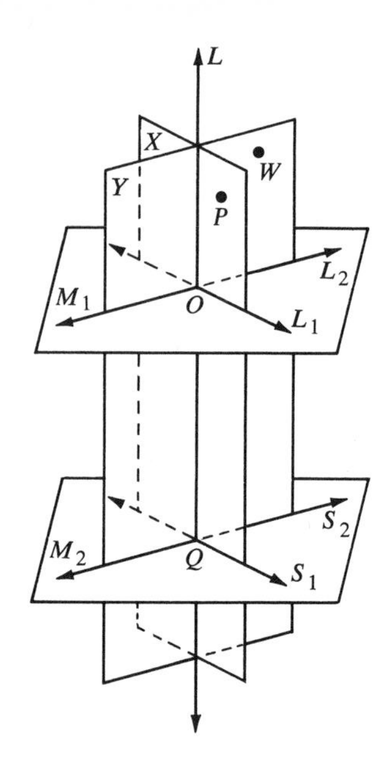

a plane X that is different from either M_1 and M_2 and intersects each of them. Therefore, by Theorem 9-5, plane X intersects planes M_1 and M_2 in parallel lines L_1 and S_1, respectively. Let W be any point not in plane X. Then W is not in line L; hence point W and line L determine a plane that we shall call plane Y. Y is different from any of the planes X, M_1, and M_2 and intersects each of them. Therefore, plane Y intersects parallel planes M_1 and M_2 in parallel lines L_2 and S_2, respectively. Since L is perpendicular to plane M_1, it is perpendicular to line L_1. Moreover, L, L_1, and S_1 are coplanar. Therefore, $L \perp S_1$ at Q. Similarly, $L \perp S_2$ at Q because it is perpendicular to line L_2, which is parallel to S_2, and all three lines are coplanar. Clearly, S_1 and S_2 are distinct lines whose point of intersection is Q. According to Theorem 9-1, the three underlined statements imply that $L \perp M_2$. Thus we have outlined a proof for the following theorem.

THEOREM 9-9 *If a line is perpendicular to one of two parallel planes, then it is perpendicular to the other.*

THEOREM 9-10 *If two distinct planes are each perpendicular to the same line, they are parallel to each other.*

COROLLARY TO THEOREM 9-10 *Through a given point not in a given plane there is one and only one plane parallel to the given plane.*

THEOREM 9-11 *If two distinct planes are each parallel to the same plane, then they are parallel to each other.*

The proofs of Theorem 9-10 and its corollary and the proof of Theorem 9-11 are requested in the exercises.

Suppose that each of the lines L_1 and L_2 is perpendicular to plane M and that $L_1 \neq L_2$. We should be able to prove that L_1 is parallel to L_2.

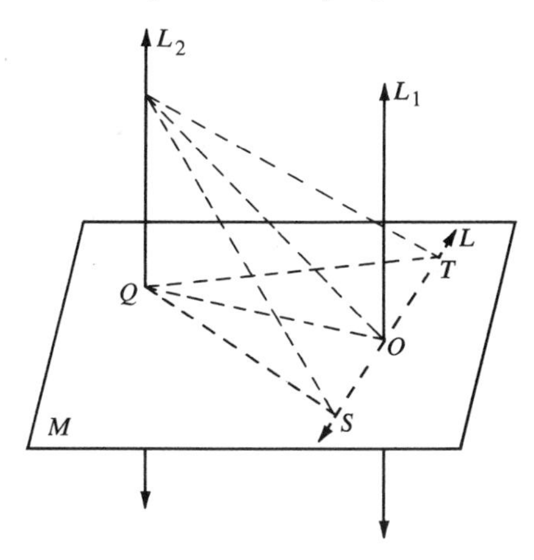

If the points of intersection of L_1 and L_2 with plane M are O and Q, respectively, then, since L_1 and L_2 are distinct, we know that $O \neq Q$. We see that $L_1 \perp M(O) \longrightarrow L_1 \perp \overleftrightarrow{OQ}$ and $L_2 \perp M(Q) \longrightarrow L_2 \perp \overleftrightarrow{OQ}$. Therefore, L_1 and L_2 are perpendicular to the same line at distinct points in the line. This, however, does not prove that L_1 is parallel to L_2 by Corollary 3 to Theorem 8-4 unless we can also prove that L_1, $\overleftrightarrow{OQ}$, and L_2 are coplanar. To do this, we let L be the line in plane M that is perpendicular to $\overleftrightarrow{OQ}$ at O. In the exercises you will be asked to complete this argument by showing that L_1, $\overleftrightarrow{OQ}$, and L_2 are contained in the plane that is perpendicular to L at O. (The drawing may suggest a way of doing this.)

We state this result as

THEOREM 9-12 *If two distinct lines are each perpendicular to the same plane, then they are parallel to each other.*

COROLLARY TO THEOREM 9-12 *If two distinct lines are respectively perpendicular to each of two parallel planes, then they are parallel to each other.*

THEOREM 9-13 *If a plane is perpendicular to one of two parallel lines, then it is perpendicular to the other.*

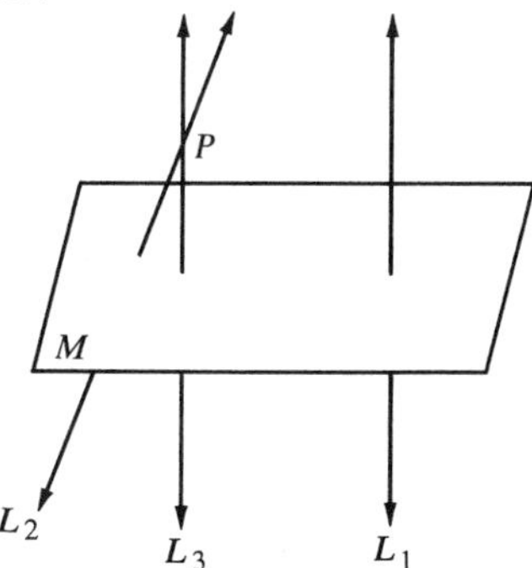

For lines L_1 and L_2 and plane M, we have

$$\left.\begin{matrix} L_1 \parallel L_2 \\ L_1 \perp M \end{matrix}\right\} \longrightarrow L_2 \perp M.$$

Consider the first contrapositive:

$$\left.\begin{matrix} M \text{ is not perpendicular to } L_2 \\ M \perp L_1 \end{matrix}\right\} \longrightarrow L_1 \nparallel L_2.$$

PROOF

Through any point P in L_2 there is a line L_3 that is perpendicular to M. From Theorem 9-12 we know that $L_1 \parallel L_3$. But $L_2 \neq L_3$, since $L_3 \perp M$ and $\sim(L_2 \perp M)$. Furthermore, $P \in L_2$ and $P \in L_3$. Thus we have $L_1 \nparallel L_2$. Why?

In the exercises you are asked to use Theorem 9-12 and 9-13 to prove Theorem 9-14. You will observe that this theorem is a generalization of Theorem 8-9 in the sense that the condition that the lines be coplanar is no longer necessary. We can now accept the transitive property of parallelism for distinct lines.

THEOREM 9-14 *If two distinct lines are each parallel to the same line, they are parallel to each other.*

COROLLARY TO THEOREM 9-14 *If two angles have their sides respectively parallel, the angles are congruent.*

Exercises

1. (a) Suppose that two lines L_1 and L_2 each contain the point P, and L_1 and L_2 are each parallel to a given plane. Can you conclude that $L_1 = L_2$? Why?
   (b) Suppose point Q is contained in each of the planes M_1 and M_2, and M_1 and M_2 are each parallel to a given plane. Can you conclude that $M_1 = M_2$? Why?
2. M_1, M_2, N_1 and N_2 are distinct planes, and L_1, L_2, and L_3 are lines. If $M_1 \parallel M_2$, $N_1 \cap M_1 = L_1$, $N_2 \cap M_2 = L_2$, $N_1 \cap M_2 = L_2$, and $N_2 \cap M_1 = L_3$, prove $L_1 \parallel L_3$.
3. M_1, M_2, and M_3 are distinct planes. If $\overleftrightarrow{AB} \perp M_1$, $\overleftrightarrow{AB} \perp M_2$, $\overleftrightarrow{CD} \perp M_1$, and $\overleftrightarrow{CD} \perp M_3$, prove $M_2 \parallel M_3$.
4. Prove:

$$\left.\begin{matrix} \overrightarrow{OQ} \text{ is the bisector of } \angle AOB \\ \overline{PQ} \perp \text{pl } AOB \text{ at } Q \\ \overline{QX} \perp \overleftrightarrow{OA} \text{ at } X \\ \overline{QY} \perp \overleftrightarrow{OB} \text{ at } Y \end{matrix}\right\} \longrightarrow \angle POY \cong \angle POX.$$

5. Given two skew lines L_1 and L_2, justify each of the following steps in the proof that there exists a common perpendicular for L_1 and L_2.
   Proof: Let P be any point in L_2. There is a line L_3 that contains P and is parallel to L_1 (1). L_3 and L_2 determine a plane we shall call X (2). Let Q be any point in L_1. There is a line L_4 that contains Q and is perpendicular to plane X at a point S (3).

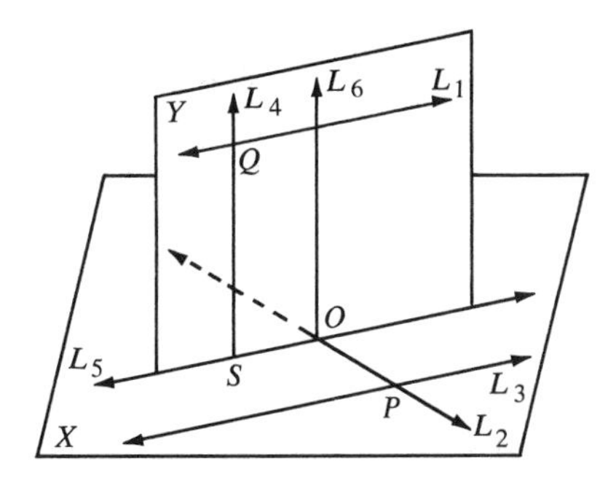

L_4 and L_1 determine a plane Y (4). Plane Y intersects plane X in a line L_5 (5). $L_4 \perp L_5$. (6). Also plane X is parallel to L_1 (7). Therefore, $L_1 \parallel L_5$ (8). It follows that $L_4 \perp L_1$ (9). L_2 and L_5 are distinct coplanar lines that are not parallel (10). Therefore, L_2 intersects L_5 in a point O (11). At O there is a line L_6 in Y that is parallel to L_4 or equal to L_4 (12). $L_6 \perp L_1$ (13). Also $L_6 \perp \text{pl}\, X$ (14). It follows that $L_6 \perp L_2$ (15). Thus we have L_6, which is perpendicular to each of the lines L_1 and L_2.

6. Prove Theorem 9-10.
7. Prove the corollary to Theorem 9-10.
8. Prove Theorem 9-11.
9. Complete the proof of Theorem 9-12.
10. Prove the corollary to Theorem 9-12.
11. Prove:

$$\left.\begin{array}{l} L_1 \times L_2 \\ L_1 \perp M_1 \\ L_2 \perp M_2 \end{array}\right\} \longrightarrow M_1 \neq M_2.$$

L 12. Combine the implication in Exercise 11 with a contrapositive of the corollary to Theorem 9-12 to construct a flow-diagram proof of the implication:

$$\left.\begin{array}{l} L_1 \times L_2 \\ L_1 \perp M_1 \\ L_2 \perp M_2 \end{array}\right\} \longrightarrow M_1 \cap M_2 (L).$$

13. Prove Theorem 9-14.
14. Prove the corollary to Theorem 9-14.
15. Prove:

$$\left.\begin{array}{l} \overrightarrow{OP} \downharpoonleft\upharpoonright \overrightarrow{QR} \\ \overrightarrow{OP} \cap \text{pl}\, M = \{O\} \\ \overrightarrow{QR} \cap \text{pl}\, M = \{Q\} \end{array}\right\} \longrightarrow \overset{\circ\!-\!\circ}{PR} \times M.$$

Perpendicular Planes

Before we can define what it means for two planes to be perpendicular, we need to define a dihedral angle.

DEFINITION A dihedral angle *is the union of two closed half planes that have a common edge and do not lie in the same plane.*

Notation: $W\text{-}\underline{XY}\text{-}Z = \overleftrightarrow{XY}/W \cup \overleftrightarrow{XY}/Z$. We use this notation to obtain the following equivalences.

$W\text{-}\underline{XY}\text{-}Z$ is a dihedral angle $\longleftrightarrow$ pl $WXY \neq$ pl $XYZ \longleftrightarrow Z \notin$ pl $WXY \longleftrightarrow W \notin$ pl XYZ.

If $W\text{-}\underline{XY}\text{-}Z$ is a dihedral angle, we denote it by the symbol $\angle W\text{-}XY\text{-}Z$.

DEFINITIONS *The common edge of the two closed half planes of a dihedral angle is called the* edge of the dihedral angle. *Each of the closed half planes is called a* face of the dihedral angle.

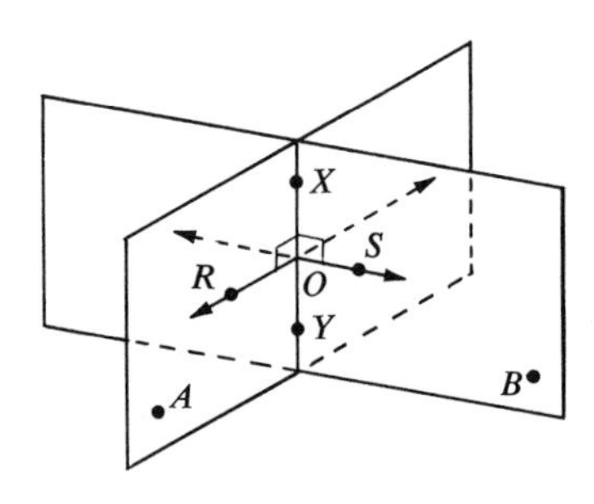

Each dihedral angle has many angles called plane angles associated with it. In our drawing $\angle ROS$ is a plane angle for $\angle A\text{-}XY\text{-}B$ according to the following

DEFINITION *A* plane angle of a dihedral angle *is the union of two rays, one in each face of the dihedral angle and each perpendicular to the edge of the dihedral angle at the same point.*

The next theorem is a direct consequence of these definitions.

THEOREM 9-15 *Any two plane angles of a dihedral angle are congruent.*

DEFINITIONS Measure of a dihedral angle. *The measure of a dihedral angle is the measure of any of its plane angles.*

Congruent dihedral angles. *Two dihedral angles are congruent if their plane angles are congruent.*

Right dihedral angles. *A right dihedral angle is a dihedral angle whose measure is* 90.

Perpendicular planes. *Two planes are perpendicular to each other if their union contains a right dihedral angle.*

COROLLARY TO THE DEFINITION OF PERPENDICULAR PLANES *If two planes are perpendicular, their intersection is a line.*

THEOREM 9-16 *If a line is perpendicular to a plane, then any plane which contains that line is perpendicular to the given plane.*

$$\left.\begin{array}{l} L \perp M_2(O) \\ L \subseteq M_1 \end{array}\right\} \longrightarrow M_1 \perp M_2.$$

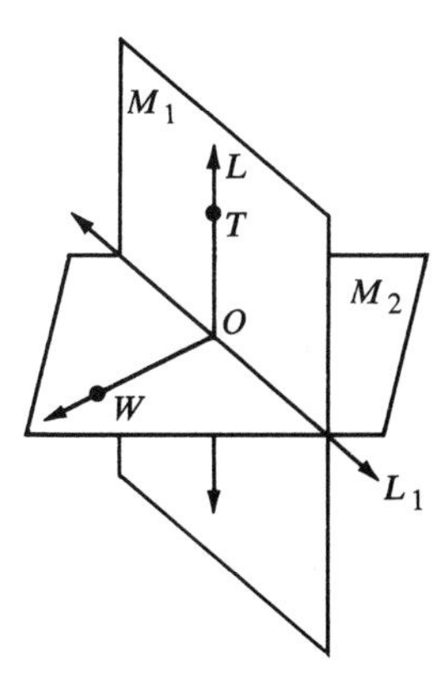

In order to prove this theorem, we observe that there exists a line OW that lies in plane M_2 and is perpendicular to line L_1, where L_1 is the line of intersection of planes M_1 and M_2. If $T \in L$ and $T \neq O$, then we can prove that $\angle TOW$ is a plane angle for the dihedral angle formed by planes M_1 and M_2. Since $\angle TOW$ is a right angle, it follows that $M_1 \perp M_2$.

THEOREM 9-17 *If two planes are perpendicular, then any line that is perpendicular to their line of intersection and lies in one of the planes is perpendicular to the other.*

Referring to the drawing for Theorem 9-16, we wish to prove

$$\left.\begin{array}{l} M_1 \perp M_2 \\ M_1 \cap M_2 = L_1 \\ L \subseteq M_1 \\ L \perp L_1(O) \end{array}\right\} \longrightarrow L \perp M_2.$$

The essential idea of the proof is to show that L is perpendicular to $\overleftrightarrow{OW}$. This is true because $\angle WOT$ is a plane angle of a right dihedral angle.

COROLLARY TO THEOREM 9-17 *If two planes are perpendicular, then any line that is perpendicular to one of them and contains a point in the other, lies in the other.*

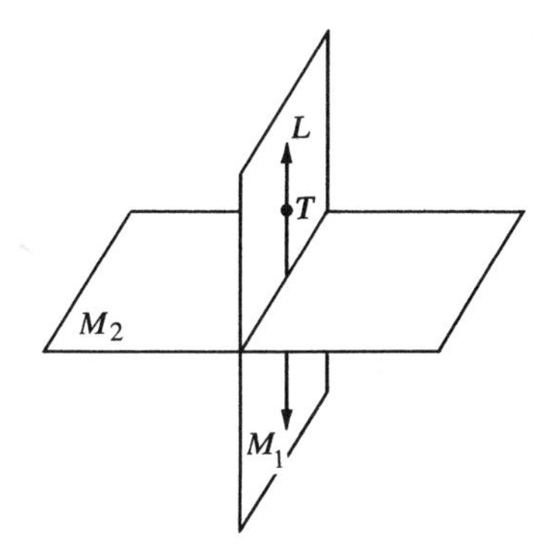

$$\left.\begin{array}{l} M_1 \perp M_2 \\ T \in M_1 \\ T \in L \\ L \perp M_2 \end{array}\right\} \longrightarrow L \subseteq M_1.$$

DEFINITIONS *P′ is the* projection *of point P on plane M if and only if P′ is the intersection of M and a line L that is perpendicular to M and contains P. If S is the set of projections of points in line L on plane M, then S is the projection of line L on plane M.*

Having defined the projection of a line on a plane, we now state the following theorem, which describes the projection of a line on a plane to which the line is not perpendicular. Suggestions for the proof of this theorem are contained in the exercises.

THEOREM 9-18 *Let P and Q be distinct points whose projections on plane M are P′ and Q′, respectively. If* $\overleftrightarrow{PQ}$ *is not perpendicular to M, then the projection of* $\overleftrightarrow{PQ}$ *on M is* $\overleftrightarrow{P'Q'}$.

From the drawing it would appear that a plane that is perpendicular to each of two distinct intersecting planes must be perpendicular to their line of intersection. In order to prove this, we state the following lemma, whose proof is considered in the exercises.

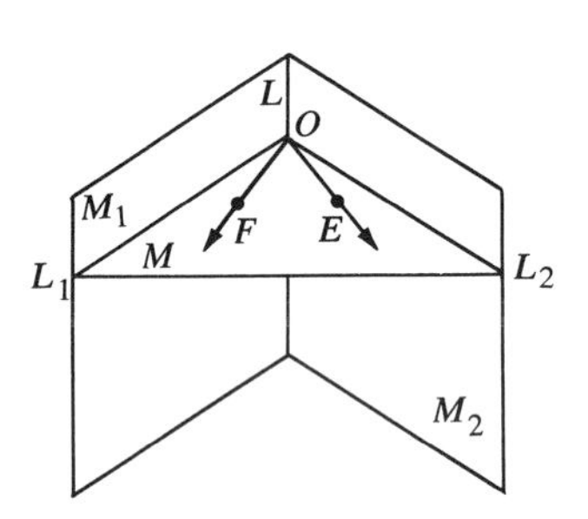

LEMMA 1 *If a plane is perpendicular to each of two distinct intersecting planes, then it intersects their line of intersection in a single point.*

THEOREM 9-19 *If two planes whose intersection is a line are each perpendicular to a third plane, then their line of intersection is perpendicular to that plane.*

We must prove:

$$\left.\begin{array}{l} M_1 \perp M \\ M_2 \perp M \\ M_1 \cap M_2 = L \end{array}\right\} \longrightarrow L \perp M.$$

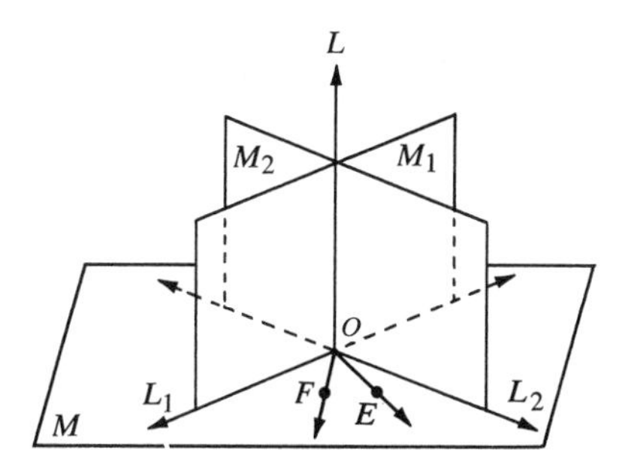

OUTLINE OF PROOF

Since $M_1 \perp M$ and $M_2 \perp M$, we know that M intersects M_1 in a line L_1 and M intersects M_2 in a line L_2. According to Lemma 1 $L \times M(O)$. Therefore, O is in each of the lines L_1 and L_2. There is a line OE in M perpendicular to L_1 and a line OF in M perpendicular to L_2. Then $\overleftrightarrow{OE} \perp M_1$ and $\overleftrightarrow{OF} \perp M_2$ by Theorem 9-17. Furthermore, $\overleftrightarrow{OE} \neq \overleftrightarrow{OF}$ because $M_1 \neq M_2$ (Theorem 9-4). Since $\overleftrightarrow{OE} \perp M_1$, it is perpendicular to any line in M_1 that contains O. Therefore, we know $\overleftrightarrow{OE} \perp L$. Similarly, we have $\overleftrightarrow{OF} \perp L$. According to Theorem 9-1, the underlined statements imply that $L \perp M$.

Combining Theorem 9-19 with the statement proved in **L**-Exercise 12, page 203, we have the following important theorem.

THEOREM 9-20 *If two planes are respectively perpendicular to two lines that intersect in a single point, then these planes intersect in a line that is perpendicular to the plane determined by the given lines.*

Exercises

1. $A\text{-}\underline{XY}\text{-}B$ is a dihedral angle and $M \perp \overleftrightarrow{XY}$ at X. $M \cap \underline{\overleftrightarrow{XY}/A} = \overrightarrow{XP}$ and $M \cap \underline{\overleftrightarrow{XY}/B} = \overrightarrow{XQ}$. Explain why $P\underline{X}Q$ is a plane angle of the dihedral angle.
2. $A\text{-}\underline{XY}\text{-}B$ is a dihedral angle. Plane $M \parallel \overleftrightarrow{XY}$, $M \cap \underline{\overleftrightarrow{XY}/A} = \overleftrightarrow{PQ}$, and $M \cap \underline{\overleftrightarrow{XY}/B} = \overleftrightarrow{RS}$. Prove $\overleftrightarrow{PQ} \parallel \overleftrightarrow{RS}$.
3. Prove Theorem 9-15.
4. Prove the corollary to the definition of perpendicular planes.
5. Prove Theorem 9-16.
6. Prove Theorem 9-17.
7. Prove: If M is the midpoint of the common base of noncoplanar isosceles triangles XAB and YAB and T is any point in $\overline{XY}$, then $\overline{AB} \perp \overline{MT}$.
8. Prove: If a plane M intersects one of two parallel planes in a line, then it intersects the other in a line.
9. Prove: If a plane is perpendicular to one of two parallel planes, then it is perpendicular to the other.
10. For planes M_1 and M_2 and lines L and T, prove

$$\left.\begin{array}{l} L \perp M_1(O) \\ M_1 \cap M_2 = T \\ M_1 \perp M_2 \\ O \notin T \end{array}\right\} \longrightarrow L \parallel M_2.$$

11. Prove: If a segment is not perpendicular to a plane, then there is a unique plane that contains that segment and is perpendicular to the given plane.
12. Prove:

$$\left.\begin{array}{l}\text{Plane } S \parallel \text{plane } T\\ \text{Line } L \parallel S\\ L \nsubseteq T\end{array}\right\} \longrightarrow L \parallel T.$$

L 13. Prove:

$$\left.\begin{array}{l}\text{Line } L \perp \text{pl } M_1\\ L \parallel \text{pl } M_2\end{array}\right\} \longrightarrow M_1 \perp M_2.$$

14. (a) Prove: If two planes are perpendicular, any line that is perpendicular to one of them and contains a point in their line of intersection lies in the other.
    (b) Prove: If two planes are perpendicular, any line that is perpendicular to one of them and contains a point in the other that is not in the line of intersection lies in the other.
    (c) Explain why the conjunction of the statements proved in parts (a) and (b) implies the truth of the corollary to Theorem 9-17.
15. Prove Theorem 9-18. (Suggestion: Use the definition of projection on a plane and the corollary to Theorem 9-17.)
16. Prove Lemma 1:

$$\left.\begin{array}{l}M \perp M_1\\ M \perp M_2\\ M_1 \neq M_2\\ M_1 \cap M_2 = L\end{array}\right\} \longrightarrow L \times M.$$

    (Suggestion: If we suppose $L \parallel M$ or $L \subseteq M$, we can reach a contradiction of Theorem 9-4. Thus $L \nparallel M$ and $L \nsubseteq M$, and we conclude that $L \times M$.)
17. Prove that there is only one common perpendicular to two given skew lines.

Trihedral Angles

If AOB, BOC, and AOC are angles, we shall use the symbol O-ABC to represent the union of angles AOB, BOC, and AOC and the interiors of these angles. If ABC is a triangle and $O \notin \text{pl } ABC$, we know that AOB, BOC, and AOC are angles. We state the following definition.

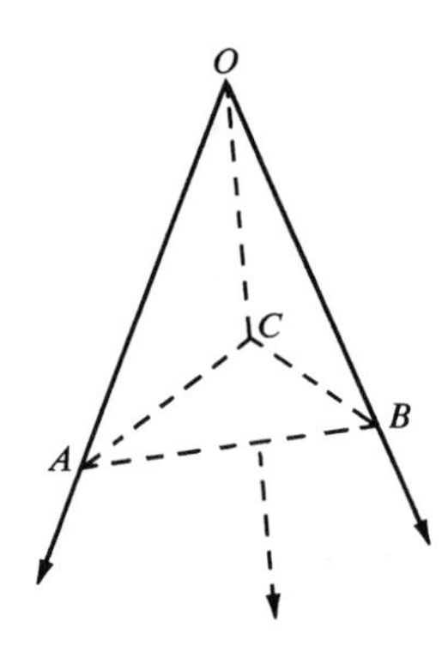

DEFINITION *O-ABC is a* trihedral angle *if and only if ABC is a triangle and O is not in its plane. In this trihedral angle, O is the* vertex, *$\overrightarrow{OA}$, $\overrightarrow{OB}$, and $\overrightarrow{OC}$ are the* edges, *each of the angles AOB, BOC, and AOC is a* face angle, *and the union of each face angle and its interior is a* face *of the trihedral angle. The triangle ABC is called a* plane section *of the trihedral angle.*

Notation: If O-ABC is a trihedral angle, we denote it by $\angle O$-ABC.

THEOREM 9-21 *The sum of the measures of any two face angles of a trihedral angle is greater than the measure of the third face angle.*

Suggestions for the proof of Theorem 9-21 are found in the exercises.

THEOREM 9-22 *Planes that are respectively perpendicular to the edges of a trihedral angle intersect in a single point that is the vertex of a second trihedral angle whose edges are respectively perpendicular to the planes that contain the faces of the first.*

In terms of the drawing below, we have trihedral angle O-ABC and planes X, Y, and Z perpendicular respectively to $\overrightarrow{OA}$, $\overrightarrow{OB}$, and $\overrightarrow{OC}$. We must prove that the planes X, Y, and Z intersect in a single point—that is, $X \cap Y \cap Z = Q$. Furthermore, if L_1, L_2, and L_3 are the lines of intersection of planes Y and Z, X and Z, and X and Y, respectively, we must show $L_1 \perp \text{pl } BOC$, $L_2 \perp \text{pl } AOC$ and $L_3 \perp \text{pl } AOB$. Finally we must prove that Q-$A'B'C'$ is a trihedral angle, where A', B' and C' are points, other than Q, in L_1, L_2 and L_3, respectively.

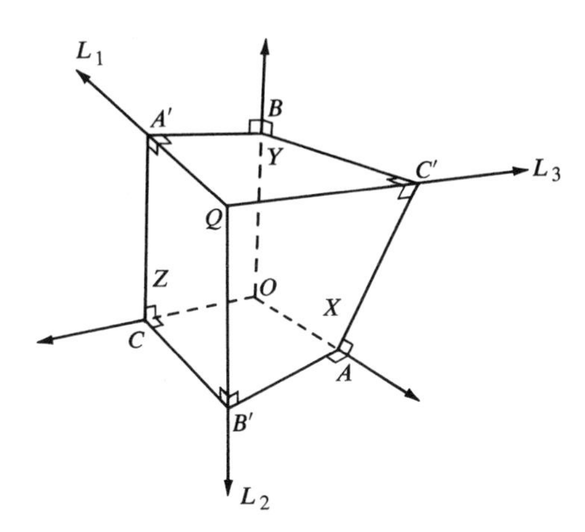

We outline a proof that planes X, Y, and Z intersect in a single point Q. The rest of the proof is requested in the exercises.

O-ABC is a trihedral angle $\overset{(1)}{\dashv}$

$$\vdash\!\!\longrightarrow \left.\begin{array}{r} AQB \text{ is an angle} \xrightarrow{(2)} \overleftrightarrow{AO} \times \overleftrightarrow{OB} \\ \text{pl } X \perp \overrightarrow{OA} \xrightarrow{(3)} X \perp \overleftrightarrow{AO} \\ \text{pl } Y \perp \overrightarrow{OB} \xrightarrow{(3)} Y \perp \overleftrightarrow{OB} \end{array}\right\} \xrightarrow{(4)} \left\{\begin{array}{l} X \cap Y = \text{line } L_3 \\ L_3 \perp \text{pl } AOB \end{array}\right.$$

$$Q\text{-}ABC \text{ is a trihedral angle } \overset{(5)}{\dashv}$$

$$\vdash\!\!\longrightarrow O \notin \text{pl } ABC \xrightarrow{(6)} C \notin \text{pl } AOB \xrightarrow{(7)} \left.\begin{array}{r} \overleftrightarrow{OC} \nsubseteq \text{pl } AOB \\ \text{pl } Z \perp \overleftrightarrow{OC} \xrightarrow{(3)} Z \perp \overleftrightarrow{OC} \end{array}\right\} \xrightarrow{(8)} \sim(Z \perp \text{pl } AOB) \quad \dashv$$

$$\vdash \xrightarrow{(9)} \left.\begin{array}{l} L_3 \nparallel Z \xrightarrow{(11)} L_3 \cap Z \neq \emptyset \\ \xrightarrow{(10)} L_3 \nsubseteq Z \end{array}\right\} \xrightarrow{(12)} L_3 \times Z(Q).$$

Thus we have:

$$\left.\begin{array}{l} X \cap Y = L_3 \\ L_3 \cap Z = Q \end{array}\right\} \xrightarrow{(13)} X \cap Y \cap Z = Q.$$

Although reasons can be supplied for each of the implications, we list only the following:

Reasons:

(8) Contrapositive of the corollary to Theorem 9-17.

(9) Contrapositive of the implication in **L**-Exercise 13, page 208, which states:

$$\left.\begin{array}{l} L \perp \text{pl } M_1 \\ L \parallel \text{pl } M_2 \end{array}\right\} \longrightarrow M_1 \perp M_2.$$

(10) Contrapositive of Theorem 9-16.

We now state a corollary that plays an important role in the development of a "three-dimensional" coordinate system. The proof of this corollary is requested in the exercises.

COROLLARY TO THEOREM 9-22 *Three planes that are respectively perpendicular to three mutually perpendicular lines intersect in a point and are mutually perpendicular to each other.*

Exercises

1. Prove Theorem 9-21. (Suggestions: Let Q-ABC be a trihedral angle. If all three face angles have the same measure, then we can easily prove the theorem. If this is not the case, then there is an angle ($\angle AOC$) whose measure is greater than or equal to the measures of each of the other two angles and greater than one of them ($\angle AOB$). If we can prove $m\angle AOB + m\angle BOC > m\angle AOC$, the theorem is easily proved. Let X be any point in $\overset{\circ}{OB}$; then there is a ray OY such that $\overset{\circ}{OY}$ is in the interior of $\angle AOC$, $m\angle AOY = m\angle AOB$, and $|OY| = |OX|$. According to Theorem 8-13, there is a line that intersects $\overset{\circ}{OA}$ in Z and $\overset{\circ}{OC}$ in W so that Z-Y-W. We apply certain of our triangle inequalities to show $m\angle XOW > m\angle YOW$. The conclusion follows.)

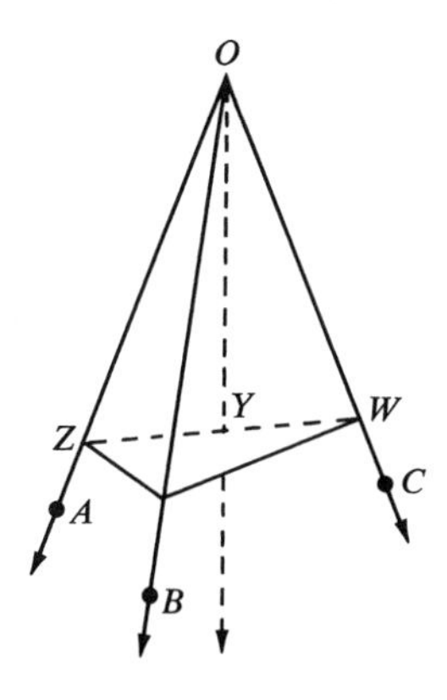

2. Complete the proof of Theorem 9-22.
3. Prove the corollary to Theorem 9-22.

References

Prenowitz and Jordan, *Basic Concepts of Geometry*
Ringenberg, *College Geometry*

Chapter 10

Introduction to Non-Euclidean Geometry

Euclidean plane geometry is characterized by the famous parallel postulate, which we stated as Postulate 17 in Chapter 8 and which we denote here as statement p.

> p: For any point in space there is at most one line that contains the point and is parallel to a given line.

Euclid's statement of the "postulate of parallels," which was his fifth and final postulate, is denoted here as statement e.

> e: That, if a straight line falling on two straight lines makes the interior angles on the same side less than two right angles, the two straight lines, if produced indefinitely, meet on that side on which are the angles less than two right angles.

Let N denote a set consisting of the first sixteen postulates in this text. Then, as we noted in Chapter 6, N is a set of postulates for neutral geometry, and statements that can be deduced from these postulates are theorems in neutral geometry.

If we make due allowances for differences in language, we observe that in Theorem 8-12 we proved that $(N \wedge p) \longrightarrow e$. It is also possible to prove that $(N \wedge e) \longrightarrow p$. (See Exercise 1, page 226.) Thus we have $N \longrightarrow (p \longleftrightarrow e)$. In other words, statements p and e are equivalent in terms of our set of postulates for neutral geometry.

As we continue our study, we shall find that, judged by our postulates for neutral geometry, many other statements are equivalent to our parallel postulate. Some of these are quite surprising. For example, Postulate 17 is equivalent to the assertion that a rectangle exists!

Now it turns out that statement p is undecidable in terms of postulate set N. This means that we can annex either p or $\sim p$ to N and thus obtain a consistent set of postulates. Let P be any consistent set of geometric postulates that contains N and p. Then P is a set of postulates for Euclidean geometry. Statements that are undecidable on the basis of N but provable on the basis of P are theorems in Euclidean geometry.

Our assertion that p is undecidable in terms of N is far from obvious. For over 2000 years generations of mathematicians made determined efforts to deduce the parallel postulate from the other, seemingly simpler postulates. They often devised elaborate and highly ingenious indirect arguments to prove that the denial of "Euclid's fifth" would produce a contradiction. None of these arguments stood up under logical analysis.

It was finally conceded that the parallel postulate is indeed undecidable when, in the nineteenth century, John Bolyai (1802–1860), a Hungarian army officer, and Nicholas Lobachevsky (1793–1856), a Russian professor of mathematics, independently introduced consistent theories of geometry based on a contradiction of Euclid's parallel postulate.

If H is a consistent set of geometric postulates that contains N and $\sim p$, then H is a set of postulates for the geometry of Lobachevsky and Bolyai, and a statement that is undecidable on the basis of N and provable on the basis of H is a theorem of this geometry.

Let p' be a statement about the number of lines through a point that are parallel to a given line that is decidable false in terms of postulate set N. Then any set of statements that contains N and p' is an inconsistent set. For example, suppose that p' is the statement "There are no parallels to any line through any point." Then p' is decidable false in terms of N, because according to Theorem 8-5 there exists at least one parallel to any line through a given external point. However, it is possible to modify the postulates in N so as to obtain a set of postulates N' for which the statement p' is undecidable. This means that $N' \cup p'$ is a consistent set of postulates. If E is a consistent set of geometric postulates that contains N' and p', then the geometry based on E is called Riemannian geometry after the great German mathematician Bernhard Riemann (1826–1866), who laid the foundation for this geometry in a lecture delivered before the Philosophical Faculty of Göttingen in 1854.

In one type of Riemannian geometry, a region can be bounded by two lines and the sum of the measures of the angles (angle measure sum) of a triangle is greater than 180. A model for this type of Riemannian geometry can be obtained by considering the surface of a sphere as a plane in which great circles play the role of lines.

The great geometer Felix Klein (1849–1925) suggested calling the geometries of Euclid, Bolyai and Lobachevsky, and Riemann respectively *parabolic, hyperbolic,* and *elliptic,* and this terminology has been almost universally accepted. Thus parabolic, hyperbolic, and elliptic geometries are based on postulate sets of types P, H, and E, respectively.

We shall not consider elliptic geometry. In the remaining sections of this text we show that neutral geometry serves as an introduction to hyperbolic geometry. To do this we merely assume the postulates of neutral geometry (Postulates 1–16) and consider the logical consequences of supposing that our parallel postulate is false.

Some Theorems in Neutral Geometry

The following statement is a contradiction of Postulate 17.

$\sim p$: There exist a line L and a point P such that there is more than one line through P parallel to L.

If we accept $\sim p$ as a postulate, any theorem we can prove on the basis of a consistent set of postulates that contains N and $\sim p$ is a theorem in hyperbolic geometry. If, on the other hand, a theorem whose hypothesis contains $\sim p$ is provable on the basis of postulate set N, then that theorem is a theorem of neutral geometry.

At this point we restate as our first theorem the statement that was proved in **L**-Exercise 14, page 187.

THEOREM 10-1 *If point P is not in line L and if L_1 and L_2 are distinct lines that contain P and are each parallel to L, then L is wholly contained in the interior of an angle that is a subset of $L_1 \cup L_2$.*

Observe that Theorem 10-1 is a theorem in neutral geometry. On the other hand, if we accept $\sim p$ as a postulate, the statement "A line is wholly contained in the interior of some angle" is a theorem in hyperbolic geometry.

One might say that Theorem 10-1 is true in an abstract sense but that it does not correspond to any situation found in the "real world." A person who makes such a statement is beginning to tread the path of the non-Euclidean geometers. Although no logical flaw could be found in some of the geometries based on a denial of the parallel postulate, these geometries were rejected for many years because it was believed that they did not correspond to reality. Finally these geometries gained acceptance by mathematicians on the grounds that to think mathematically one needs only a set of precisely stated assumptions (postulates) from which conclusions (theorems) can be derived by logical reasoning. According to this view, questions about "correspondence to reality" are in the domain of physicists, astronomers, and surveyors and are not the concern of the mathematician. Although non-Euclidean geometry was accepted originally solely on the basis of its logical consistency and in spite of the fact that it defies intuition, it is now considered by physicists to be a better description of our universe than Euclidean geometry.

Our first theorem indicated how certain incidence relations in non-Euclidean geometry differ from our Euclidean expections. We now show how the angle-measure sum of a triangle is affected when we change the parallel postulate. We shall prove the following remarkable theorem.

THEOREM 10-2 *If there are two lines through some point P parallel to line L, then there exists at least one triangle whose angle-measure sum is less than* 180.

To facilitate the proof of Theorem 10-2, we state three lemmas.

LEMMA 1 *If the sum of the angle measures of a triangle is greater than or equal to* 180, *then the measure of an exterior angle is less than or equal to the sum of the measures of the two angles of the triangle that are not adjacent to it.*

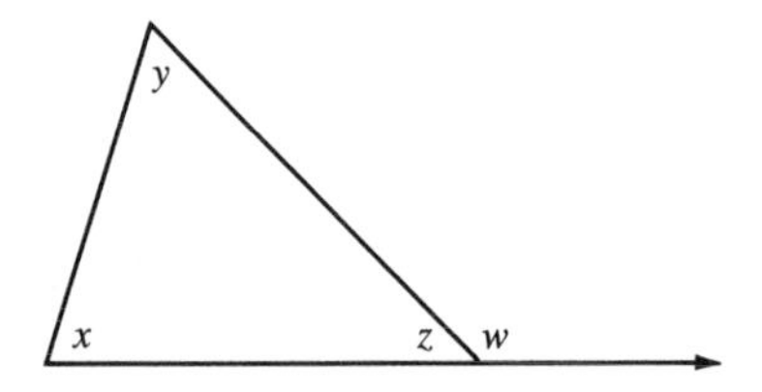

PROOF

$$\left.\begin{aligned} x + y + z &\geqq 180 \\ z + w &= 180 \end{aligned}\right\} \longrightarrow x + y + z \geqq z + w \longrightarrow x + y \geqq w.$$

LEMMA 2 *If there are two lines through point P parallel to line L and $\overrightarrow{PQ} \perp L(Q)$, then one of these lines must make an acute angle with $\overrightarrow{PQ}$.*

The proof of Lemma 2 is requested in the exercises.

LEMMA 3 *(a) If c and d are positive real numbers and n is a counting number, then there exists a value of n such that $c/2^n < d$.*
(b) If $c > 0$ and the statement $x < d + (c/2^n)$ is true for all counting numbers n, then $x \leqq d$.

The proof of Lemma 3 is considered in Appendix B.

We now outline the proof of Theorem 10-2. Suppose the theorem is false. Then the angle-measure sum of every triangle is greater than or equal to 180. Let P be a point such that there are two lines through P parallel to L and let $\overleftrightarrow{PR_0}$ be perpendicular to L at R_0. According to Lemma 2 one of the two lines through P parallel to L must make an acute angle with $\overrightarrow{PR_0}$. Let this line be called $\overleftrightarrow{PX}$, so that $\overleftrightarrow{PX}$ is parallel to L and $\angle R_0PX$ is an acute angle.

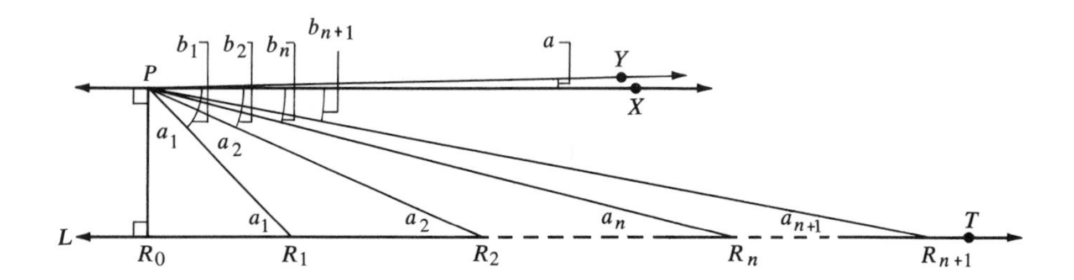

In $\overline{\overleftrightarrow{PR_0}/X}$ there is a ray PY such that $m\angle R_0PY = 90$. Let $m\angle YPX = a$. Then $0 < a < 90$. Why? Let T be a point in $L \cap \overleftrightarrow{PR_0}/X$. In $\overrightarrow{R_0T}$ there is a point R_1 such that $|R_0R_1| = |PR_0|$. PR_0R_1 is an isosceles triangle and $m\angle R_0PR_1 = m\angle R_0R_1P = a_1$. Since the exterior angle of $\triangle PR_0R_1$ at R_0 is a right angle, Lemma 1 implies that

$$a_1 + a_1 = 2a_1 \geqq 90 \quad \text{and} \quad a_1 \geqq 45.$$

Let $m\angle YPR_1 = b_1$. Then $b_1 + a_1 = 90$, so that

$$\left.\begin{aligned} b_1 + a_1 &= 90 \\ a_1 &\geqq 45 \end{aligned}\right\} \longrightarrow b_1 \leqq 45.$$

Moreover, $b_1 > a$.

Now we repeat this argument with a new triangle. In $\overset{\circ}{\overrightarrow{R_0T}}$ there is a point R_2 such that $|R_1R_2| = |PR_1|$ and R_0-R_1-R_2. Triangle PR_1R_2 is isosceles, so that $m\angle R_1PR_2 = m\angle PR_2R_1 = a_2$. By Lemma 1,

$$a_2 + a_2 = 2a_2 \geqq a_1 \quad \text{and} \quad a_2 \geqq \frac{a_1}{2}.$$

Also

$$\left.\begin{aligned} a_2 &\geqq \frac{a_1}{2} \\ a_1 &\geqq 45 \end{aligned}\right\} \longrightarrow a_2 \geqq \frac{45}{2}.$$

Let $m\angle YPR_2 = b_2$. Then $a_2 + b_2 = b_1$ and

$$\left.\begin{aligned} a_2 + b_2 &= b_1 \\ b_1 &\leqq 45 \\ a_2 &\geqq \frac{45}{2} \end{aligned}\right\} \longrightarrow b_2 \leqq \frac{45}{2}.$$

It is also true that $b_2 > a$.

Continuing in this way, we obtain a sequence of real numbers $b_1, b_2, \ldots, b_n, \ldots,$ which are respectively less than or equal to

$$45, \frac{45}{2}, \ldots, \frac{45}{2^{n-1}}, \ldots,$$

and all of which are greater than the fixed positive number a. In other words, if our supposition is true, we must have

$$\frac{45}{2^{n-1}} > a$$

for all positive integers n. But this is impossible, because, according to Lemma 3(a), repeated halving of 45 must eventually produce a number less than a. So our supposition is false and the theorem is proved.

The phrase "continuing in this way" suggests that mathematical induction is involved in the proof of Theorem 10-2. Observe that in $\overrightarrow{R_0T}$ we have a sequence of points $R_0, R_1, R_2, \ldots, R_n, \ldots,$ such that R_{n-1}-R_n-R_{n+1} and $|R_nR_{n+1}| = |PR_n|$. It follows that the points $R_0, R_1, \ldots, R_n, \ldots$ appear in $\overrightarrow{R_0T}$ in the order indicated by their subscripts. If now we let $m\angle PR_nR_{n-1} = m\angle PR_nR_0 = m\angle R_{n-1}PR_n = a_n$ and let $m\angle YPR_n = b_n$, we can prove that $a_n + b_n = b_{n-1}$ and $b_n > a$. Proofs for these statements are required in the exercises. We have shown that

$$a_1 \geqq \frac{45}{2^0}, \quad b_1 \leqq \frac{45}{2^0}, \quad a_2 \geqq \frac{45}{2} \quad \text{and} \quad b_2 \leqq \frac{45}{2}.$$

It is possible to complete the proof that $b_n \leqq 45/2^{n-1}$ by mathematical induction by proving that

$$\left.\begin{array}{l} a_n \geqq \dfrac{45}{2^{n-1}} \\ \\ b_n \leqq \dfrac{45}{2^{n-1}} \end{array}\right\} \longrightarrow b_{n+1} \leqq \frac{45}{2^n}.$$

This proof, which makes use of the fact that $b_{n+1} + a_{n+1} = b_n$, is requested in the exercises.

We state here for ready reference some previously proved theorems of neutral geometry that are useful in this discussion. Observe that each of the next four theorems was proved prior to the acceptance of Postulate 17.

THEOREM 10-3 *The measure of an exterior angle of a triangle is greater than the measure of either of the angles of the triangle that are nonadjacent to it. (Theorem 7-4.)*

THEOREM 10-4 *The sum of the measures of any two angles of a triangle is less than* 180. *(Corollary 1 to Theorem 7-4.)*

THEOREM 10-5 *If two given distinct lines are perpendicular to the same line and all three lines are coplanar, then the two given lines are parallel. (Corollary 3 to Theorem 8-4.)*

THEOREM 10-6 *If L is a line and point $P \notin L$, then there is at least one line that contains P and is parallel to L. (Theorem 8-5.)*

Recall that our proof that the sum of the measures of the angles of a triangle is equal to 180 made use of the parallel postulate. However, without the parallel postulate it is possible to prove that the sum of the measures of the angles of a triangle is *less than or equal to 180.* This remarkable theorem, due to Legendre (1752–1833), makes an essential contribution to this discussion. In order to prove it, we introduce Lemmas 4 and 5.

LEMMA 4 *For real numbers a, b, and c, $a = b + c \longrightarrow (b \leqq \frac{1}{2}a \vee c \leqq \frac{1}{2}a)$.*

The proof of Lemma 4 is left as an exercise.

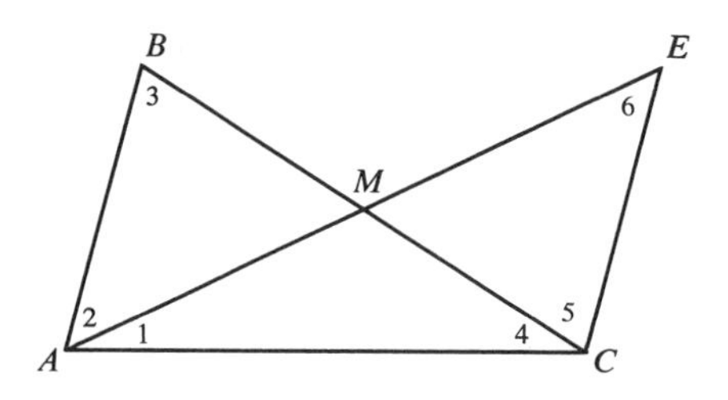

LEMMA 5 *Given $\triangle ABC$, there exists a triangle $A_1B_1C_1$ with the same angle-measure sum as $\triangle ABC$ and such that $m\angle A_1 \leqq \frac{1}{2}m\angle A$.*

OUTLINE OF PROOF

Let M be the midpoint of $\overline{BC}$ and let E be a point in $\overrightarrow{AM}$ such that $|AE| = 2|AM|$. Then $\triangle ABM \cong \triangle ECM$. Why? Labeling the angles as shown in the figure, we have

$$\begin{aligned} m\angle A + m\angle B + m\angle C &= m\angle 1 + m\angle 2 + m\angle 3 + m\angle 4 \\ &= m\angle 1 + m\angle 6 + m\angle 5 + m\angle 4 \\ &= m\angle EAC + m\angle AEC + m\angle ACE. \end{aligned}$$

Also, $m\angle A = m\angle 1 + m\angle 6$. Therefore, by Lemma 4, we have either (1) $m\angle 1 \leqq \frac{1}{2}m\angle A$ or (2) $m\angle 6 \leqq \frac{1}{2}m\angle A$. If (1), then $\triangle A_1B_1C_1$ with $A_1 = A$, $C_1 = C$, and $B_1 = E$ is the required triangle. If (2), then $\triangle A_1B_1C_1$ with $A_1 = E, C_1 = C$, and $B_1 = A$ is the required triangle.

THEOREM 10-7 (LEGENDRE) *The angle-measure sum of any triangle is less than or equal to* 180.

PROOF

Let S denote the angle-measure sum of $\triangle ABC$. According to Lemma 5, there is a triangle $A_1B_1C_1$ with angle-measure sum S and with $m\angle A_1 \leqq \frac{1}{2}m\angle A$. Applying Lemma 5 repeatedly, we obtain a sequence of triangles $A_1B_1C_1$, $A_2B_2C_2, \ldots, A_nB_nC_n, \ldots$, each with angle-measure sum S and such that

$$m\angle A \geqq 2m\angle A_1,\ m\angle A_1 \geqq 2m\angle A_2,\ \ldots,\ m\angle A_n \geqq 2m\angle A_{n-1},\ \ldots.$$

It follows by mathematical induction that $m\angle A \geqq 2^n m\angle A_n$ or $m\angle A_n \leqq m\angle A/2^n$ for all $n \in C$. The rest of the argument can be presented as follows.

$$\left.\begin{aligned} &m\angle A_n \leqq \frac{m\angle A}{2^n} \\ &m\angle B_n + m\angle C_n < 180 \end{aligned}\right\} \dashv$$

$$\left.\begin{aligned} &\leftrightarrow m\angle A_n + m\angle B_n + m\angle C_n < 180 + \frac{m\angle A}{2^n} \\ &m\angle A_n + m\angle B_n + m\angle C_n = m\angle A + m\angle B + m\angle C \end{aligned}\right\} \dashv$$

$$\leftrightarrow m\angle A + m\angle B + m\angle C < 180 + \frac{m\angle A}{2^n} \dashv$$

$$\leftrightarrow m\angle A + m\angle B + m\angle C \leqq 180.$$

Observe that the reason for the last implication is Lemma 3(b).

Consider now convax quadrilateral $ABCD$. Its angle-measure sum is the sum of the angle-measure sums of $\triangle ACB$ and $\triangle ACD$. Why? Therefore, the angle-measure sum of quad $ABCD$ is less than or equal to $2 \cdot 180$ or 360. Thus we have the following theorem.

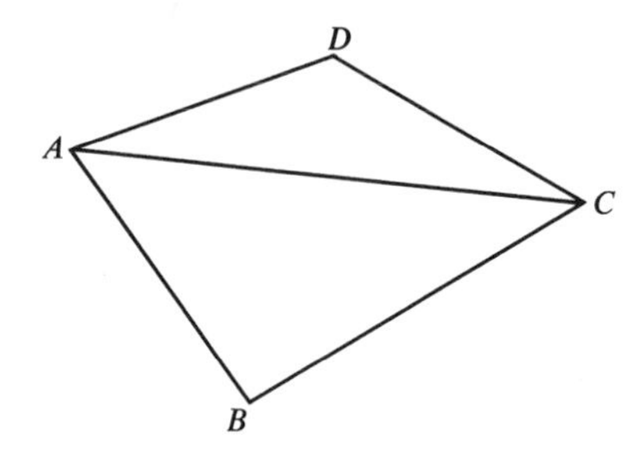

THEOREM 10-8 *The angle-measure sum of any convax quadrilateral is less than or equal to* 360.

Rectangles in Neutral Geometry

Girolamo Saccheri (1667–1733) was prominent among those mathematicians who sought to vindicate Euclid by proving the fifth postulate. Saccheri's principal tool in his investigations was an "isosceles quadrilateral," which is now known as a Saccheri quadrilateral.

DEFINITIONS *Quadrilateral $ABCD$ is a* Saccheri quadrilateral *with* base $\overline{AB}$ *and* summit $\overline{CD}$ *if* $\overline{AD} \cong \overline{BC}$, $\overline{AD} \perp \overline{AB}$, $\overline{BC} \perp \overline{AB}$, *and* $C \in \overleftrightarrow{AB}/D$. *Angles DAB and ABC are the* base angles *of this Saccheri quadrilateral, and angles ADC and DCB are the* summit angles.

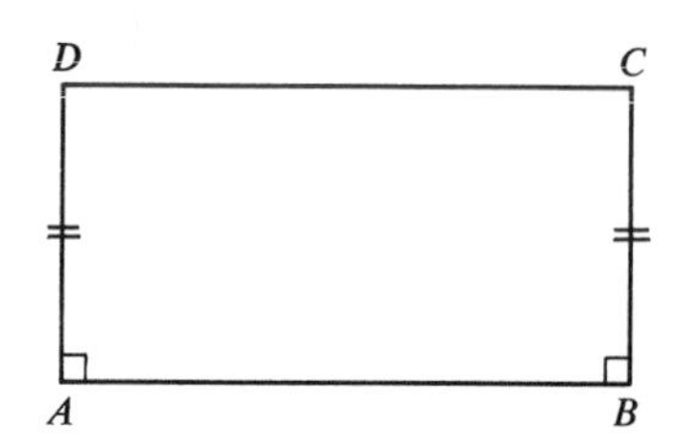

When we name a Saccheri quadrilateral, the first two letters will always denote the base. Thus if $XYZW$ is a Saccheri quadrilateral, then $\overline{XY}$ is the base and $\overline{ZW}$ is the summit.

Proofs of the next two theorems about Saccheri quadrilaterals are requested in the exercises.

THEOREM 10-9 *A Saccheri quadrilateral is a convax quadrilateral.*

THEOREM 10-10 *The summit angles of a Saccheri quadrilateral are congruent.*

Having proved Theorem 10-10, Saccheri considered three possibilities for the summit angles:

(1) Each of the summit angles is obtuse (obtuse-angle hypothesis).
(2) Each of the summit angles is a right angle (right-angle hypothesis).
(3) Each of the summit angles is acute (acute-angle hypothesis).

He sought to establish the right-angle hypothesis by eliminating the other two. He was able to prove that the obtuse-angle hypothesis is impossible in neutral geometry. (Note that we have accomplished this with Theorem 10-8.) His efforts to obtain a contradiction on the basis of the postulates of neutral geometry by assuming the acute-angle hypothesis were unsuccessful, because, as we shall see, no such contradiction exists.

For the purposes of this section, we adopt a definition of a rectangle that differs somewhat from the definition given in Chapter 8.

DEFINITION *A quadrilateral is a* rectangle *if each angle which contains a pair of its consecutive sides is a right angle.* [$ABCD$ *is a rectangle* $\longrightarrow$ $ABCD$ *is a quadrilateral and* $A\underline{B}C$, $B\underline{C}D$, $C\underline{D}A$ *and* $D\underline{A}B$ *are right angles.*]

We know that defining a thing does not guarantee its existence. In the next sequence of theorems, we consider some of the logical consequences that follow from the assumption that a rectangle exists.

First we note that (i) the opposite sides of a rectangle are parallel and (ii) a rectangle is a convax quadrilateral. Now we shall prove that if one rectangle exists, then there exists a rectangle with an arbitrarily large side. To do this, we need the following lemma.

LEMMA 6

$$\left.\begin{array}{l} ABB_0A_0 \text{ is a rectangle} \\ B_0 \text{ is the midpoint of } \overline{BB_1} \\ A_0 \text{ is the midpoint of } \overline{AA_1} \end{array}\right\} \longrightarrow ABB_1A_1 \text{ is a rectangle.}$$

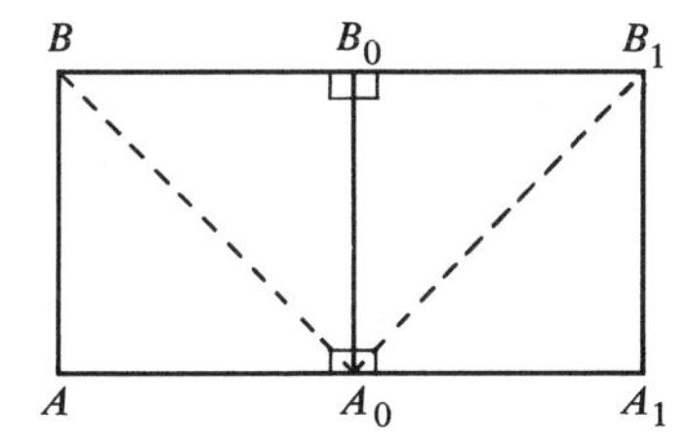

OUTLINE OF PROOF

It follows from SAS that $\triangle B_0BA_0 \cong \triangle B_0B_1A_0$. Then $\overline{BA_0} \cong \overline{B_1A_0}$ and $\underline{m\angle BA_0B_0 = m\angle B_1A_0B_0}$. We can prove that $\underline{B \text{ is in the interior of } \angle AA_0B_0}$ and $\underline{B_1 \text{ is in the interior of } \angle B_0A_0A_1}$. Also we have $\underline{m\angle AA_0B_0 = m\angle A_1A_0B_0}$.

The underlined statements imply that $m\angle AA_0B = m\angle A_1A_0B_1$. Thus we have $\angle AA_0B \cong \angle A_1A_0B_1$, $\overline{A_0A} \cong \overline{A_0A_1}$, and $\overline{BA_0} \cong B_1A_0$. Hence $\triangle A_0AB \cong \triangle A_0A_1B_1$. Therefore, $m\angle A_0A_1B_1 = m\angle AA_1B_1 = 90$. Why? By a similar argument, we can show that $m\angle BB_1A_1 = 90$. Therefore, AA_1B_1B is a rectangle.

Using Lemma 6 and an induction argument, we can prove

LEMMA 7 *If AA_0B_0B is a rectangle, there exists a rectangle AA_nB_nB with $|AA_n| = |AA_0| \cdot 2^n$.*

Since, by Lemma 3, there exists a value of n such that $|AA_0| \cdot 2^n > p$, where p is any given positive number, we have the following theorem.

THEOREM 10-11 *If there exists a rectangle AA_0B_0B and p is any positive number, then there exists a rectangle AA_nB_nB with $|AA_n| > p$.*

THEOREM 10-12 *If there exists a rectangle and if p and q are any two positive numbers, then there exists a rectangle $ABCD$ such that $|AB| > p$ and $|AD| > q$.*

The proof of Theorem 10-12 is considered in the exercises.

THEOREM 10-13 *If one rectangle exists, then there exists a rectangle with two adjacent sides of preassigned lengths p and q.*

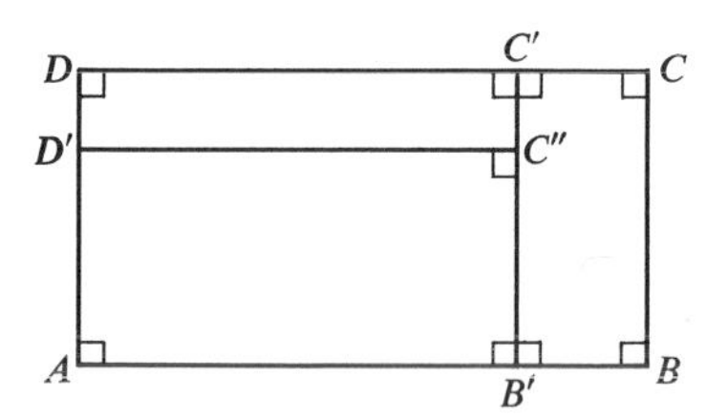

PROOF

By Theorem 10-12, there exists a rectangle $ABCD$ with $|AB| > p$ and $|AD| > q$. There is a point B' in $\overset{\circ\!\!-\!\!\circ}{AB}$ such that $|AB'| = p$. Let C' be the foot of the perpendicular from B' to $\overleftrightarrow{DC}$. Also there is a point D' in $\overset{\circ\!\!-\!\!\circ}{AD}$ such that $|AD'| = q$. Let C'' be the foot of the perpendicular from D' to $\overleftrightarrow{B'C'}$. We intend to prove that $AB'C''D'$ is the required rectangle. In order to do this, we shall first prove that $\angle AB'C'$ is a right angle.

Suppose $m\angle AB'C' > 90$; then the angle-measure sum of quadrilateral $AB'C'D$ is greater than 360, and this contradicts Theorem 10-8 because $AB'C'D$ is a convax quadrilateral. (How do we know this?) Suppose $m\angle AB'C' < 90$; then the angle-measure sum of quadrilateral $B'BCC'$ is greater than 360, and again we have a contradiction of Theorem 10-8 because $B'BCC'$ is also a convax quadrilateral. It follows that $m\angle AB'C' = 90$.

Now $AB'C''D'$ has right angles at A, B', and C''. By an argument similar to the one given above, we can prove that $m\angle AD'C'' = 90$. Therefore, $AB'C''D'$ is a rectangle and, since $|AB'| = p$ and $|AD'| = q$, our proof is complete.

THEOREM 10-14 *If one rectangle exists, then every right triangle has an angle-measure sum of* 180.

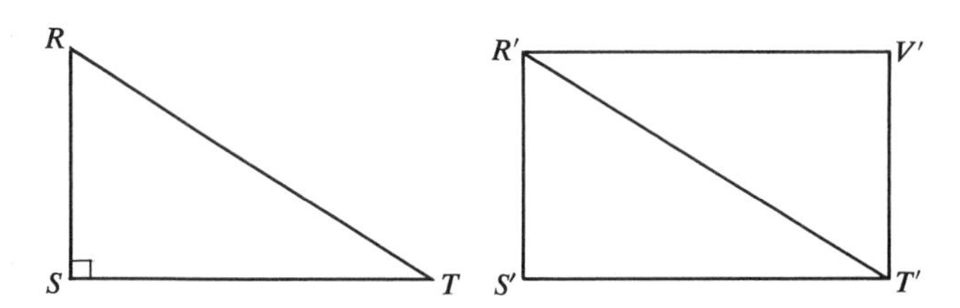

PROOF

Let RST be a right triangle with right angle at S. By Theorem 10-13, there exists a rectangle $R'S'T'V'$ with $|R'S'| = |RS|$ and $|S'T'| = |ST|$. Then $\triangle RST \cong \triangle R'S'T'$, and these two triangles have the same angle-measure sum, call it s. Let t be the angle-measure sum of $\triangle R'T'V'$. $\underline{\text{Now } s + t = 360}$, because 360 is the angle-measure sum of rectangle $R'S'T'V'$. Also $\underline{s \leqq 180}$ and $\underline{t \leqq 180}$ by Legendre's theorem. The underlined statements imply that $s = 180$ (see Exercise 18, page 227), and our proof is complete.

THEOREM 10-15 *If one rectangle exists, then every triangle has an angle-measure sum of* 180.

The proof of Theorem 10-15 is requested in the exercises.

THEOREM 10-16 *If one rectangle exists, then Euclid's parallel postulate holds.*

PROOF

Let p represent a statement that is equivalent to Euclid's parallel postulate. Then Theorem 10-2 has the form: $\sim p \longrightarrow$ there exists at least one triangle whose angle-measure sum is less than 180.

The contrapositive is: Every triangle has an angle-measure sum greater than or equal to 180 $\longrightarrow p$. Theorem 10-15 asserts that: One rectangle exists $\longrightarrow$ every triangle has an angle-measure sum of 180. We have the deductive sequence:

One rectangle exists $\longrightarrow$ every triangle has an angle-measure sum of 180 $\longrightarrow$ every triangle has an angle-measure sum greater than or equal to 180 $\longrightarrow p$.

Therefore, one rectangle exists $\longrightarrow p$, and our proof is complete.

In the exercises you are asked to prove

THEOREM 10-17 *If Euclid's parallel postulate holds, then a rectangle exists.*

The conjunction of Theorems 10-16 and 10-17 implies that Euclid's parallel postulate and the assertion that a rectangle exists are equivalent statements in neutral geometry.

THEOREM **10-18** *If there exists one triangle with angle-measure sum of* 180, *then there exists a rectangle.*

Our plan of proof involves showing that there exists a right triangle whose angle-measure sum is 180. Then we use this fact to prove that there exists a rectangle.

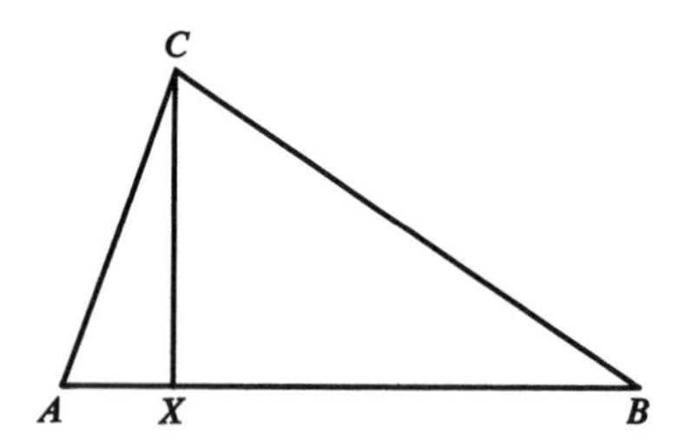

PROOF

If ABC is a triangle, then one of its sides is at least as long as either of the other two sides. Suppose this side is $\overline{AB}$. Then by Corollary 5 to Theorem 7-6, X, the foot of the perpendicular from C to $\overleftrightarrow{AB}$, is in $\mathring{A}\mathring{B}$. It follows that X is in the interior of $\angle ACB$ and, hence, that $m\angle ACX + m\angle XCB = m\angle ACB$. Let p and q represent the angle-measure sums of $\triangle AXC$ and $\triangle CXB$, respectively. Then

$$\begin{aligned} p + q &= m\angle AXC + m\angle A + (m\angle ACX + m\angle BCX) + m\angle B + m\angle BXC \\ &= 90 + \text{angle-measure sum of } \triangle ABC + 90 \\ &= 2\cdot 90 + 180 = 360. \end{aligned}$$

We can prove that $p = 180$. By Legendre's theorem $p \leqq 180$. If $p < 180$, then $q > 180$, contrary to Legendre's theorem as applied to $\triangle CXB$. Thus $p = q$, and $\triangle AXC$ is a right triangle whose angle-measure sum is 180.

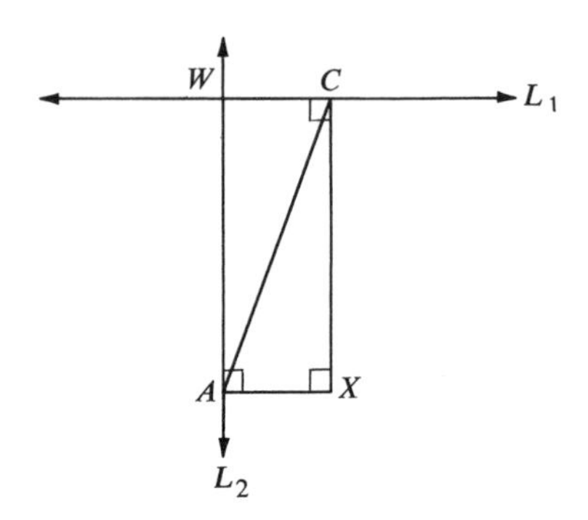

By Theorem 6-21, there is, in the plane of $\triangle AXC$, a line $L_1 \perp \overleftrightarrow{XC}(C)$ and a line $L_2 \perp \overleftrightarrow{AX}(A)$. According to Corollary 4 to Theorem 8-8, L_1 and L_2 intersect at a single point, which we call W. By Theorem 10-5, $L_1 \parallel \overleftrightarrow{AX}$ and $L_2 \parallel \overleftrightarrow{CX}$. Now W and C are distinct points in L_1, and W and A are distinct points in L_2. Why? Therefore, $L_1 = \overleftrightarrow{WC}$ and $L_2 = \overleftrightarrow{WA}$, and we have $\overleftrightarrow{WC} \parallel \overleftrightarrow{AX}$ and $\overleftrightarrow{WA} \parallel \overleftrightarrow{CX}$. It follows by Corollary 2 to Theorem 8-1 that $C \in \overleftrightarrow{WA}/X$ and $C \in \overleftrightarrow{AX}/W$. Hence C is in the interior of $\angle WAX$. This implies that $\underline{m\angle WAC + m\angle CAX = 90}$. By similar reasoning, we have A in the interior of $\angle WCX$ and, hence, $\underline{m\angle WCA + m\angle ACX = 90}$.

Since $\triangle ACX$ is a right triangle whose angle-measure sum is 180, we have $m\angle CAX + m\angle ACX + 90 = 180 \longrightarrow \underline{m\angle CAX + m\angle ACX = 90}$. The underlined statements imply that $m\angle CAX = m\angle WCA$ and $m\angle WAC = m\angle ACX$. By ASA, we have $\triangle AWC \cong \triangle CXA$. Therefore, $m\angle AWC = m\angle CXA = 90$. Quadrilateral $AXCW$ is a rectangle, because each of its angles is a right angle. Our proof is complete.

THEOREM 10-19 *If one triangle has an angle-measure sum of* 180, *then every triangle has an angle-measure sum of* 180.

THEOREM 10-20 *If one triangle has an angle-measure sum of* 180, *then Euclid's parallel postulate holds.*

The proofs of Theorems 10-19 and 10-20 are considered in the exercises.

THEOREM 10-21 *If one triangle has an angle-measure sum that is less than* 180, *then every triangle has an angle-measure sum of less than* 180.

PROOF

Let $\triangle RST$ be the triangle whose angle-measure sum is less than 180 and let $\triangle XYZ$ be any other triangle. If p is the angle-measure sum of $\triangle XYZ$, Legendre's theorem states that $p = 180$ or $p < 180$. If $p = 180$, then, by Theorem 10-19, $\triangle RST$ has an angle-measure sum of 180, contrary to our hypothesis.

Observe that Theorems 10-21 and 10-19 tell us that in neutral geometry exactly one of the following statements is true:

(1) Every triangle has an angle-measure sum of 180.
(2) Every triangle has an angle-measure sum less than 180.

The neutral geometry for which the first statement is true is Euclidean geometry. The neutral geometry that is characterized by the second statement corresponds to the geometry of Bolyai and Lobachevsky (hyperbolic geometry), which is considered in the next section.

THEOREM 10-22 *Euclid's parallel postulate is equivalent to the assertion that a rectangle exists.*

The proof of Theorem 10-22 is requested in the exercises.

THEOREM 10-23 *If there is a point P and a line L not containing it such that there is not more than one line containing P parallel to L, then Euclid's parallel postulate holds.*

Restatement:

$P \notin L$.
There is not more than one line containing P and parallel to L. $\longrightarrow$ Euclid's parallel postulate holds.

Our plan of proof consists of showing that a rectangle exists. Then, according to Theorem 10-22, it follows that Euclid's parallel postulate holds.

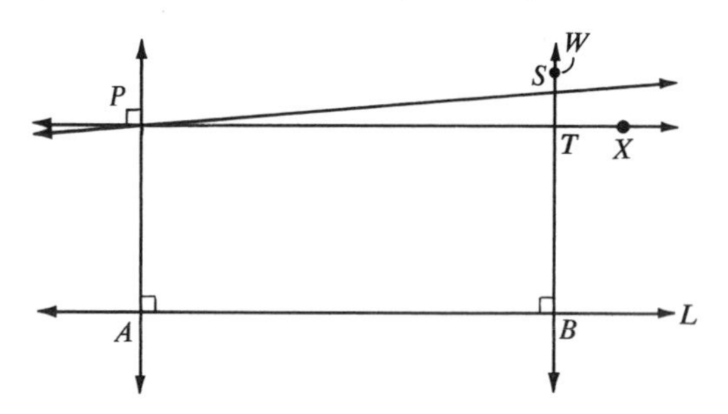

OUTLINE OF PROOF

There exists a line through P perpendicular to L at point A. Let B be any point in L other than A. There is a line BW in plane PAB such that $\overleftrightarrow{BW} \perp \overleftrightarrow{AB}$. Also, there is a line PX in plane PAB such that $\overleftrightarrow{PX} \perp \overleftrightarrow{AP}$. According to Corollary 4 to Theorem 8-4, $\overleftrightarrow{PX}$ and $\overleftrightarrow{BW}$ intersect in a single point, which we shall call T. We can prove that $ABTP$ is a quadrilateral in which $P\underline{A}B$, $A\underline{B}T$, and $A\underline{P}T$ are right angles. It remains for us to prove that $P\underline{T}B$ is a right angle.

According to Theorem 10-5, $\overleftrightarrow{PX} \parallel \overleftrightarrow{AB}$. By Theorem 7-15, there is a line PS that is perpendicular to $\overleftrightarrow{BW}$ at S. Now $\overleftrightarrow{PS}$ and $\overleftrightarrow{AB}$ are distinct lines in plane PAB and each is perpendicular to $\overleftrightarrow{BW}$. It follows that $\overleftrightarrow{PS} \parallel \overleftrightarrow{AB}$. If $S \neq T$, we have two distinct lines through P parallel to $\overleftrightarrow{AB}$ in violation of the second condition in the hypothesis. Therefore $S = T$ and, hence, $P\underline{S}B = P\underline{T}B$. Accordingly, $P\underline{T}B$ is a right angle. The underlined statements imply that $ABTP$ is a rectangle. The existence of this rectangle implies that Euclid's parallel postulate holds.

Exercises

1. Let e represent Theorem 8-12 and let p represent the parallel postulate. Prove that $e \longrightarrow p$ on the basis of our first sixteen postulates. (To do this, you may use any theorem or postulate that precedes Postulate 17.)

L 2. Prove that the statement "If a line intersects one of two parallel lines in a single point, then it intersects the other in a single point provided all three lines are coplanar." (Theorem 8-7) is equivalent to the parallel postulate on the basis of the first sixteen postulates.

3. Prove that the statement "If two distinct lines are parallel to the same line, then they are parallel to each other, provided all three lines are coplanar" (Theorem 8-9) is equivalent to the parallel postulate on the basis of the first sixteen postulates.

4. Prove Lemma 2.

5. Prove the statements

$$a_n + b_n = b_{n-1} \quad \text{and} \quad b_n > a,$$

which appear in the outlined proof of Theorem 10-2.

6. Complete the proof of Theorem 10-2 by mathematical induction.

7. Prove Lemma 4.

8. Complete the proof of Theorem 10-7 by mathematical induction.

9. Prove Theorem 10-9.

10. Prove Theorem 10-10.

11. Prove:

$$\left.\begin{array}{l} ABC \text{ is a Saccheri quadrilateral} \\ M \text{ is the midpoint of } \overline{AB} \\ N \text{ is the midpoint of } \overline{CD} \end{array}\right\} \longrightarrow \left\{\begin{array}{l} \overleftrightarrow{MN} \perp \overleftrightarrow{AB} \\ \overleftrightarrow{MN} \perp \overleftrightarrow{CD}. \end{array}\right.$$

L 12. Prove:

$$\left.\begin{array}{l} ABCD \text{ is a convax quadrilateral} \\ \angle DAB \text{ is a right angle} \\ \angle \mathrm{ABC} \text{ is a right angle} \\ |\mathrm{AD}| < |\mathrm{BC}| \end{array}\right\} \longrightarrow m\angle BCD < m\angle ADC.$$

L 13. Prove the fourth converse of the implication stated in Exercise 12.

14. Prove:

$$\left.\begin{array}{l} ABCD \text{ is a convax quad} \\ \angle A \text{ and } \angle B \text{ are right angles} \\ \angle C \cong \angle D \end{array}\right\} \longrightarrow ABCD \text{ is a Saccheri quad.}$$

15. Using our new definition of "rectangle," prove
    (a) The opposite sides of a rectangle are parallel.
    (b) A rectangle is a convax quadrilateral.

16. Prove Lemma 7.

17. Prove Theorem 10-12.

18. If $a, b, c \in R$, prove:

$$\left.\begin{array}{l} a + b = c \\ a \leqq \frac{1}{2}c \\ b \leqq \frac{1}{2}c \end{array}\right\} \longrightarrow a = \tfrac{1}{2}c.$$

19. Prove Theorem 10-15.

20. Prove Theorem 10-17.

21. Prove Theorem 10-19.

22. Prove Theorem 10-20.

23. Prove Theorem 10-22.

24. Prove that the line containing the summit of a Saccheri quadrilateral does not intersect the line containing the base.

25. Prove: If there exists a line L and a point P such that there are two distinct lines through P parallel to L, and if L_1 is any line 'and P_1 is any point not in L_1, then there are two distinct lines through P_1 parallel to L_1.

26. Prove that the statement "Through any point in the interior of an angle there exists at least one line that intersects both sides of the angle and does not contain the vertex of the angle" is equivalent to the parallel postulate.

Hyperbolic Geometry

We are now in a position to convert some of the theorems we have already proved into theorems in hyperbolic geometry by the simple process of accepting $\sim p$ as a postulate and annexing it to our set of postulates for neutral geometry. Thus we have the following theorems in hyperbolic geometry from Theorems 10-2 and 10-22.

THEOREM **10-24** *There exists a triangle with angle-measure sum less than* 180.

THEOREM **10-25** *There exist no rectangles.*

Combining Theorems 10-21 and 10-24, we have the following important theorem and corollary.

THEOREM **10-26** *The angle-measure sum of any triangle is less than* 180.

COROLLARY TO THEOREM **10-26** *The angle-measure sum of any convax quadrilateral is less than* 360.

In Euclidean geometry, two triangles having their corresponding angles congruent may have the same shape (that is, are similar), but we were able to prove in Chapter 8 that they are not necessarily congruent. (See Exercise 11, page 192.) In other words, AAA does not represent a congruence theorem in Euclidean geometry. However, AAA is a congruence theorem in hyperbolic geometry, as indicated by the following theorem.

THEOREM **10-27** *Two triangles are congruent if their corresponding angles have equal measures.*

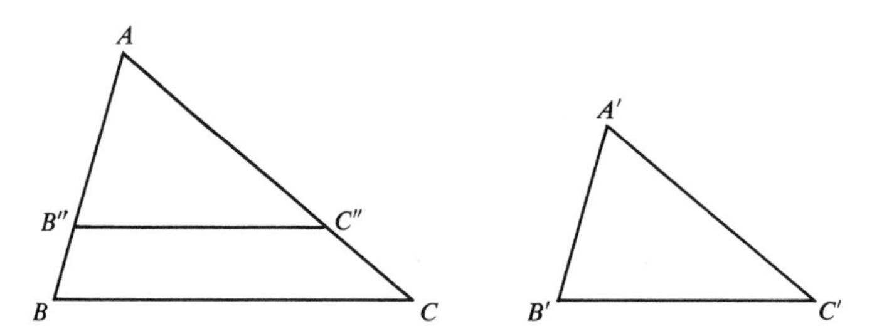

OUTLINE OF PROOF

Suppose the theorem is false. Then there exist two triangles $\triangle ABC$ and $\triangle A'B'C'$ that are not congruent and yet $m\angle A = m\angle A'$, $m\angle B = m\angle B'$, and

$m\angle C = m\angle C'$. We must have $|AB| \neq |A'B'|$ because, if not, the triangles would be congruent by ASA. Similarly, we must have $|BC| \neq |B'C'|$ and $|CA| \neq |C'A'|$.

Consider the two triples (1) $|AB|$, $|BC|$, $|AC|$ and (2) $|A'B'|$, $|B'C'|$, $|A'C'|$. One of these triples must contain two numbers that are greater than the corresponding numbers in the other triple. Therefore, there is no loss of generality in supposing that $|AB| > |A'B'|$ and $|AC| > |A'C'|$. Accordingly, there is a point B'' in $\overset{\circ\!\!-\!\!\circ}{AB}$ such that $|AB''| = |A'B'|$, and there is a point C'' in $\overset{\circ\!\!-\!\!\circ}{AC}$ such that $|AC''| = |A'C'|$. It follows by SAS that $\triangle AB''C'' \cong \triangle A'B'C'$, so that

$$m\angle AB''C'' = m\angle B' = m\angle B$$

and

$$m\angle AC''B'' = m\angle C' = m\angle C.$$

But

$$m\angle AB''C'' + m\angle C''B''B = 180$$

and therefore,

$$m\angle B + m\angle C''B''B = 180.$$

Similarly,

$$m\angle C + m\angle CC''B'' = 180.$$

The last two statements imply that the angle-measure sum of convax quadrilateral $BCC''B''$ is 360. Since this contradicts the corollary to Theorem 10-26, our indirect proof is complete.

If we accept the idea that triangles having their corresponding angles congruent are the same shape, we see from Theorem 10-27 that in hyperbolic geometry any two triangles having the same shape must be congruent. In fact, any two "similar" figures would be congruent and so the same size. In a hyperbolic world all replicas such as pictures and statues would have to be life size to avoid distortion.

DEFINITION *The* defect δ of a triangle *ABC is given by the formula*

$$\delta = 180 - (m\angle A + m\angle B + m\angle C).$$

We conclude this introductory discussion of hyperbolic geometry by noticing the close relationship that exists between the defect of a triangle and statements about the area of the "triangular region" bounded by the triangle. In order to do this we first define a triangular region as the union of a triangle and its interior; then we list three statements about the areas of triangular regions that are intuitively evident and that would be postulated in a formal discussion of area. These statements are

(1) The area of a triangle (triangular region) is a uniquely determined positive real number.
(2) Congruent triangles have equal areas.

(3) If triangular region ABC is "split" into two triangular regions ABX and BXC with $X \in \overset{\circ\!-\!\circ}{AC}$, then $|ABC| = |ABX| + |BAC|$, where the symbol $|ABC|$ represents the area of triangular region ABC.

THEOREM 10-28 *The defect of a triangle satisfies properties* (1), (2), *and* (3) *above.*

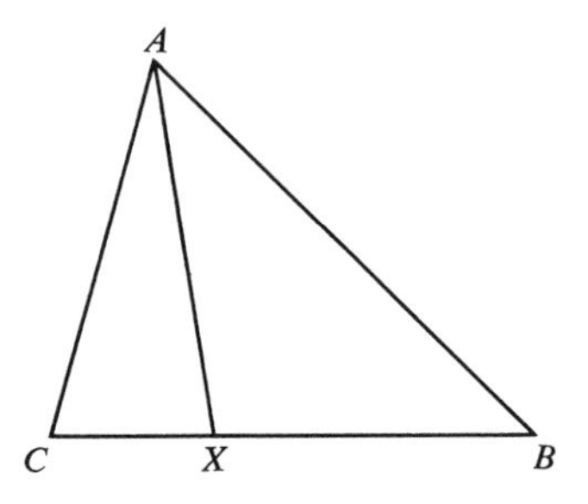

PROOF

Since we have agreed in Chapter 6 on the unit for measuring angles, property (1) is satisfied because the defect of a triangle is a uniquely determined positive number. Property (2) holds because congruent triangles have equal angle sums and, hence, equal defects. To prove property (3), we consider $\triangle ABC$ with $X \in \overset{\circ\!-\!\circ}{BC}$, so that $\triangle ABC$ is "split" into $\triangle ACX$ and $\triangle AXB$. Let δ, δ_1, and δ_2 represent the defects of $\triangle ABC$, $\triangle ACX$, and $\triangle AXB$, respectively. Then

$$\delta_1 = 180 - (m\angle C + m\angle CAX + m\angle AXC),$$

$$\delta_2 = 180 - (m\angle B + m\angle AXB + m\angle XAB),$$

$$\begin{aligned}\delta_1 + \delta_2 &= 360 - (m\angle CXA + m\angle AXB) - [(m\angle C + m\angle B) + \\ &\quad (m\angle CAX + m\angle XAB)] \\ &= 360 - 180 - (m\angle C + m\angle B + m\angle CAB) \\ &= 180 - (m\angle C + m\angle B + m\angle A).\end{aligned}$$

Therefore, $\delta_1 + \delta_2 = \delta$.

It is interesting to observe that in Euclidean spherical geometry the angle-measure sum of any triangle is greater than 180 and that its "excess," which is the amount by which its angle-measure sum exceeds 180, also satisfies properties (1), (2), and (3) above.

We have seen that P, the postulates of Euclidean geometry, and H, the set of postulates of hyperbolic geometry, are identical except for the "parallel postulate." The logical structure called hyperbolic geometry that has been erected on H is fully as extensive and complex as Euclidean geometry. One might argue that this structure is based on a false premise and that it will eventually yield false theorems. It was this belief that motivated mathematicians for over 2000 years in their efforts to prove Euclid's parallel postulate. As we noted in

Chapter 3, there is no absolute test for the consistency of any set of postulates. But it can be proved that Euclidean (parabolic) geometry and hyperbolic geometry stand or fall together on the question of consistency. If either is inconsistent, so is the other.

Exercises

1. Prove the corollary to Theorem 10-26.
2. Which of the three properties referred to in Theorem 10-28 hold for the defect of a triangle in Euclidean geometry?
3. Prove:

$$\left.\begin{array}{l} ABC \text{ is a triangle} \\ X \in \overleftrightarrow{AB} \end{array}\right\} \longrightarrow \left\{\begin{array}{l} \text{The defect of } \triangle AXC \text{ is less than the} \\ \text{defect of } \triangle ABC \end{array}\right.$$

4. Prove:

$$\left.\begin{array}{l} ABC \text{ is a triangle} \\ D \in \overleftrightarrow{AB} \\ E \in \overleftrightarrow{BC} \\ F \in \overleftrightarrow{CA} \end{array}\right\} \longrightarrow \left\{\begin{array}{l} \text{The defect of } \triangle ABC \text{ is the sum of the} \\ \text{defects of triangles } ADF, BED, CFE, \\ \text{and } DEF \end{array}\right.$$

5. If points X, Y, and Z are three noncollinear points in the interior of $\triangle ABC$, prove that $\triangle ABC$ has a larger defect than $\triangle XYZ$.
6. The defect δ of a convax quadrilateral $ABCD$ is defined by the equation $\delta = 360 - (m\angle A + m\angle B + m\angle C + m\angle D)$. If δ_1 and δ_2 are the defects of $\triangle ABC$ and $\triangle ACD$, respectively, prove that $\delta = \delta_1 + \delta_2$.
7. Prove: If three angles of a quadrilateral are right angles, then the fourth angle is acute.
8. Definition: A *Lambert quadrilateral* is a quadrilateral with three and only three right angles. Prove:
   (a) If $ABCD$ is a Lambert quadrilateral with right angles at D, A, and B, then $|CB| > |AD|$.
   (b) If $ABCD$ and $AB'C'D$ are Lambert quadrilaterals with right angles at A, B, B', and D, and $B \in \overleftrightarrow{AB'}$, then $|BC| < |B'C'|$.
   [*Note:* The statements in parts (a) and (b) imply that, in hyperbolic geometry, two parallel lines having a common perpendicular diverge continuously on both sides of this perpendicular.]

References

Alder, *Modern Geometry*
Eves, *A Survey of Geometry*
Golos, *Foundations of Euclidean and Non-Euclidean Geometry*
Kay, *College Geometry*
Klein, *Elementary Mathematics from an Advanced Viewpoint*
Moise, *Elementary Geometry from an Advanced Standpoint*
Prenowitz and Jordan, *Basic Concepts of Geometry*
Wolfe, *Introduction to Non-Euclidean Geometry*

Appendix A
Some Algebraic Properties Useful in Geometry

Properties of Equality

The statement $a = b$ says that a and b name the same thing (the same set, the same real number, and so on).

E_1: $a = a$.	Reflexive property of equality.
E_2: If $a = b$, the truth or falsity of any statement in which a appears is not changed if b is used instead.	Substitution property of equality.
E_3: If $a = b$, then $b = a$.	Symmetric property of equality.
E_4: If $a = b$ and $b = c$, then $a = c$.	Transitive property of equality.
For real numbers a, b, c, d	
E_5: If $a = b$ and $c = d$, then $a + c = b + d$.	Addition property of equality.
E_6: If $a = b$ and $c = d$, then $ac = bd$.	Multiplication property of equality.

Properties of Real Numbers

R_1: $a + b$ is a real number.	Closure property of addition.
R_2: ab is a real number.	Closure property of multiplication.
R_3: $a + b = b + a$.	Commutative property of addition.
R_4: $ab = ba$.	Commutative property of multiplication.
R_5: $(a + b) + c = a + (b + c)$.	Associative property of addition.
R_6: $(ab)c = a(bc)$.	Associative property of multiplication.
R_7: There is a unique real number 0 such that $a + 0 = a$.	Addition property of zero.
R_8: There is a unique real number 1 such that $a \cdot 1 = a$.	Multiplication property of one.
R_9: For each $a \neq 0$, there is a multiplicative inverse $\frac{1}{a}$ such that $a(1/a) = 1$.	Existence of multiplicative inverse.

R_{10}: For each real number a, there is an additive inverse $-a$ such that $a + (-a) = 0$.	Existence of additive inverse.
R_{11}: $a(b + c) = ab + ac$.	Distributive property.
R_{12}: $a + b + c = (a + b) + c$, $a + b + c + d = (a + b + c) + d, \ldots$	Definition of addition of more than two numbers.
R_{13}: $abc = (ab)c$, $abcd = (abc)d, \ldots$.	Definition of multiplication of more than two numbers.
R_{14}: There are infinitely many real numbers.	
R_{15}: Between any two real numbers there is another real number.	Density property of real numbers.
R_{16}: There is no greatest real number and no least real number.	
For R_{17}, we state the following definitions: Let S be any nonempty set of real numbers, and suppose that y is a real number such that the statement $x \leqq y$ is true for every real number x in S. We call y an *upper bound* of S, and we say that S is *bounded above*. If no real number less than y is an upper bound of S, then y is called the *least upper bound* of S.	
R_{17}: If S is a nonempty subset of the real numbers, R, that has an upper bound in R, then it has a least upper bound in R.	Completeness property
R_{18}: If a and b are positive real numbers such that $a < b$, there is a counting number n such that $na > b$.	Archimedean property.

Order Properties of Real Numbers

O_1: For any two real numbers the statements $a > b$ and $b < a$ are either both true or both false. $(a > b \longleftrightarrow b < a.)$	
O_2: Exactly one of the following statements is true: $a < b$, $a = b$, $a > b$.	Trichotomy property (comparison property).
O_3: If $a > b$, then $a + c > b + c$.	Addition property of order.
O_4: If $a > b$ and $b > c$, then $a > c$.	Transitive property of order.
O_5: (1) If $a > b$ and $c > 0$, then $ac > bc$. (2) If $a > b$ and $c < 0$, then $ac < bc$.	Multiplication property of order.
O_6: If $a + b = c$ and $b > 0$, then $a < c$.	

[*Note:* True statements are obtained from statements O_3–O_6 when the symbol $>$ is replaced wherever it occurs by the symbol $\geqq$ and the symbol $<$ is replaced wherever it occurs by the symbol $\leqq$.]

Theorems about Real Numbers

RT_1 (1) *If $a, b, c \in R$, then $a + c = b + c \longleftrightarrow a = b$.*
(2) *If $a, b, c \in R$ and $c \neq 0$, then $ac = bc \longleftrightarrow a = b$.*

RT_2 *If $a \in R$ and $a \neq 0$, then a has only one multiplicative inverse.*

COROLLARY TO RT_2 *If $a \in R$ and $a \neq 0$, then the multiplicative inverse of the multiplicative inverse of a is a.*

RT_3 *If $a \in R$, then $a \cdot 0 = 0$.*

RT_4 *If $a, b \in R$ and $ab = 0$, then $a = 0$ or $b = 0$.*

DEFINITION *If $a, b, c \in R$ and $b \neq 0$, then*

$$\frac{a}{b} = c \longleftrightarrow a = cb.$$

RT_5 *If $a, b \in R$ and $b \neq 0$, then*

$$\frac{a}{b} = \frac{1}{b} \cdot a.$$

RT_6 *If $a, b \in R$ and $ab \neq 0$, then*

$$\frac{1}{a} \cdot \frac{1}{b} = \frac{1}{ab}.$$

RT_7 *If $a, b, c, d \in R$ and $bd \neq 0$, then*

$$\frac{a}{b} \cdot \frac{c}{d} = \frac{ac}{bd}.$$

RT_8 *If $a, b, c \in R$ and $bc \neq 0$, then*

$$\frac{a}{b} = \frac{ac}{bc}.$$

RT_9 *If $a \in R$, then a has only one additive inverse.*

COROLLARY 1 TO RT_9 *If $a \in R$, then $-(-a) = a$.*

COROLLARY 2 TO RT_9 $-0 = 0$.

DEFINITION *If $a, b, c \in R$, then $a - b = c \longleftrightarrow a = b + c$.*

RT_{10} *If $a, b \in R$, then $a - b = a + (-b)$.*

RT_{11} *If $a \in R$, then $(-1)a = -a$.*

RT_{12} *If $a, b \in R$, then $(-a)b = -(ab)$.*

RT_{13} *If $a, b \in R$, then $(-a)(-b) = ab$.*

RT_{14} *If $a, b \in R$ and $b \neq 0$, then*

$$\frac{-a}{b} = \frac{a}{-b} = -\frac{a}{b}.$$

RT_{15} *If $a, b \in R$ and $b \neq 0$, then*

$$\frac{-a}{-b} = \frac{a}{b}.$$

RT_{16} *If $a, b, c \in R$ and $b \neq 0$, then*

$$\frac{a}{b} + \frac{c}{b} = \frac{a + c}{b}.$$

RT_{17} *If $a, b, c, d \in R$ and $bd \neq 0$, then*

$$\frac{a}{b} + \frac{c}{d} = \frac{ad + bc}{bd}.$$

RT_{18} *If $a, b, c, d \in R$ and $bd \neq 0$, then*

$$\frac{a}{b} = \frac{c}{d} \longleftrightarrow ad = bc.$$

RT_{19} *If $a, b, c, d \in R$ and $bcd \neq 0$, then*

$$\frac{\frac{a}{b}}{\frac{c}{d}} = \frac{ad}{bc}.$$

RT_{20} *If $a, b \in R$, then $a < b \longleftrightarrow a + c < b + c$.*

RT_{21} *If $a, b \in R$, then $a > b \longleftrightarrow a - b > 0$ and $a < b \longleftrightarrow a - b < 0$.*

COROLLARY TO RT_{21} *A real number is negative if and only if its additive inverse is positive, and a real number is positive if and only if its additive inverse is negative.*

RT_{22} *If $a, b, c, d \in R$, then*

$$\left.\begin{array}{l} a > b \\ c > d \end{array}\right\} \longrightarrow a + c > b + d.$$

RT_{23} *If $a, b \in R$, then $a > b \longleftrightarrow -a < -b$ and $a < b \longleftrightarrow -a > -b$.*

RT_{24} *If $a, b \in R$, then $(ab > 0) \longleftrightarrow$ (a and b have like signs) and $(ab < 0) \leftrightarrow$ $\leftrightarrow$ (a and b have opposite signs).*

COROLLARY 1 TO RT_{24} *(1) If $a \in R$, then $a^2 \geqq 0$. (2) If $a \in R$ and $a \neq 0$, then $a^2 > 0$.*

COROLLARY 2 TO RT_{24} *If $a \in R$, then a and $1/a$ have like signs when $a \neq 0$.*

RT_{25} *If $a, b, c \in R$ and $a > 0$, then $b > c \longleftrightarrow ab > ac$.*

RT_{26} *If $a, b, c \in R$ and $a < 0$, then $b > c \longleftrightarrow ab < ac$.*

RT_{27} *If $a, b \in R$, $a > 0$, and $b > 0$, then*

$$a > b \longleftrightarrow a^2 > b^2,$$
$$a = b \longleftrightarrow a^2 = b^2.$$

COROLLARY TO RT_{27} *If $a, b \in R$, then $a = b \longrightarrow a^2 = b^2$.*

Notation: If $x \in R$, then the absolute value of x is denoted by the symbol $|x|$.

DEFINITION *For $x \in R$, $|x| = x$ if $x \geqq 0$ and $|x| = -x$ if $x < 0$.*

RT_{28} *(1) If $x \in R$, then $|x| \geqq 0$. (2) If $x \in R$ and $x \neq 0$, then $|x| > 0$.*

RT_{29} *If $a, b \in R$, then $a \geqq b \longleftrightarrow |a - b| = a - b$ and $a \leqq b \longleftrightarrow |a - b| = b - a$.*

RT_{30} *If $x \in R$, then $\sqrt{x^2} = |x|$.*

RT_{31} *If $x \in R$, then $|x| = |-x|$.*

RT_{33} *If $x \in R$, then $-|x| \leqq x \leqq |x|$.*

RT_{34} *If $x, y \in R$, then (1) $|x||y| = |xy|$, (2) $|x|/|y| = |x/y|$, provided $y \neq 0$.*

RT_{35} *If $a, x \in R$ and $a > 0$, then $|x| < a \longleftrightarrow -a < x < a$.*

COROLLARY TO RT_{35} *If $a, x \in R$ and $a \geqq 0$, then $|x| \leqq a \longleftrightarrow -a \leqq x \leqq a$.*

RT_{36} *If $x, y \in R$, then $|x + y| \leqq |x| + |y|$.*

Notation: We use the symbol $a\text{-}b\text{-}c$ to indicate that the real number b is between the real numbers a and c. Thus

$$a\text{-}b\text{-}c \longleftrightarrow (a < b < c \vee a > b > c).$$

RT$_{37}$ *If $a, b, c \in R$, then*

$$a\text{-}b\text{-}c \longleftrightarrow (|b - a| + |c - b| = |c - a| \wedge a, b, c \text{ are distinct}).$$

RT$_{38}$ *If $a, b, c \in R$, then $a\text{-}b\text{-}c \longleftrightarrow (b - a)(c - b) > 0$.*

RT$_{39}$ *If a, b, c are distinct real numbers, exactly one of them is between the other two.*

Use of the real number properties in proof is illustrated by the following outline proof for part of RT$_{37}$ and RT$_{39}$.

Consider RT$_{37}$:

$$a\text{-}b\text{-}c \longleftrightarrow (|b - a| + |c - b| = |c - a| \wedge a, b, c \text{ distinct}).$$

This theorem is equivalent to the conjunction of the following two statements:

$$\text{I:} \quad \left.\begin{array}{l} |b - a| + |c - b| = |c - a| \\ a, b, c \text{ distinct} \end{array}\right\} \longrightarrow a\text{-}b\text{-}c,$$

$$\text{II:} \quad a\text{-}b\text{-}c \longrightarrow \left\{\begin{array}{l} |b - a| + |c - b| = |c - a| \\ a, b, c \text{ distinct.} \end{array}\right.$$

Outline proof of I:

$$\begin{array}{l}
|b - a| + |c - b| = |c - a| \dashv \\
\leftrightarrow (|b - a| + |c - b|)^2 = (|c - a|)^2 \dashv \\
\leftrightarrow |b - a|^2 + 2|b - a||c - b| + |c - b|^2 = |c - a|^2 \dashv \\
\leftrightarrow (b - a)^2 + 2|b - a||c - b| + (c - b)^2 = (c - a)^2 \dashv \\
\leftrightarrow b^2 - 2ab + a^2 + 2|b - a||c - b| + c^2 - 2bc + b^2 = c^2 - 2ac + a^2 \dashv \\
\leftrightarrow 2|b - a||c - b| = 2bc - 2b^2 - 2ac + 2ab \dashv \\
\leftrightarrow |b - a||c - b| = bc - b^2 - ac + ab \dashv \\
\leftrightarrow |b - a||c - b| = b(c - b) - a(c - b) \dashv \\
\leftrightarrow |b - a||c - b| = (b - a)(c - b) \dashv \\
\leftrightarrow |(b - a)(c - b)| = (b - a)(c - b) \dashv \\
\left.\begin{array}{l}
\leftrightarrow (b - a)(c - b) \geqq 0 \\
\quad \left.\begin{array}{l} \underline{a \neq b} \longrightarrow (b - a) \neq 0 \\ \underline{c \neq b} \longrightarrow (c - b) \neq 0 \end{array}\right\} \longrightarrow (b - a)(c - b) \neq 0
\end{array}\right\} \dashv \\
\leftrightarrow (b - a)(c - b) > 0 \dashv \\
\leftrightarrow [(b - a > 0) \wedge (c - b > 0)] \vee [(b - a < 0) \wedge (c - b < 0)] \dashv \\
\leftrightarrow [b > a \wedge c > b] \vee [b < a \wedge c < b] \longrightarrow (a < b < c) \vee (c < b < a) \dashv \\
\leftrightarrow a\text{-}b\text{-}c.
\end{array}$$

To establish RT$_{39}$, we must prove (1) at least one of the numbers a, b, and c is between the other two and (2) not more than one of them is between the other two.

In order to prove (1), we observe that according to RT_{38},

$$a\text{-}b\text{-}c \longleftrightarrow (b-a)(c-b) > 0.$$

Accordingly,

$$\sim(a\text{-}b\text{-}c) \longleftrightarrow (b-a)(c-b) \leqq 0.$$

We can prove that at least one of the numbers a, b, c is between the other two by proving that the conjunctive statement

$$\sim(a\text{-}b\text{-}c) \wedge \sim(b\text{-}a\text{-}c) \wedge \sim(a\text{-}c\text{-}b)$$

is false—that is, that it leads to a contradiction. We arrange our argument as follows:

$$\begin{array}{ll}
\left.\begin{array}{l}\sim(a\text{-}b\text{-}c \longrightarrow (b-a)(c-b) \leqq 0 \\ \sim(b\text{-}a\text{-}c) \longrightarrow (a-b)(c-a) \leqq 0\end{array}\right\} \dashv & \\
\qquad \mapsto (b-a)(c-b)(a-b)(c-a) \geqq 0 \dashv & \\
\left.\begin{array}{l}\qquad \mapsto (a-b)^2(c-a)(b-c) \geqq 0 \\ \sim(a\text{-}c\text{-}b) \longrightarrow (c-a)(b-c) \leqq 0\end{array}\right\} \dashv & \\
\qquad \mapsto (a-b)^2(c-a)^2(b-c)^2 \leqq 0 \dashv & \\
\qquad \mapsto [(a-b)(c-a)(b-c)]^2 \leqq 0. &
\end{array}$$

However, since a, b, and c are distinct real numbers, none of the factors $a-b$, $c-a$, $b-c$ is zero. Therefore,

$$(a-b)(c-a)(b-c) \neq 0.$$

It follows that the statement

$$[(a-b)(c-a)(b-c)]^2 \leqq 0$$

is false because the square of a nonzero real number is positive. Hence, by the trichotomy property, it cannot be 0 or negative. Thus we have a contradiction that resulted from supposing that all three of the statements a-b-c, a-c-b, and c-a-b are false. Therefore, at least one of these statements is true.

We now prove (2) by using RT_{36} to prove that no two of these statements can be true at the same time.

$$\left.\begin{array}{l} a\text{-}b\text{-}c \longrightarrow |b-a| + |c-b| = |c-a| \\ a\text{-}c\text{-}b \longrightarrow |c-a| + |b-c| = |b-a| \end{array}\right\} \longrightarrow 2|c-b| = 0 \longrightarrow c = b.$$

However, $c \neq b$, because all three numbers are distinct by hypothesis. It follows that the statement $a\text{-}b\text{-}c \wedge a\text{-}c\text{-}b$ is false. It follows that $a\text{-}b\text{-}c \longrightarrow \sim(a\text{-}c\text{-}b)$. In like manner we can prove $a\text{-}b\text{-}c \longrightarrow \sim(c\text{-}a\text{-}b)$. Thus if one of the statements a-b-c, a-c-b, c-a-b is true, the other two are false, and our proof of RT_{39} is complete.

Reference

Katz, *Axiomatic Analysis*

Appendix B
Mathematical Induction

Consider the following statements:

P_1: $8 - 3 = 1 \cdot 5.$
P_2: $8^2 - 3^2 = 11 \cdot 5.$
P_3: $8^3 - 3^3 = 97 \cdot 5.$
P_4: $8^4 - 3^4 = 803 \cdot 5.$

From these statements we might formulate the conjecture that $8^n - 3^n$ is a multiple of 5 for all counting numbers n. Since this conjecture has been verified for some but not all of the set of counting numbers, it is merely a guess based on inductive reasoning. We noted in Chapter 2 that some conjectures obtained by inductive reasoning are false; hence, we cannot use induction to prove statements. However, there is a form of deductive reasoning, known as *mathematical induction,* that can be used to establish the truth of this conjecture. We now state the *principle of mathematical induction* and accept it as a postulate.

PRINCIPLE OF MATHEMATICAL INDUCTION Let $P_1, P_2, \ldots, P_n, \ldots$ be a sequence of statements and let H be the assertion that P_n is true for all counting numbers greater than or equal to the counting number t. The assertion H will be accepted as proved if (1) there is a special proof that P_t is true, and (2) there is a general proof that $P_k \longrightarrow P_{k+1}$ for $k \in C$ and $k \geqq t$.

The special proof is sometimes referred to as the intial verification, and the general proof is called the auxiliary theorem.

Returning to our example, we see that the statements P_1, P_2, P_k, P_{k+1}, and P_n are, respectively,

$$8 - 3 = 1 \cdot 5, \qquad 8^2 - 3^2 = 11 \cdot 5, \qquad 8^k - 3^k = a \cdot 5,$$

$$8^{k+1} - 3^{k+1} = a_{k+1} \cdot 5 \quad \text{and} \quad 8^n - 3^n = a_n \cdot 5,$$

where a_k, a_{k+1}, and a_n are integers. The first of these statements, P_1, is true, and this fact provides the initial verification for the assertion that $8^n - 3^n$ is a multiple of 5 for all $n \in C$. In other words, the statement "$8^n - 3^n$ is a multiple of 5" is true for $n = 1$. Thus $t = 1$ for this example.

Now we must attempt to construct the general proof that $P_k \longrightarrow P_{k+1}$ as required by our statement of the principle of mathematical induction. Therefore we must prove

$$8^k - 3^k = a_k \cdot 5 \longrightarrow 8^{k+1} - 3^{k+1} = a_{k+1} \cdot 5,$$

where a_k and a_{k+1} are integers. In order to do this, we try to express $8^{k+1} - 3^{k+1}$ in a form that involves $8^k - 3^k$. One way to do this is to carry out one stage of the division indicated by

$$\frac{8^{k+1} - 3^{k+1}}{8^k - 3^k}.$$

We have

$$\begin{array}{r|l} & \quad 8 \\ 8^k - 3^k & 8^{k+1} - 3^{k+1} \\ & \underline{8^{k+1} - 8 \cdot 3^k} \\ & \qquad 8 \cdot 3^k - 3^{k+1} \end{array}$$

Thus

$$\begin{aligned} 8^{k+1} - 3^{k+1} &= 8(8^k - 3^k) + 8 \cdot 3^k - 3^{k+1} \\ &= 8(8^k - 3^k) = 3^k(8 - 3) \\ &= 8(8^k - 3^k) + 5 \cdot 3^k. \end{aligned}$$

Since $8^k - 3^k = a_k \cdot 5$ by hypothesis, we have

$$\begin{aligned} 8^{k+1} - 3^{k+1} &= 8a_k \cdot 5 + 3^k \cdot 5 \\ &= (8a_k + 3^k)5. \end{aligned}$$

Since a_k and k are integers, we know that $8a_k + 3^k$ is some integer, which we denote by a_{k+1}. Therefore, $8^{k+1} - 3^{k+1} = a_{k+1} \cdot 5$. Thus we have proved the auxiliary theorem:

$$8^k - 3^k = a_k \cdot 5 \longrightarrow 8^{k+1} - 3^{k+1} = a_{k+1} \cdot 5,$$

where a_k, $a_{k+1} \in I$. In other words, if $8^k - 3^k$ is a multiple of 5, so is $8^{k+1} - 3^{k+1}$.

Now that we see how to proceed, we can arrange our argument as follows:

$8^k - 3^k$ is a multiple of 5 $\dashv$
$\leftrightarrow 8^k - 3^k = a_k \cdot 5$, where a_k is an integer $\dashv$
$\leftrightarrow 8(8^k - 3^k) = 8a_k \cdot 5 \dashv$
$\leftrightarrow 8(8^k - 3^k) + 5 \cdot 3^k = 8a_k \cdot 5 + 5 \cdot 3^k \dashv$
$\leftrightarrow 8(8^k - 3^k) + 3^k(8 - 3) = (8a_k + 3^k)5 \dashv$
$\leftrightarrow 8^{k+1} - 3^{k+1} = (8a_k + 3^k)5 \dashv$
$\leftrightarrow 8^{k+1} - 3^{k+1} = a_{k+1} \cdot 5$, where $a_{k+1} = 8a_k + 3^k$ and $a_{k+1} \in I \dashv$
$\leftrightarrow 8^{k+1} - 3^{k+1}$ is a multiple of 5.

We apply the principal of mathematical induction to two other examples.

EXAMPLE 1

If S_n is the sum of the first n odd numbers, find a formula for S_n.

DISCUSSION

We see that

$$\begin{aligned}
S_1 &= 1, \\
S_2 &= 1 + 3 = 4, \\
S_3 &= 1 + 3 + 5 = 9, \\
&\vdots \\
S_k &= 1 + 3 + \cdots + (2k - 1) && (k \text{ terms}), \\
S_{k+1} &= 1 + 3 + \cdots + (2k - 1) + (2k + 1) && (k + 1 \text{ terms}), \\
&\vdots \\
S_n &= 1 + 3 + \cdots + (2n - 1) && (n \text{ terms}).
\end{aligned}$$

Considering the values of S_1, S_2, and S_3 it would seem reasonable to conjecture that $S_n = n^2$. However, it is equally reasonable to suggest that $S_n = n^3 - 5n^2 + 11n - 6$. In each case, the formula "checks out" for $n = 1$, $n = 2$, and $n = 3$. Thus $S_1 = 1$, $S_2 = 4$, and $S_3 = 9$ for either formula. However, when we try to prove the second formula by mathematical induction, we find that the general proof of the auxiliary theorem fails. We must try to prove that $P_k \longrightarrow P_{k+1}$, where P_k is the statement

$$S_k = k^3 - 5k^2 + 11k - 6$$

and P_{k+1} is the statement

$$S_{k+1} = (k + 1)^3 - 5(k + 1)^2 + 11(k + 1) - 6.$$

Clearly, $S_{k+1} = S_k + (2k + 1)$. If $S_k = k^3 - 5k^2 + 11k - 6$, we have

$$S_{k+1} = k^3 - 5k^2 + 11k - 6 + (2k + 1).$$

Therefore, if our formula is correct, we must have

$$(k + 1)^3 - 5(k + 1)^2 + 11(k + 1) - 6 = k^3 - 5k^2 + 13k - 5$$

for all $k \in I$. It is easy to show that this statement is false. Therefore, the conjecture that $S_n = n^3 - 5n^2 + 11n - 6$ must be abandoned.

It is possible to present a general proof for the conjecture that $S_n = n^2$. We must prove that $P_k \longrightarrow P_{k+1}$, where P_k is the statement $S_k = k^2$ and P_{k+1} is the statement $S_{k+1} = (k + 1)^2$. We arrange our argument as follows:

$$\left.\begin{aligned} S_k &= k^2 \\ S_{k+1} &= S_k + (2k + 1) \end{aligned}\right\} \longrightarrow S_{k+1} = k^2 + 2k + 1 \longrightarrow S_{k+1} = (k + 1)^2.$$

This establishes our general proof and assures us that the formula $S_n = n^2$ is correct.

EXAMPLE 2

Prove that $2^m > m$ for all $m \in C$.

SPECIAL PROOF

When $m = 1$, we have $2^1 > 1$.

GENERAL PROOF

To establish the general proof we must prove the auxiliary theorem $P_k \longrightarrow P_{k+1}$, where P_k is the statement $2^k > k$. Thus we must prove the implication $2^k > k \longrightarrow 2^{k+1} > k + 1$ for $k \in C$. This is easily accomplished as follows:

$$\left.\begin{array}{l} 2^k > k \longrightarrow 2^{k+1} > 2k \\ k \in C \longrightarrow k \geqq 1 \longrightarrow 2k \geqq k + 1 \end{array}\right\} \longrightarrow 2^{k+1} > k + 1.$$

We are now in a position to use the Archimedean property to prove that for any positive real number a there exists a number $m \in C$ such that $2^m > 1/a$.

PROOF

If $a \geqq 1$, then $1/a \leqq 1$, and any counting number may be assigned to m with the assurance that $2^m > 1 \geqq 1/a$. If $0 < a < 1$, then $1/a > 1$. By the Archimedean property there exists a counting number m such that $m \cdot 1 > 1/a$. Since by Example 2 we have $2^m > m$, we see that, for such a value of m, $2^m > 1/a$.

Now that we have proved that if a is a positive real number there exists a counting number m such that $2^m > 1/a$, we are ready to prove Lemma 3 in Chapter 10.

LEMMA 3 (CHAPTER 10) *(a) If c and d are positive real numbers, then there exists a counting number n such that $c/2^n < d$. (b) If $c > 0$ and the statement $x < d + (c/2^n)$ is true for all counting numbers n, then $x \leqq d$.*

PROOF OF PART (a)

$$\left.\begin{array}{l} c > 0 \\ d > 0 \end{array}\right\} \longrightarrow \frac{d}{c} > 0 \longleftarrow$$

$$\longleftrightarrow \text{there is a counting number } n \text{ such that } 2^n > \frac{1}{\frac{d}{c}} \longleftarrow$$

$$\longleftrightarrow \text{there is a counting number } n \text{ such that } d > \frac{c}{2^n}.$$

PROOF OF PART (b)

We restate part (b) as follows:

$$\left.\begin{array}{l} c > 0 \\ x < d + \dfrac{c}{2^n} \text{ for all } n \in C \end{array}\right\} \longrightarrow x \leqq d.$$

The second contrapositive is

$$\left.\begin{array}{l} c > 0 \\ x > d \end{array}\right\} \longrightarrow \sim \left(x < d + \frac{c}{2^n} \text{ for all } n \in C \right).$$

Observe that our conclusion asserts that the statement

$$x < d + \frac{c}{2^n} \qquad \text{for all } n \in C$$

is false. Thus there must exist at least one counting number n such that $x \geqq d + (c/2^n)$. In other words, we have the following equivalence:

$$\sim\left(x < d + \frac{c}{2^n} \text{ for all } n \in C\right) \leftrightarrow$$

$$\leftrightarrow \text{there exists a counting number } n \text{ such that } x \geqq d + \frac{c}{2^n}.$$

We must prove the following implication:

$$\left.\begin{matrix} c > 0 \\ x > d \end{matrix}\right\} \longrightarrow \text{there exists a counting number } n \text{ such that } x \geqq d + \frac{c}{2^n}.$$

$$\left.\begin{matrix} x > d \longrightarrow x - d > 0 \\ c > 0 \end{matrix}\right\} \longrightarrow \text{there exists a counting number } n \text{ such that } \frac{c}{2^n} < x - d \dashv$$

$$\leftrightarrow \text{there exists a counting number } n \text{ such that } x \geqq d + \frac{c}{2^n}.$$

This completes our proof of Lemma 3.

Exercises

1. If S_n represents the sum of the first n counting numbers, prove that
$$S_n = \frac{n(n-1)}{2}.$$
2. If S_n represents the sum of the squares of the first n counting numbers, prove that
$$S_n = \frac{n(n+1)(2n+1)}{6}.$$
3. If $0 < |x| < 1$, prove that $(1 + x)^n > 1 + nx$, provided n is a counting number greater than or equal to 2.
4. Prove that the number of subsets of a set consisting of exactly n elements is 2^n.
5. Prove that $a^n - b^n$ is a multiple of $a - b$ for $n \in C$. [*Note:* $a^n - b^n$ is a multiple of $a - b$ if and only if $a^n - b^n = (a - b) \cdot f(a, b)$, where $f(a, b)$ is a polynomial in a and b with integral coefficients.]
6. If A_n is the amount of money accumulated when one dollar is invested at compound interest for a period of n years at i percent with interest compounded annually, prove
$$A_n = (1 + i)^n.$$

Reference

Katz, *Axiomatic Analysis*

Bibliography

Alder, Claire Fisher. *Modern Geometry.* New York: McGraw-Hill Book Company, 1967.

Birkhoff, George David and Beatley, Ralph. *Basic Geometry.* Chicago: Scott, Foresman and Company, 1941.

Carroll, Lewis. *Pillow Problems and A Tangled Tale.* New York: Dover Publications, Inc., 1958.

Carroll, Lewis. *Symbolic Logic and the Game of Logic.* New York: Dover Publications Inc., 1958.

Courant R. and Robbins, H. *What is Mathematics.* Fair Lawn, N.J.: Oxford University Press, 1941.

Dubnov, Y. S. *Mistakes in Geometric Proofs.* Boston: D. C. Heath and Company, 1963.

Eves, Howard. *A Survey of Geometry.* Boston: Allyn and Bacon, Inc., 1963.

Fetisov, A. I. *Proof in Geometry.* (Topics in Mathematics Translated from the Russian) Boston: D. C. Heath and Company, 1963.

Fujii, John N. *Geometry and its Methods.* New York: John Wiley and Sons, Inc., 1969.

Golos, Ellery B. *Foundations of Euclidean and Non-Euclidean Geometry.* New York: Holt, Rinehart and Winston, Inc., 1968.

Hemmerling, Edwin M. *Fundamentals of College Geometry.* New York: John Wiley and Sons, Inc., 1970.

Katz, Robert. *Axiomatic Analysis.* Boston: D. C. Heath and Company, 1964.

Kay, David C. *College Geometry.* New York: Holt, Rinehart and Winston, Inc., 1969.

Klein, Felix, *Elementary Mathematics from an Advanced Viewpoint.* New York: The Macmillan Company, 1932.

Moise, Edwin E. *Elementary Geometry from an Advanced Standpoint.* Reading, Mass.: Addison-Wesley Publishing Company, 1963.

Prenowitz, Walter and Jordan, Meyer. *Basic Concepts of Geometry.* Waltham, Mass.: Blaisdell Publishing Company, 1965.

Ringenberg, Lawrence A. *College Geometry.* New York: John Wiley and Sons, Inc., 1968.

Stabler, E. R. *An Introduction to Mathematical Thought.* Reading, Mass.: Addison-Wesley Publishing Company, Inc., 1953.

Stoll, Robert R. *Sets, Logic and Axiomatic Theories.* San Francisco: W. H. Freeman and Company, 1961.

Suppes, Patrick. *Introduction to Logic.* Princeton, N.J.: D. Van Nostrand Company, Inc., 1960.

Wolfe, H. E. *Introduction to Non-Euclidean Geometry.* New York: The Dryden Press, Inc. 1945.

Zippin, Leo. *Uses of Infinity.* New York: Random House New Mathematical Library, 1962.

Index